Annual Reports on Analytic[al] Atomic Spectroscopy

Reviewing

1975

Volume 5

Editor: C. W. Fuller
(Tioxide International Ltd.)

Published by
THE CHEMICAL SOCIETY
BURLINGTON HOUSE, LONDON, W1V 0BN

ISBN 0 85186 757 X

ISSN 0306-1353

Printed by Billingham Press Limited, Stockton-on-Tees, Cleveland, England.

FOREWORD

With the publication of Volume 5 of *ARAAS* it is possible to view the development of the publication over the past five years and make an informed assessment of the relevance and scope of the series. At its inception some reservation was expressed concerning the need for another addition to the secondary literature of analytical spectroscopy. As one who has been associated with *ARAAS* from its conception, however, it is pleasing to look at its first five 'formative' years and be able to report that a considerable body of evidence now exists, from reviews and comments from a widespread body of practising spectroscopists and from the steadily increasing sales volume, that *ARAAS* appears to fill a real need in the provision of readily accessible information presented in a brief and concise manner.

It is also possible from the first five volumes of *ARAAS* to make a realistic survey of the present status and progress of analytical atomic spectroscopy. It is apparent that the range of application of atomic absorption spectroscopy continues to increase and that electrothermal atomization has now taken a permanent place in the arsenal of analytical techniques. The renaissance in atomic emission spectroscopy for liquid samples, heralded at the introduction of the nitrous oxide/acetylene flame, has been further stimulated by the current widespread interest in the applications of high-frequency plasma sources, particularly to simultaneous multi-element trace analysis. From a fundamental viewpoint it is also clear that our understanding of the physical and chemical processes occurring in flames, plasmas and other sources continues to increase; the application of laser probe and mass spectrometric techniques in studies of these processes shows great promise.

It is possible to conclude that both fundamental and applied analytical atomic spectroscopy continue to advance rapidly. It is difficult to know what the future will bring, and although, as stated by Alkemade*, we do not know 'what will actually be found around the corner', at least the 'corners' around which something may be found are evidentally being subjected to close examination.

Gordon Kirkbright

*C. Th. J. Alkemade, *Proc. Soc. Analyt. Chem.*, 1973, **10**, 130.

We regret to record the recent death of Mr. James B. Attrill, a member of the Chemical Society's editorial staff, who had been closely associated with the production of ARAAS since its inception.

C. W. Fuller
G. F. Kirkbright

CONTENTS

PART I FUNDAMENTALS AND INSTRUMENTATION

GLOSSARY OF ABBREVIATIONS

Wherever possible elements are referred to by their chemical symbols and compounds by their formulae throughout this volume. The following abbreviations, which conform to I.U.P.A.C. recommendations and common usage, have also been used.

(F)AAS	(flame) atomic absorption spectroscopy
(F)AES	(flame) atomic emission spectroscopy
AFS	atomic fluorescence spectroscopy
G(L)C	gas (liquid) chromatography
NAA	neutron activation analysis
EDL	electrodeless discharge lamp
GDL	glow discharge lamp
HCL	hollow cathode lamp
ICP	inductively coupled (r.f.) plasma
ADDC	ammonium diethyldithiocarbamate
APDC	ammonium pyrrolidinedithiocarbamate
DDDC	diethylammonium diethyldithiocarbamate
EDTA	ethylenediaminetetra-acetic acid
MECA	molecular emission cavity analysis
MEK	methyl ethyl ketone
MIBK	methyl iso-butyl ketone
PTFE	polytetrafluoroethylene
TBP	tri-n-butyl phosphate
TCA	trichloroacetic acid
TOPO	tri-n-octylphosphine oxide
TPP	triphenylphosphine

PART I

FUNDAMENTALS AND INSTRUMENTATION

1 Light Sources

1.1 HOLLOW CATHODE LAMPS

Much of the new work reported involving the use of HCLs has been concerned with the modification of lamp design or operational parameters for specific analytical purposes. Potential applications of new types of demountable spectral lamps, based upon the Sullivan-Walsh high-intensity HCL, have been reported, both as high-intensity sources for AFS (844) and as AES excitation sources (891) (See also section 2.3). If the booster filaments in the lamps incorporate lanthanum hexaboride (844) they may be rapidly regenerated, even after repeated exposure to air. Stephens and Flinn (772) have described a multi-cathode discharge source which contained separate, sequentially-operated cathodes fabricated from wires of appropriate metals. The useful life was ≅100h, but the emission intensity was 1~2 orders of magnitude lower than that from corresponding single-element lamps. Demountable 'see-through' HCLs have been described by Bath and Woodriff (1129), for use in conjunction with a sequentially-pulsed HCL continuum source for simultaneous automatic correction for non-atomic absorbance signals. The principal advantage of such a system is the low light loss. Sealed 'see-through' lamps have been described earlier by Strasheim and Butler. (*Appl. Spectrosc.* 1962, **16**, 109). A claim has been made (1359) that the fabrication of cathodes by compression moulding from metal particles of less than 70 μm in diameter at normal temperatures leads to 2~4-fold increases in absorption-line intensities. Interest in the routine use of pulsed operation of HCLs has not been widespread. They have been used, however, in molecular absorption and fluorescence spectrometry (710) and also as excitation sources for AFS. A mini-computer-controlled power supply (726) for the latter use has been manufactured and evaluated. Peak and average currents, d.c. level, and pulse width and period could be individually controlled with this system (see also *ARAAS* 1974, **4**, 3). It is difficult to envisage how the advantages of using HCLs operated in this mode could be regarded as sufficient to justify their widespread use in AFS at the present time, except when multi-element systems are to be developed. Piepmeier and de Galan (640, 644, 1126) have studied the variation in the properties of Cu and Ca lines emitted by HCLs operated in a pulsed mode. Pulses up to one ampere and widths as short as 10 μs were employed at repetition frequencies between 3 and 300 Hz. Temporal changes in the line profile of a pulsed lamp give rise to the observation of an apparent line profile whose shape depends upon the detector–system response characteristics, and thus influence the analytical curves. For the Ca 422.7 nm line (640), self-reversal and line width increased with radial distance from the centre of the cathode bore and the time during the pulse, and decreased with increasing modulation frequency up to 6·4 kHz. Up to 1·6 kHz, electronic modulation gave narrower and more intense lines than optically–chopped d.c. operation.

Torok and Zaray (1456) have described the construction of purpose-built twin HCLs which may be cooled with liquid air and could be used for studying excitation processes in hollow cathode discharges and in emission analysis. Wagenaar and de Galan (594) studied the influence of spectral line profiles upon analytical curves in AAS. A Fabry-Perot interferometer was used to study Ag and Cu line profiles before and after flame absorption at various HCL currents. The measured profiles were digitized and deconvoluted and the flame-shift values were applied to the calculation of analytical curves. Good agreement was obtained between the directly measured curves and the curves thus calculated. Hannaford (798) has also reviewed the theoretical aspects of the influence of spectral line profiles in AAS.

Human (845) has measured the shapes of the Ca 422.7 nm and the Cr 425.4 nm lines from boosted-output HCLs with a pressure-scanning Fabry-Perot interferometer. A two-layer model was assumed for the source and theoretical line profiles were calculated, assuming a Gaussian shape for the absorption line in the lamp. The theoretical profiles were fitted to the measured profiles and from this the Doppler halfwidths and optical densities in the two layers were obtained and temperature values calculated. Zechev and co-workers (1528) found considerable Gaussian and Lorentzian broadening of the Fe 372.0 nm line. Predictably, it was found that increasing the discharge current stimulated the Gaussian component of the spectral profile, and a correlation was found between the Gaussian component and the cathode voltage drop. An echelle spectrometer (1370) has been used to measure spectral line profiles for Ca, Ag and Al. Increases in line-width resulted in a corresponding poorer AA sensitivity.

The excessive Doppler broadening of B HCL lines in Kr-filled HCLs is avoided in Ne-filled sources; the latter (858) have therefore been used for B isotopic analysis using the 208.8/208.9 nm doublet. A further example of line overlap in AAS (137) has been reported — Re has been determined at 346.046 nm, using the neon line at 346.053 nm as a source. A detection limit of 11 μg/ml was obtained in a N_2O/C_2H_2 flame.

It is perhaps worth noting here that the performance of HCLs for a number of elements, notably As, B, Se, U and the rare earths, continues to be, at best, mediocre and no great progress has been made in this area.

Other references of interest —

Fe – Ne HCL spectrum: 302

1.2 ELECTRODELESS DISCHARGE LAMPS

Interest in the use of EDLs has continued, prompted perhaps by reports of the substantial improvement in stability of the emission from EDLs attained by external temperature control (*ARAAS* 1974, **4**, 57). Fixed temperature (1192) enables satisfactory results to be obtained from simultaneous excitation of elements with markedly different optimum temperature requirements, and the merits of EDLs as multi-element excitation sources (1192, 1122) have again been stressed. The main applications at present of EDLs are: *(a)* as intense excitation sources for Zn and Cd, where AF detection limits are vastly superior to those obtained by AA; and *(b)* as sources for As and Se in AA analysis, where the stability and intensity of the corresponding HCLs is so poor.

From a fairly rigorous study of the optimization of the experimental parameters involved in microwave-excited EDL construction (725), it was concluded that lamp shape is the most important factor. However, although the familiar 40 mm x 8 mm i.d. dimensions were found to give the most efficient coupling, an additional 'ballast' section 60 mm x 5 mm i.d. improved the lamp operating characteristics. Volatile covalent hydrides (1132) have been used to prepare EDLs for selected group IV, V and VI elements. The primary advantage of this procedure is that it allows the introduction of a reproducible amount of element into the lamp blank. Lamps prepared in this way were reported to be intense, reproducible in characteristics, stable over a wide temperature range and long-lived. The spectral characteristics of S, Se, As and P EDLs (642) beween *ca* 177 and 217 mm have also been discussed.

A detailed study of the properties of alkaline earth EDLs (1075) as light sources to be used in a strong magnetic field for Zeeman scanning of absorption profiles in flames has been reported. Temperature, line width, fill gas, unwanted line emission, stability and useful life were investigated.

1.3 LASERS

Although there has been continued speculation about the potential uses of laser excitation sources in chemistry in general (563) and analytical chemistry in particular (680, 732, 649), useful novel applications of these sources in routine analytical atomic spectroscopy are scarce. Dye lasers have again been used (see *ARAAS* 1974, **4**, 281) to monitor very low Na vapour densities (526, 1000) by AFS and studies of intracavity quenching of laser emission (445, 960, 1009) have been extended by several workers. Absorption by Na, Ba, and Eu (445) and Na, Li, Sr, Ba and Cs (1009) has been observed using air/ C_2H_2 flame atomizers; the method has also been applied to K, Rb and La salts (960) using atomization from a graphite electrode. A narrow-band tunable dye laser has also been used (536) to excite Ba atomic fluorescence in a level-crossing study on the 6s 6p $^1P_1$, level (553.5 nm) of Ba(I).

The effect of dye laser emission bandwidth (525) on absorption coefficients and resonance excitation has been studied on a theoretical basis. Generalised expressions were derived for absorption coefficients that take into account not only the atomic Lorentz and Doppler broadening, but also the line shape of the irradiations. The construction of a continuously-dumped argon-ion laser (238), suitable for phase fluorimetry has been described briefly.

Other references of interest —

Iodine laser emission: 150, 1012

1.4 CONTINUUM SOURCES

Some further results of preliminary studies (*ARAAS* 1974, **4**, 4) of the use of exploding wires (689, 711, 847) as intense ultraviolet continuum excitation sources have been described by Sacks *et al.* The wires are exploded by capacitive discharge. At 220 nm, the peak irradiance was more than five orders of magnitude greater than that obtained from a 1·6kW Xe arc lamp (689), with a RSD of 0·054 between shots. Optimum intensity was obtained in an Ar atmosphere (711) and the output increased almost linearly with pressure over the range 100 to 700 torr, becoming virtually pressure independent at above 100 torr.

A theoretical model (349) has been formulated for calculating the spectral radiant emission from a pulsed inert-gas plasma arc lamp. Continuum sources such as a 200 W Xe-Hg lamp at wavelengths below 300 nm (731) have been shown to be very promising for AAS when an echelle monochromator is employed to attain adequate dispersion. An interesting alternative approach has also been suggested, based upon selective modulation of resonance lines (1120), using the droplet-generator system of Malmstadt and Hieftje (*Anal. Chem.*, 1968, **40**, 1860) to periodically introduce into a flame droplets of metal-salt standard solution. This pulsating atomic vapour was used to effect selective modulation of resonance lines against an unmodulated, continuum background.

1.5 ZEEMAN - SPLIT SOURCES IN AAS

A great deal of interest (369, 770, 990, 991, 1176, 1235) has been shown in the analytical potential of the use of Zeeman splitting (described briefly in *ARAAS*, 1974, **4**, 18). The method has been applied, for example, to the determination of Hg in hair and NBS steel samples (369), which had received no previous chemical treatment. Much attention has been paid to the design and operation of purpose-built sources, including both EDLs (770), and d.c. discharge lamps (990) employing wire or sheet-metal electrodes. It has been claimed (991) that Zeeman splitting can be used to compensate for source drift, as well as scatter. It seems probable that interest in this technique has been promoted to a great extent by the stringent requirements for precise background correction which are imposed by a number of electrothermal atomizer systems. However, it should be remembered that the technique is not a universal panacea, because of the variability of the splitting pattern from element to element.

2 Excitation Sources and Atomizing Systems

2.1 ARCS AND SPARKS

2.1.1 Fundamental Studies

The temperature of, and the movement of materials in, the various plasmas available to spectrochemical analysis continue to be investigated. Vulcanovic and his co-workers provide valuable information that will help in the understanding of the sources of spectra. 'Liquid bullets' have been injected into the plasma of a d.c. arc in air (graphite) and the spatial intensity distribution of Na (213), CaO (666) and their residence time in the arc recorded with a high-speed camera (214). It was suggested by Petrovic *et al.* (629) that the distribution of particles in the plasma can be improved by constricting the arc in cooled metal tubes. According to Decker and McFadden (211) one should be careful where one looks for the region of maximum sensitivity since in their opinion this region will, for some elements, be found quite a distance off the axis of the plasma. The position seems to be elemental dependent and related to the inside diameter of the electrode crater. Stupp and Overhoff have used some sophisticated apparatus to study the vaporization of graphite in the d.c. arc (403). A computer programme to enable determinations of the radial distribution of temperature and electron pressure has been written (1321). Calculations of the optimal characteristics for d.c. arc operation have also been made and reported in detail (284). From such information *i.e.* the relationship between plasma characteristics, temperature and line intensity, improvements have been made by Kubota and Ishida (1296).

The work of Walters *et al*, on the subject of spatial and temporal resolution of sparks is noteworthy (1090, 1091, 1092, 1093, 1094, 1095, 1096). These papers were presented at the II F.A.C.S.S. meeting; the abstracts in the programme stimulate more than average interest in the work of this progressive school.

2.1.2 Modifications of Existing Systems

Two papers presented at Euroanalysis II, 1975 confirm that the spectra produced by a d.c. arc can be affected significantly by a magnetic field. In the first paper (1306), it was suggested that an alternating magnetic field can influence several spectro-analytical problems. The second paper (1311) was concerned only with improving and explaining the sensitivity of 9 volatile elements. Pavlovic and Mikailidi further suggest that the most marked effects occur when the cathode is the sample-carrying electrode.

There are still modifications and studies to be made of existing systems which can improve sensitivity and other analytical parameters. Time resolution of the arc (1486), often called selective distillation, has improved limits of sensitivity and line identification. Strasheim *et al* (579) have improved their method of analysis for various Al alloys. The addition of a third electrode to the well tested 'one-up, one-down', system is claimed to "increase the stability of arc ignition" (341).

A Kenotron (a switch) introduced into a spark source (950) has increased the pulse voltage by a factor of two and makes this source suitable as a "universal atomizer for AAS studies". Unipolar sparks with highly reproducible current impulses are produced with a new circuit designed by Mast and Pfeilsticker (592). Used in the mode suggested, 1000 impulses are superimposed at a repetition rate of 3 kHz. The liquid layer on solid sample spark technique has been exploited by Ohls (1303). This system when compared with other methods used in an industrial steel laboratory also produced greater sensitivity. The virtues of time resolved spark source have been expounded by Crouch (110, 646, 1347). Details are given for the determination of a number of elements.

2.1.3 Mixed Sources

Since the early 1930's there have been a number of attempts to combine arcs and sparks with flames. Recently Woodruff and Malmstadt (709) have used such a coupled system to analyse a variety of solutions. A report (174) from the Macaulay Institute — the home of "cathode layer arc spectroscopy"— described a triple-flow gas sheathed d.c. arc which it claims with typical conservative realism, "is giving better results". Fijalkowski and co-workers (626) examined three types of discharge and the 'best conditions' for the determination of some difficult non-metallic elements. Vapours produced by arcs and analysed by sparks (767) and those produced by sparks and analysed by flames (841) have been reported. Solutions, after desolvation, have been directed into arcs (371) and sparks (45). In the latter method the analysis of rare earths has been performed with better reproducibility than with the porous-cup technnique. Powder aerosols injected into d.c. arc plasmas gave surprisingly reproducible results with RSDs of ~0.05 at levels of a fraction of a ppm (591).

2.1.4 Buffers

Investigations into the nature and effect of spectroscopic buffers has continued, albeit at a slightly slower pace than of recent years. Buffering, it is claimed, offers its main advantage by improving the residence time of elements in the plasma (1392). The role of heat of evaporation has been studied by Lakatos (587) in the analysis of petroleum products. The temperature of the anode spot, responsible for the uptake of sample into the excitation areas, has been studied under buffering conditions by Decker and Kobus (576). No analytical applications were attempted.

Alkali metals still form the basis of most buffers. With the pressure for increased reproducibility of analytical information, Zadgorska *et al* (1313) found that the various halides and common acid salts of potassium they used should be selected with care. Rippetoe *et al.* (234) used KCl to improve the stability of a d.c. arc plasma. The use of 1% w/w NaCl buffer for rare-earth analysis has been studied (418); unfortunately no comparisons were given with other techniques. Work on NaCl mixed with NaF and Ga_2O has also been carried out (418). Lower determination limits for a wide range of elements were claimed. NH_4F added to samples was used in an attempt to understand plasma behaviour (665). Tripkovic (625) has investigated the effects of halogens introduced into plasmas. Iodine was mixed (5:1) with graphite and light output of the source studied. The conclusions were that without iodine, at certain levels, no lines for Te, Sb, As, Mo and Ni were observed. When iodine was added in the manner described, the lines of all the elements were seen. No analytical applications were attempted. Organic reagents containing iodine were used to suppress the interference of Mo in the determination of a wide variety of elements in molybdenum oxide (39).

The addition of Ar to the air around the plasma and Ba to the sample was found to improve both the sensitivity and detection limits of numerous elements (38, 1498). At a recent conference in Poland several workers reported that mixtures of air–Ar–N_2 could be manipulated to improve the analysis of aluminium alloys (627).

2.2 PLASMAS

2.2.1 d.c. Arc Plasmas

Within the past year there has been an explosion of interest in spectrochemical analysis using plasmas and a review of available d.c. arc plasmas has been made (1403). This expansion is logical and is an extension of the interest by analysts in AAS into multi-element determination. The temperatures (<300K) of the commonly used flames are insufficient to provide the required analytical sensitivity for many of the elements often sought. It is not surprising that electron temperatures and concentrations have been studied by several workers (206, 597). The need for multi-element determinations has led many workers to develop *'new'* devices by adapting existing units. Vickers (648) has rotated a d.c. arc on a disc electrode, thereby generating a plasma with a "hole in the middle" into which the solution may be injected. Levels of background, interference effects and detection limits have all been studied. Denton (698) attempted to use an 'old' arc plasma source but found that too much improvement was necessary to enable him to perform the required analysis. Layman and Hieftje (226) used a computer to perform the manipulations of their plasma. The computer not only controls the pulsations of the discharge but the coincidental injection into the arc of the discrete volumes of the sample solution. Sensitivity, precision, background and interference effects have been studied for some of the easier elements.

A new source for trace analysis using two rotating plasmas constrained within a horizontal graphite tube has been developed (605). Improvements in the power of detection by almost an order of magnitude were suggested. Solutions have been studied and the injection of powders is to be attempted in the future.

2.2.2 r.f. and Microwave Plasmas

The inductively-coupled plasma has, beyond all doubt, invoked the most interest in instrumental analysis for several years. The decision "which system is best ?"—will no doubt cause many emotive arguments both within and without conference halls, lecture theatres, offices and laboratories for some time to come. The first application of this excitation system to analysis was by Greenfield and co-workers in the early 1960's. A demonstration of a 'low power' system was given at the XII C.S.I. in Exeter, England in 1965. For those intending to use or with an interest in this source unit the new ICP Newsletter, published by The Department of Chemistry of the University of Massachusetts, Amherst, Massachusetts 01002 U.S.A., under the editorship of Barnes provides a worthwhile source of information. Greenfield has published several excellent reviews on the application of ICP's to analysis (104, 173, 701). The systems used in all these reports have been built "in-house". The obvious advantages of multi-element analysis with precision and accuracy, claimed by many workers, have quickly been appreciated by many instrument manufacturers. Dahlquist (615, 854) and Davison (741) have attempted to relate the experience gained in pioneering work to industrial analytical requirements. Brech (17, 735) described the application of a new source unit to similar problems. The modern proliferation of abbreviations has extended into this subject, Boumans has used the initials SMEA (Simultaneous-multi-element-analysis) to describe the application of his equipment to the analysis of a variety of solutions. Fassel *et al* have reported studies which serve to bridge the gap between academic information and investigation and the practical application of ICP to SMEA (484, 738, 906). The initial promise of an interference-free system claimed in early reports is evaporating like the morning dew in the warming sun as is illustrated by the work of Mermet (182, 1462) Laison (228) and Koirtyohann (885). The first paper notes the deleterious effect of Na on Cr and P determinations, in the second, the failure to attain local thermal equilibrium in the plasma causes concern and also contributes conflicting information for the interaction of the same matrix element (Na) on the determination of Cr with addition of data on Ca and Cd. The third paper reports that mineral acid concentrations, in this case HNO_3, can alter the results obtained with an ICP.

The theoreticians have also been hard at work on this new system. Barnes *et al* (856) have undertaken computer simulation of processes occuring in the ICP and came to conclusions concerning the plasmas and their behaviour patterns. The shape of lines in the various regions of the plasma tailflame has been investigated and the possibility of ionization interferences considered. Some of the effects observed in the ICP cannot be explained in terms of the Saha equation.

Mermet and Robin (182, 619) have investigated plasmas at two frequencies (619). A microwave helium plasma operating at 2450 MHz has been developed (671) for the analysis of metalloenzymes and up to 10 nm KCL is used to enhance the line intensities. A high-frequency, high-power source unit at 3000 MHz has been developed by Van Calker and Hollenberg (643). To date, only the invention of the source is described. An unambiguous comparison of an ICP and a CMP (Capacitively-coupled microwave plasma) has recently been made by Boumans *et al* (607). It was found that the ICP was a 'better' source for practical analysis. The same author found that the most suitable operating conditions for an ICP (1457, 1465) were a compromise between maximum sensitivity with poor reproducibility and good reproducibility but reduced sensitivity.

2.3 GLOW DISCHARGES

Hollow-cathode lamps with and without demountable cathodes continue to be used in emission spectrochemical analysis. Very little new work appears to be necessary on developing or improving the lamps for AAS. From the series of papers presented at a recent conference one obtains a comprehensive review of the state of the art. Berneron (588) gives a review of the value of the Grimm Lamp in the analysis of alloys and an account (589) of the erosion of metal surfaces by the glow discharge. Lowe (846) has modified the original Grimm lamp by adding a second discharge at a low voltage by way of an electron-emitting cathode. Aluminium alloys as well as pure metals were investigated. Laqua has provided another excellent review of the work in Dortmund (793) and expresses the opinion that no other single optical emission method has yet offered such a combination of possibilities for analysis. Baudin and Remy (842) suggest that the future for the lamp lies in its value as a atom reservoir for AAS and used a modified Grimm lamp to study diffusion profiles in flat specimens (849). Gough (838) suggested that the use of the lamp in AAS is best restricted to the determination of minor constituents in alloys. Cathodic sputtering in AAS studies and observations of the Hanle effect (860) suggest that new developments are possible with these discharges. Surface collisions from low-energy heavy-particle bombardment of gases were studied by Folk (794). Measurements of emission radiation from the Grimm lamp have been made by Butler (843). Fabry-Perot interferometers were used by two independent workers to study line-profiles (567, 1312). In the first case, to study the processes taking place within the lamp and in the second, isotopic studies of Pb.

2.4 LASERS

In spite of the fact that the advantages of laser microspectral analysis have again been stressed for quantitative microanalysis in general (641) and forensic analysis in particular (1330), progress in this field continues to be slow. Generally laser atomizers remain an academic curiosity to many of those engaged in routine analysis. Their high cost and limited range of application relative to other systems has remained an inhibiting factor. It should be emphasised that this may not always be the case if further improvements in laser technology can be achieved.

A number of groups (835, 836, 1016) have investigated the use of laser atomization in AAS. Matousek and Orr (835) employed a pulsed infra-red CO_2 laser with 0.1 J pulses with a width of less than 1 μs and a repetition rate of 5 Hz. Solid, pure-metal samples were contained in a graphite furnace, but it was found necessary to heat the furnace to a temperature just below the threshold for continuous perceptible atom production to obtain satisfactory reproducibility. Others (610, 836) have overcome the problem of inefficient ground state atom production by employing a pulse repetition rate of 100 kHz, which resulted in efficient, quasi-continuous vaporization. Theoretical calculations (1016) carried out for the Cu and Cr lines at 327.4 and 428.9 nm respectively, indicated that detection limits lower than 10^{-3}% were unattainable using the total-absorption method and conventional procedures for AAS analysis with a laser atomizer.

Lacqua and co-workers (836) have pointed out that laser atomization AAS avoids some of the shortcomings of spark cross-excitation laser emission methods, such as contamination and dilution problems. However, the same group have investigated the use of microwave cross excitation (610, 612, 837), which also serves to overcome these problems. It was found (612) that incomplete evaporation, short residence time, and rarifaction of the plasma by cross-excitation discharge prevented the detection of masses below the picogram level. Using single-shot atomization (837) and photoelectric detection, an RSD of about 0.3 was obtained, with detection limits for a number of elements in Zn in the ppm range.

An improved laser micro-spectral analyser (1042) has been described. Cross excitation may be delayed with respect to the laser pulse, or synchronised to it, and a pulse repetition rate of 4 min^{-1} may be used. Briggs and Kraft (720) have discussed the effect of spectrograph optical characteristics on laser microprobe emission detection limits. They reported that best results were obtained with the largest aperture instrument used, a 0.75 m f/5 spectrograph. A comparison (688) has been made of the relative signal-to-noise ratios of time differentiated photelectric and photographic densitometric recordings of AE from laser-generated plasmas. The superiority of the photoelectric technique was less than anticipated.

Other references of interest —

AE from laser-produced plasmas: 403, 404, 528, 658
Evaporation of solids by laser pulses: 1069, 1070

2.5 FLAMES

2.5.1 Theoretical Studies

The results of a number of basic theoretical studies concerned with flame systems which are pertinent to analytical spectroscopy are reported here. Gaydon (795) has reviewed the current status of spectroscopic studies on equilibria in flame gases. Frank and Krauss (873) have used a short-duration spark between pure Ca electrodes in a pure oxygen atmosphere, to study the origin of the green bands of CaO(H). They concluded, from time-resolved experiments which showed that the intensity ratio of the green and orange bands of CaO remains unchanged over a wide temperature range, that hydrogen cannot be a constituent of the emitting species for the green bands which appear, therefore, to originate from CaO. Van Hurk, Hollander and Alkemade (123, 942) have determined excitation energies of SrOH and BaO bands measured in flames. Excitation energy differences were derived directly from the ratio of thermal-band intensities measured as a function of temperature. Absorption excitation energies were derived from the temperature dependence of the ratio of thermal-band to atomic line intensity under thermal equilibrium conditions using assumed values for the dissociation energy of SrOH and BaO. Flames with temperatures between 1907 and 2886 K were employed.

Laser Raman spectrometry is finding increasing application in the investigation of flame systems. Setchell (140) has examined the air/CH_4 flame by this technique and reported accurate axial and transverse temperature-profile measurements from recorded nitrogen spectra. Q-branch bands of CO_2, O_2 and H_2 in the flame were recorded; band intensities qualitatively agreed with the concentrations predicted from equilibrium calculations. Absolute CO concentrations were determined for rich flames. Arden and co-workers (360) have utilized laser Raman spectroscopy in a study of the gaseous combustion products of flames produced on standard burners for O_2/H_2, O_2/C_2H_2, O_2/CH_4 and O_2/C_3H_8. Omenetto (797) has reviewed the possibility of local sensing of physical parameters such as temperature and species concentrations in flames using laser sources.

L'Vov and co-workers (1167) calculated the temperature values and composition of N_2O/C_2H_2 flames over a wide range of fuel/oxidant ratios. The effects of introducing water, changing the pressure and the atomization efficiencies (β) of some refractory elements were also calculated. These studies predict full dissociation of all monoxides except those with dissociation energies (Do) in the range 180–200 kcal mol^{-1} (B, Ce, Hf, La, Nb, Pr, Ta, Th, U and Zr). Low atomization efficiencies in the air/C_2H_2 flame for elements whose oxides have Do between 130 and 150 kcal mol^{-1} may be explained by incomplete vaporization of aerosol particles. Stephens and Stevenson have derived a relationship between CN radical concentration and β of elements introduced into the N_2O/C_2H_2 flame; the relationship has been examined under various experimental conditions and β values estimated for a number of elements (216). The possibility of complex vapour phase oxide formation by U was discussed.

The collisional quenching of electronically-excited Sn atoms has been studied by time-resolved AAS (946). Deactivation-rate constants have been reported for quenching of $Sn(5^1D_2)$ with a range of collision partners and the resulting data compared with those for analogous states within Group IV (carbon 2^1D_2 and lead 6^1D_2). Laniepce (958) has reported measurements of the intensity of the fluorescence radiation (6s 6d)$\rightarrow$(6s, 6p), emitted by optically-excited Hg atoms, in the presence of nitrogen. These indicate the order of magnitude of the cross-sections for excitation transfer between the (6s, 6d) levels of Hg and for quenching induced by collisions with the nitrogen molecules.

An interferometric method based on synchronous scanning of a Fabry-Perot interferometer and a monochromator tuned to select one free spectral range has been developed to measure atomic line profiles in absorption and used to determine the profiles of Ag 328 nm and 338 nm and Cu 325 and 327 nm in an air/C_2H_2 flame (207). The results have been interpreted in terms of Voigt a-parameters and peak absorption coefficients. The same technique has been applied to determine profiles for these lines from HCLs for calculation of analytical growth curves in AAS in the air/C_2H_2 flame (1416); the effect of lamp current on sensitivity and linearity was studied. L'Vov and co-workers (1340) have measured the shift of the Ca, Sr and Ba lines at 422.7, 460.7 and 553.5 nm respectively emitted by air/C_2H_2 flames relative to the same lines emitted by HCLs. In agreement with theory the average ratio of shift to the Lorentzian line-width was 0.38.

A method for electronic-excitation temperature measurements based on a two-line atomic absorption technique utilizing Pb lines of different lower levels has been proposed (1294). The absorption of the Pb 283.3 nm resonance line and the Pb 261.37/261.42 doublet is measured; the method has been applied successfully to the determination of Pb electronic-excitation temperature in an air/C_2H_2 flame and with Ta and glassy carbon strip atomization devices.

Held and Stephens (771) have evaluated the effect of photon trapping on line intensities observed in AFS and FES. Repeated self-absorption and re-emission of resonance radiation at high atom concentration results in trapping, causing a decrease in resonance fluorescence (RF) and an increase in direct-line fluorescence (DLF). Equations relating the intensity ratio of DLF and RF in a three-level system to atom concentration are derived for atomic vapours at low pressure. The effect of photon trapping on the fluorescence of the Cu resonance line at 325 nm is illustrated.

Other references of interest —

Low pressure flames: 124, 947.
Flame temperature measurements: 120, 1335, 1533.
Atomization mechanisms in air/C_2H_2 flame: 242, 435, 867, 1231, 1534.
Matrix isolation of reactive metal atoms: 574.
Spectra and burning velocity of C_2N_2/F_2 flame: 463.
Organic solvent flames: 343.
Use of AAS in high temperature evaporation and diffusion studies: 308.
Mass spectrometry of flame species: 955.
Technique for measurement of lifetimes of excited molecular states: 859.
Schlieren photography of flames for study of effect of organic solvents: 1302.
Role of metastable Na_2 dimers in theory of induced AF: 468.
Flame Chemiluminescence of alkali and alkaline earth halides: 65, 97.

2.5.2 Flame Types and General Studies

Koirtyohann and Lichte (1103) have reviewed the current status of flame spectrophotometric methods for trace analysis and comment that the low cost and ease of operation of these techniques are particularly attractive features and will ensure their continuing popularity. Safety practices for AAS developed by the Scientific Apparatus Manufacturers Association of the USA have been described (49); the items covered include specifications for flame-exhaust systems, rules for safe handling of cylinder gases, conditions for the safe operation of premixed flame burners, precautions in the use of volatile organic solvents and miscellaneous hazards such as UV radiation and solutions containing cyanide.

The application of the premixed N_2O/H_2 flame to the determination, by AAS and FES, of easily-atomized elements in troublesome organic solvents at trace levels has been recommended (890); solvents such as benzene, xylene and gasoline were directly aspirated into this flame with little background signal. In other commonly used (hydrocarbon) flames the critical C/O ratio, at which yellow carbon luminescence appears, is easily exceeded with these solvents and background interference is observed. The use of a H_2/Cl_2 flame for AA and emission spectroscopy has been discussed (1123); it is claimed that the exclusion of O_2 from the flame can facilitate the atomization of some elements which produce refractory oxides in conventional flames and detection limits for elements such as Al, As, B, Cs, Cu, Ba, Fe and Si have been presented. Owing to the nature of the products of combustion, HCl, it is doubtful whether this flame would be viable for routine analysis.

Held *et al.* have described a novel flame "atom trap" system which permits appreciable sensitivity enhancements in AAS for a number of elements (907). In this technique solutions are nebulized into an air/C_2H_2 flame and the vaporized sample is trapped on a cooled surface created by a water-cooled tube placed within the interconal zone; after a suitable period (up to 5 minutes) the sample is released back into the flame by stopping the water flow in the tube and allowing it to be heated by the flame gases. The efficiency, sensitivity and atomization kinetics of the system have been discussed.

L'Vov and co-workers (416, 417) calculated the lateral distribution of dry aerosol particles in premixed flames from slot burners by the application of Stokes Law to the transport of dry particles in the lateral flow of the flame gases. The effect on lateral flame profiles of the presence of excess concomitants was also discussed. According to the authors the 'lateral flame profiles' are more plausibly explained by the transport of aerosol particles

(Koirtyohann and Pickett, *Anal. Chem.*, 1968, **40**, 2068) than by diffusion of the vaporized element (West, Fassel and Kniseley, *Anal. Chem.*, 1973, **45**, 1586). The same workers have also given a mathematical model describing the lateral distribution of atoms in laminar flames from slot burners. The model allows quantitative explanation of observed effects in AAS and the estimation of the maximum possible enhancement by addition of foreign species to the analyte solution. The spatial distributions of the atoms of 16 elements in the N_2O/C_2H_2 flame have been determined (367) and discussed in relation to the thermodynamic properties of their oxides. Chakrabarti and McNeil (916) have undertaken a similar investigation of the determination of V by AAS in N_2 and Ar-shielded N_2O/C_2H_2 flames. Sutton and Lowe (866) have studied the interference effects in AAS produced by a series of mineral acids on B, Zr, Ti, Mo, Al and V in N_2O/C_2H_2 flames. The results yield further information on lateral diffusion effects.

Hieftje and co-workers have continued their extensive practical studies concerned with flame spectrometry. The diffusion of atoms and ions from individual solute particles vaporizing in a laminar flame (752, 1118) has been studied. Direct time-resolved measurement of the spatial atomic-concentration gradient surrounding an individual vaporizing particle is possible. Comparison of these concentration profiles at various times after the onset of vaporization enables the effects of diffusion, vaporization and reactions of analyte with flame gas species to be determined. A flame emission spectrometer with a droplet-generator sample-introduction system (Hieftje and Malmstadt, *Anal. Chem.*, 1968, **40**, 1860) has been interfaced to a PDP–12/40 computer for data logging, data reduction and real-time optimization of the instrument (1117). The instrument was adaptively optimized to minimize matrix and ionization interferences and to maximise the signal-to-noise ratio. Saturday and Hieftje (1119) reported the development of a modified burner system for use with a premixed $H/O_2/C_2H_2$ flame (see *ARAAS*, 1974, **4**, 8) which overcomes earlier difficulties caused by flashback and thermal deformation of burner components. Hieftje and Bystroff (638) examined the noise power spectra of both background (OH band) and analyte (Na) emission signals from two locations in shielded air/C_2H_2 flames. Low in the flame, inert gas shielding serves best to minimise flame instability; high in the flame, a quartz separator proves superior. With the burners employed, although no useful high-frequency minima were observed in any of the noise spectra, the presence of strong fluctuations at discrete low frequencies (around 25 Hz) high in the unshielded and Ar-shielded flames indicated that these frequency regions should be avoided whenever possible.

The existence of an interference mechanism in the N_2O/C_2H_2 flame, whereby elements are reduced to carbides so that their total vaporization is prevented, has been postulated by Rubeska (1544) and by L'Vov and Orlov (1534). The enhancements observed by refactory oxide-forming elements, such as Al on absorption of Mo or V, are then explained by their prevention of this reduction so that more complete vaporization is attained. Rusnakova studied interference effects on Ca, Al and Si in the N_2O/C_2H_2 flame (1545). The determination of B by atomic absorption and flame emission spectrometry in N_2O/H_2 and N_2O/C_2H_2 flame has been studied (618). Using the B line at 249.8 nm, in AAS the detection limits obtained with these flames were 0.2 μg ml^{-1} and 2.5 μg ml^{-1} respectively, while in AES the detection limit in the N_2O/H_2 flame was 0.03 μg ml^{-1}; high flame background prevented emission measurements in the N_2O/C_2H_2 flame. The successful determination of B in alloy steels by FES without prior solvent extraction with the N_2O/H_2 flame was reported. Rains and Menis (34) have studied the distribution of the Al emission at 396.2 nm in the N_2O/C_2H_2 flame using repetitive optical scanning in the derivative mode. This system enabled interferences from CN, CH, and C_2 band emission to be overcome. A detection limit of 0.01 μg ml^{-1} Al was reported and the application of the method to the determination of Al in steel, iron, ferrosilicon and phosphate rock samples described. Urbain and Desquesnes (624, 853) have described a differential scanning technique for multi-element analysis by atomic emission spectrometry using two N_2O/C_2H_2 flames.

The influence of the absorption and emission spectra of the air-C_2H_2 flame on the AA determination of Fe has been investigated (1216). The importance of diffuse parasitic light and the true flame background (band and continuum) on the observed spectra of Na-rich laminar premixed air/C_2H_2, air/H_2 and N_2O/C_2H_2 flames has been discussed (217). The sensitivity differences in the AAS determination of Cr at 357.87 nm in air/C_2H_2 and air/H_2 flames for different Cr compounds have been investigated in detail (383). Other workers have also undertaken a detailed study of the determination of Cr in the air/C_2H_2 flame (1564, 1570). Nakahara and Musha have described the determination of In in diffusion flames (1478); the addition of Mg halides was found to be very effective for elimination of interferences (except from Si and V). The same authors have described a similar study for Pb (1378).

Fuwa and Haraguchi (656) have recorded the absorption spectra of SO_2, PO, InF, InCl, InBr, InO, AlF and AlO in flames when acid or salt solutions were nebulized. It is claimed that some of these spectra can be used for analysis and others are useful for elucidation of atomization mechanisms in the flames employed.

Human and Zeegers (636) have observed molecular fluorescence for CaOH, SrOH and BaCl in a $O_2/Ar/H_2$ flame using a continuum source of radiation. Fluorescence and emission spectra were identical; the spectral distribution of the fluorescence emission within a band is independent of the wavelength distribution of the irradiating light, indicating that the absorbed energy is redistributed before the fluorescence process. The fluorescence efficiencies were determined and found to be of the same order of magnitude as the atomic fluorescence efficiencies.

Other references of interest—

Solid propellant flames: 969.
Atom formation review: 796.
Chemiluminescence in FES; review: 498.
Effect of acetone vapour in AAS: 166.
Absorption tube technique: 91.
Atomization systems, reviews: 1056, 1057.
Structure of turbulent flames: 1047.
Diffusion of atoms in flames: 1166.
Non-selective absorption by metal halides: 345.
Flame emission and AAS detectors for chromatography: 72, 135, 224, 889.
GC–AAS system for det. of alkyl lead and Se Compounds: 831.
Terminology for cool flames: 679, 1252.
Al atomization in $O_2/N_2/H_2$ flame: 868.
Sn atomization in H_2/air flame: 1391.

2.5.3 Burners and Nebulizers

A safe practice procedure has been recommended to minimize hazards from flashback of N_2O/C_2H_2 flames utilizing a commercially-available burner system and a manual gas-control system; an automatic gas-control unit was also described (312). A number of papers have reported modified burner designs to improve safety and analytical performance, especially for the N_2O/C_2H_2 flame (319, 584, 888). In a comparison of the performance of single and triple-slot air/C_2H_2 burners for AAS it has been demonstrated that only elements with oxides of high dissociation energy show an enhanced sensitivity with the triple-slot burner; for elements with oxides of low dissociation energy requiring oxidising flames for analysis, lower sensitivities are frequently observed with the triple-slot burner, possibly due to dilution in its larger flame volume (673).

Steiner and co-workers (426) have utilized a modified nebulizer-burner system to study the reactions occurring during the passage of the gas/sample mixture from the nebulizer exit to the burner orifice. Thermometric, gas chromatographic, mass spectrometric and infra-red spectroscopic techniques were employed. The results show differences in temperature and pressure of the gas mixture, which cause change in its composition, at various locations in the system. They confirm that the present arrangement of the nebulizer-burner system employed makes it impossible to prevent or make adequate adjustments to stabilize these variables.

In a Popular Science article entitled 'Its Superspray' a method of nebulization using a 'Babington Nebulizer' has been described (164). In this device air is passed into a hollow sphere in whose surface there is a small hole or slot. A fine aerosol is produced when sample liquid is passed over the outer surface of the sphere. The air operating pressure is in the range 5 to 20 psi and an aerosol particle size of 2 to 50μl is claimed; larger droplets can be removed by placing a shroud or impact ball over the outlet of the nebulizer. Further data relating to the analytical performance of this system would be of interest.

Rubeska *et al* (11) have described a branched capillary system which allows buffer solutions ($LaCl_3$, KCl etc.) to be introduced simultaneously with the sample solution to a conventional nebulizer. The ratio of the solution flowrates is controlled by the length of the capillaries employed. Ward and Biechler (12) have utilized a similar device to mix a 2000 μg ml^{-1} Na solution with sample solutions in the determination of Ca in natural waters by AAS using a N_2O/C_2H_2 flame. Taylor (881) has proposed the use of a dual nebulizer to introduce Na solutions separately to overcome the ionization interference of the alkali metals on the determination of K in a premixed air/C_2H_2 flame.

Manning (553, 765) has described the advantages of nebulization of small volume samples (100μl) into flames *via* conventional nebulizer-burner systems. Repeatability was better than that of electrothermal systems and was of the order of 0·01 RSD. Samples of high dissolved solid content (e.g. serum) may be handled with little or no dilution without clogging of the burner slot. The technique has been demonstrated by the determination of Cd and Zn in serum and of 11 elements in NBS standard orchard leaves. Thompson and Godden (584) have similarly employed pulsed nebulization of 200μl sample solutions in conjunction with use of a high solids N_2O/C_2H_2 burner. Other workers (1332) have optimized the injection method described by Sebastiani *et. al* (*Z. Anal. Chem.*, 1973, **264**, 110) to operate with increased sensitivity using 40–100μl samples. A simple capillary device for reproducible introduction of microlite solution samples into a nebulizer/flame system has been described (1481).

Willis has made a detailed study of the feasibility of direct introduction of solids into flames as suspensions of finely ground powder (887, 1239). In this study it was shown that only particles below *ca.* 12μm in diameter contribute significantly to the observed absorption and that the atomization efficiency increases rapidly with decrease of particle size. The atomization efficiency, usually only 20 to 40% of that observed with a solution sample, has been found, for a given element, to vary only by a factor of *ca.* 2 between rocks and minerals of very different types. Where results correct to within a factor of this order are acceptable the technique avoids the need for time-consuming acid digestion procedures currently used to prepare rock, soil and sediment samples.

The use of ultrasonic nebulization techniques continues to attract some interest. In a series of papers, Isaaq and Morgenthaler have described the performance and application of a nebulizer and desolvation system consisting of a commercial ultrasonic nebulizer connected to a temperature controlled heater in series with a condenser and burner head (1229, 1230, 1238). At optimum conditions of temperature and flow rate the proposed system was found to have a sample efficiency of 85% and a desolvation efficiency of 72%. Sensitivities in AAS comparable to, or better than, those observed with other heated chamber-condenser systems were obtained for a range of elements. The system has good tolerance for solutions of high salt content (1 to 3%); no burner-slot clogging and minimal memory effects were observed. Denton and Gutzler (244) have continued work in FES with a pre-mixed O_2/H_2 flame and have described the use of an ultrasonic nebulizer with this flame. A filter flame emission photometer for the determination of P in air and natural waters *via* measurement of HPO emission at 526 nm in an air/H_2 flame has been reported (223); ultrasonic nebulization and an improved burner design in this system are claimed to facilitate a significant improvement in detection limit for P (0.003μg ml^{-1} H_3PO_4 in aqueous solution).

Other references of interest —

Automated sample preparation: 51.
Effect of viscosity of petroleum products during FES: 962.
Comparison of burners for Zn and Cd AFS: 179.
Patents: 75, 296, 970.
Burner liquid fuels: 1540.

2.5.4 Other Sample-introduction Devices

Aldous, Mitchell and co-workers have continued their studies utilizing a modified Delves microsampling cup technique, in which a molybdenum cup and a N_2O/C_2H_2 flame is employed (see *ARAAS*, 1974, **4**, 11). The use of this system has been shown (766, 1228) to minimize interferences observed in the determination of elements such as Ag, Cd, Cu, Mn, Pb and Zn; instrument response was independent of sample matrix for both peak absorbance and integrated absorbance measurements. The technique has been successfully applied directly to determination of these elements in a variety of matrices. Reports from the same laboratory (573, 761) have also described the use of a similar system for the determination of less volatile elements such as Co, Cr, Mn and Ni. The N_2O/C_2H_2 flame gives a high rate of cup heating and cup temperatures high enough (*ca.* 1800K) to volatilize the less volatile metal compounds.

Katskov, Kruglikova and L'Vov (415) have developed an AAS technique for solid samples in which the sample (50mg) is placed into a porous graphite capsule. The capsule is closed with graphite powder, placed horizontally into an air/C_2H_2 or N_2O/C_2H_2 flame, and heated electrically. The atomic vapour diffuses through the graphite and its absorption is measured above the capsule; particulate material does not diffuse and hence no scatter from this source is observed. Detection limits for a wide range of elements between 10 and 100 ppb have been reported for 40 mg samples. Prudnikov (414) has reported further studies with a graphite rod sampling system in which a T-shaped adaptor and air/C_2H_2 or N_2O/C_2H_2 flame is employed (see *ARAAS*, 1973, **3**, 12).

Thermal vaporization of powdered solids, using a d.c. arc between graphite electrodes, and transport of the vapour to premixed flames has been described (1307). Laser-beam vaporization of nickel-base alloys samples, corundum plates and solution residue on photographic plates was also employed similarly in conjunction with an air/C_2H_2 flame.

Skogerboe and co-workers (237) have described an interesting technique in which sample injection into flames and microwave plasmas is affected by conversion of analyte elements to their volatile chlorides which are easily dissociated thermally. 2 μl samples are placed in a cuvette within a furnace and dried at 110°; the furnace is then heated to 850° and HCl gas passed over the sample. The volatile chloride produced is then introduced into a flame or plasma and the transient absorption or emission signal measured. Detection limits between 0.05 and 3μg ml^{-1} level for Sn, Te, Zn, Bi, Cd, Ge, Mo and Pb have been obtained.

The work of Belcher, Townshend and co-workers with molecular emission cavity analysis (MECA) (see *ARAAS,* 1974, **4**, 11) has continued. Particular attention has been devoted to extension of the capability of the technique for the determination of sulphur (434). The interference from metal ions on the sulphur determination has been shown to be removed by the addition of excess phosphoric acid (921). The technique may be used to determine the composition of mixtures of inorganic sulphur compounds, such as sulphide, thiocyanate, sulphite, thiosulphate and sulphate, by the differing time required to achieve maximal S_2 emission at 384 nm for the different anionic species (872, 1364). Despite the availability of commercial equipment for this technique no other reports of its application have been brought to our attention.

Other references of interest—

Screw feed for powder sample: 1261.
MECA: 309, 363, 501, 953.
Determination of Bi by candoluminescence: 494.

2.6 ELECTROTHERMAL ATOMIZERS

There has been a further decline during 1975 in the number of references describing novel electrothermal atomization systems. Emphasis has been on improving the scope, reliability and precision of the technique using various instrumental and procedural modifications.

A considerable amount of work has been performed on the minimisation of interference effects observed in the analysis of samples; with some impressive results. The emergence for the first time, of a relatively large number of review papers (109, 220, 446, 565, 753, 792, 801, 975, 989, 1011, 1019, 1057, 1171, 1173, 1174, 1195, 1521) would indicate that electrothermal atomizataion has survived infancy and should now progress towards full and long-lasting maturity.

The Zn and Fe contamination of six brands of disposable pipette tips was studied (13) before and after thorough washing with 1% V/V HCl. One brand still exhibited appreciable Fe contamination even after washing. Unfortunately the brands were not named. Further work on automatic sample injection (803, 1322) has resulted in an RSD of 0·005 for most elements and is thus comparable with the precision obtainable for flame techniques. (See also *ARAAS,* 1972, **2**, Ref 385; 1974, **4**, Ref 1530).

Massmann and El Gohary (611) have stressed that the measurement of the non-specific background absorption of samples can cause problems when the background absorption is not caused by a dissociation continuum, but results from the complex fine structure of the electronic absorption spectrum of certain species (*e.g.* C, OH etc.), (*ARAAS,* 1973, **3**, 17). Other workers (248, 378, 1060) have plotted the absorption spectra of many common species (NaCl, KCl, KI etc.). The spectra were found to be similar to those observed in cool flames; they exhibited more than one peak and were markedly dependent on wavelength. The non-absorbing adjacent line method for correction of background absorption was not recommended. It would be most useful if all of this type of work on background absorption spectra could be tabulated in some form.

Various instrumental modifications to improve the precision of background correction using a D_2 or H_2 lamp have been described (74, 800, 826, 827, 971, 1010, 1221, 1367).

Further work (770, 799, 1040) has been reported on the use of the Zeeman effect to split resonance lines into pairs of non-absorbing lines in order to correct for background absorption in electrothermal atomization (*ARAAS*, 1973 **3**, 4; *ARAAS*, 1974, **4**, 18).

An ingenious method of overcoming or considerably minimising the problem of non-specific background absorption was to simply modify the sample matrix by the addition of a suitable reagent (749, 754, 762, 824, 829, 1066, 1468, 1547). For example, if excess NH_4NO_3 is added to a sea-water matrix then, during the dry-ashing phase, NH_4Cl is volatilized and the residual $NaNO_3$ results in a much lower background absorption signal than the original NaCl; the addition of excess Ni lessens volatility of As and Se, thus allowing the use of much higher dry-ashing temperatures; the addition of NH_4F, $(NH_4)_2$, SO_4 or $(NH_4)_2$ HPO_4 markedly decreases Cd volatility; the addition of 10% V/V $HClO_4$ prevents the formation and premature volatilisation of GeO.

Further work has been reported on the formation of pyrolitic graphite coatings by adding a small quantity of CH_4 to the Ar (or N_2) purge gas (83, 110, 121, 134, 177). CH_4 was preferred to C_2H_4, C_3H_8 or C_2H_2 (177). These pyrolitic coatings reduce the graphite porosity, improve the precision and sensitivity for a large number of elements and considerably improve the lifetime of the graphite element (up to 3–4 times). Considering that this simple technique was reported over two years ago (*ARAAS*, 1973, **3**, 16). it is surprising that some manufacturers have not yet modified their equipment or handbooks to allow the use of this technique. Soaking graphite tubes with a 5% m/V solution of Na_2WO_4 was found to improve the sensitivity for Si (702). This improvement was thought to be caused by preferential formation of tungsten carbide (*ARAAS*, 1973, **3**, 17).

Other references of interest—

Degree of atomization: 317.
Drift compensation integrator: 692.
Programmed heating: 222.
Sample decomposition techniques. 959.

2.6.1 Graphite Rod Devices

Torsi and Tessari (246, 247) have reported further theoretical and experimental work on the time resolved distribution of atoms in electrothermal atomization from a graphite surface (*ARAAS*, 1974, **4**, 13). A theoretical model of the release and transport of atoms from a graphite surface, under specified thermal perturbation, was developed and found to give good agreement with experimental results. West *et al* (682) have developed an equation showing the rate of evaporation of atoms from a graphite filament. Agreement for this discrepancy were postulated. Winefordner and Chuang (1382) have described a simple AF spectrophotometer consisting of a graphite rod atomizer and an Eimac Xe between the observed and calculated signal profiles was not always observed and reasons line noise) for Ag, Cd, Pb and Zn were shifted towards higher temperatures in the presence continuum source. The concentration detection limits for Ag, Cd, Co, Cr, Cu, Fe, Mg, Mn, Ni, Pb, Sn, Tl and Zn varied from 2 x 10^{-4} to 10^{-1} $\mu g\ ml^{-1}$. The various calibration graphs were linear up to concentrations of between 85 and 1000 times the detection limits.

Other references of interest—

Atomic fluorescence: 121.
Interelement effects: 1536.
Temperature measurement: 1294.

2.6.2 Graphite Miniature Furnaces

Posma *et al.* (1247) have considered the fundamental aspects of various noise sources associated with graphite miniature furnaces and it was concluded that the measuring system was shot noise limited. Further work (600) from the same laboratory has described a model for the formation of analytical signals which combined the volatilization model proposed by Torsi and Tessari (246, 247) and the basic ideas of L'Vov (*ARAAS*, 1971, **1**, Ref 360). The influence of concentration, heating rate, sheathing gas and the geometrical dimensions of the device were investigated. It was concluded that there was no correlation between the observed volatility and either the saturated vapour pressure of an element or the dissociation energy of the oxide, chloride or sulphate of the element. The rate determining step was thought to be influenced by interactions between the atoms of the element and the graphite surface.

Another study (815) on the "appearance" temperatures of a number of elements under various conditions has been reported. These 'appearance' temperatures (in this case defined as the temperature at which the peak height was twice the standard deviation of the base of certain anions (e.g. SO_4, PO_4), however these anions had no effect on the appearance temperatures of Cr, Cu, Ni and Sn.

Again the addition of H_2 to the inert gas (757, 816, see *ARAAS*, 1971 **1**, Ref 259), was found to improve the sensitivity and reduce the background absorption for a number of elements in a variety of matrices. Johnson and Skogerboe (85) have described a useful device for renewing the contact surfaces of the supporting electrodes on the Varian model 63 flameless atomizer.

Other references of interest—

Background correction: 429, 825, 971.
Peak height/area: 683.
Solid samples: 1164, 1393.

2.6.3 Graphite Tube Furnaces

A further study (395) on the development of a kinetic theory of atomization for graphite tube furnaces (*ARAAS*, 1974, **4**, 14) has been reported. Equations were derived for losses occurring during the dry-ashing stage. The advantages of operation under stopped gas-flow conditions and the relative merits of peak height versus peak area measurement were also considered. It was calculated that stopped gas-flow operation could improve peak height by a maximum factor of 2.7 for Cu and that peak area measurement is preferred to peak height measurement when atomization is slow.

The optimisation of the instrumental and working conditions for graphite tube furnaces is of prime importance (10, 582, 595, 748, 817) if optimum sensitivity and precision are to be obtained. Unfortunately the degree of improvement was not expressly specified in most cases.

A number of papers describing emission measurements for graphite tube furnaces have appeared. Such studies were first reported by King (*Astrophys J.*, 1905, **21**, 236). Massmann and Gucer (*Spectrochim. Acta.*, 1974, **29B**, 283) have observed the emission spectra of some rare-earth elements emitted along the optical axis of a graphite tube furnace. Ottaway and Shaw (497, 828) have similarly detected resonance line emission from Ag, Al, Ca, Cu, Cr, K, Li, Mg, Na and Ti. The quoted K an Na detection limits were 1.6×10^{-6} and 2.6×10^{-6} $\mu g\ ml^{-1}$ respectively. However the problem of blanks for these two determinands is a limiting factor. Winefordner *et al.* (1333) have also detected resonance line emission from Cr, K, Li and Sr using both single pulse and continuous nebulization (in conjunction with a desolvation chamber) into a graphite tube furnace. However it was concluded that the absorption technique was superior.

It was found (746, 1234) that much lower dry-ashing temperatures could be used for the analysis of biological samples if two rectangular holes (10 x 7 mm) were introduced at both ends of the graphite tube (each hole located 9 mm from the 2.5 mm diameter sample introduction hole). Adams *et al.* (756) have used a reduced size graphite tube furnace with a N_2 purged optical path and vacuum monochromator for the AA determination of I, P and S with some impressive results.

The advantages of constant-temperature electrothermal atomizers, where the sample is directly introduced into the heated graphite tube atomizer, have been stressed by various workers (31, 110, 160, 800, 929, 1121, 1195. See also Woodriff, R., Stone, R. W., and Held, A. M., *Appl. Spectrosc.*, 1968, **22**, 408). In one case (110) the sample was continuously nebulized into the heated graphite tube. It was claimed that with continuous nebulization of the sample the precision was improved severalfold, mainly because of a non-transient output signal. The power requirements are somewhat greater than a furnace operated in the conventional transient mode. Herrmann *et al.* (1464) have determined F by mixing the sample with SiO_2 and H_2SO_4. The liberated SiF_4 was passed into a graphite tube furnace and the Si absorption monitored.

Various methods of calibration for the direct analysis of a wide variety of solid samples have been tested (822, 823). Reference materials with a similar matrix to the sample, synthetic solid reference materials, graphite-based standards, solution standards and standard addition were all considered. Gries and Norval (184) have extended their earlier work (*ARAAS*, 1973, **3**, 17) on the preparation of solid calibration standards. These were prepared by direct ion implantation into a metal matrix or by incorporation of various substances into a urea matrix.

Layer by layer analysis of solids by bombardment of the sample with an ion beam followed by adsorption of the sputtered products on the inner surface of a graphite tube was found to be a feasible technique (1006). For a Cu alloy containing 0.01% Ag, a 9 nm layer was the minimum sputtered layer thickness required to detect the Ag.

Other references of interest —

Atomization mechanisms: 1001.
Collisional broadening: 133.
Interference studies: 813, 814, 1205, 1507.
Patents: 315, 458, 1210.
Peak height/area: 632, 684, 813, 818, 819, 908, 910.
Rapid heating rate: 1447.
Standard addition: 813, 964.
Thermal diffusion: 64.
Vapour-pressure measurement: 344.

2.6.4 Metal Filament Devices

Ohta and Suzuki (695) have described an inexpensive Mo microtube atomizer which required a low operating power (4V, 50A). Some results for the analysis of trace metals in rocks after solvent extraction were given. The device would appear to offer no advantage, other than cost, over similar graphite devices, (see *ARAAS*, 1974, **4**, 16).

A patent (70) has been granted for a device in which the atomized sample from the electrically-heated metal filament is directed into a silica tube. The device has a low thermal capacity and almost instantaneous response. For Cd, Mg and Zn AA detection limits of less than 50 fg were observed using sample volumes of up to 1 μl. A similar type of device (1242) has been evaluated for the analysis of 19 elements in a variety of matrices. For complex matrices (e.g. seawater) electrodeposition was recommended in order to overcome matrix effects (*ARAAS*, 1974, **4**, 16).

Interelement effects observed using a tantalum ribbon system have been evaluated (87) and were found to be very dependent on the operating conditions.

Other references of interest—

Tantalum ribbon: 44, 356.

2.7 OTHER EXCITATION AND ATOMIZING SYSTEMS

The analysis of metals and alloys by cathodic sputtering in conjunction with either AA or AF does not appear to have progressed very far despite many favourable claims for this technique (*ARAAS*, 1973, **3**, 19). Amos *et al* (1145), stressed that the main advantages of the technique lay in the examination of thin coatings and alloys, and for the determination of metals that either require complex dissolution procedures or are poorly atomized in conventional flames.

The only new developments in the cold-vapour mercury technique were improvements in the detection limit (1–20 pgHg) using a long path (600 mm) absorption tube (239) and use of the AF technique (310, 633, 923). The main advantages of the fluorescence technique are:—simpler instrumentation using single-beam operation and linear amplifiers (310, 633); reduction in the effect of background absorption; a linear response over a greater range of concentrations. Epstein, Rains and Barnes (1142) have made an interesting study of the factors influencing precision and accuracy of mercury determinations. Parameters such as sample preparation, stabilizing agents, analysis time, matrix interference and peak height versus peak area measurement were all considered

There have been few new developments in the hydride generation technique. By using a N_2/H_2 diffusion flame supported on a 10 mm i.d. Pyrex tube, and AFS, significant improvements in detection limits and linear ranges of the calibration graphs were observed for As, Sb, Se and Te (400, 586). As (III) could be selectively determined in the presence of As (V) by buffering the sample to a pH of between 3 and 5 prior to adding the $NaBH_4$ (878). Smith (399) has performed a comprehensive interelement study using the sodium borohydride reduction technique. He concluded that group I, II, IIIA and IVA elements do not interfere, Ag, Au, Co, Cu, Ni, Pd, Pt, Rh, Ru always interfere and that hydride-forming elements mutually interfere.

The detection of F compounds from a GLC column has been achieved (20) by mixing the GC effluent with a constant bleed of Na vapour at 800°C and monitoring the Na absorption in a heated silica tube. The presence of F compounds was indicated by a decrease in the Na absorption signal. An inexpensive combined GLC column and T-shaped heated silica absorption tube (830) have been used to detect Se compounds transpired by Se accumulator plants.

Other references of interest—

Hydride generation: 428, 691, 880, 919, 1217, 1271, 1274, 1508.

Mercury: 570, 911, 1317, 1319.

RF sputter/AA: 963, 967.

3 Optics

3.1 BEAM MANIPULATION

Several workers have considered the use of wave length-modulation techniques in emission spectrometry (*ARAAS*, 1973, **3**, 20; 1974, **4**, 18). O'Haver *et al.* (219) have written a detailed review of the use of derivative spectroscopy and have paid particular attention (407, 493) to the use of repetitive scanning techniques for the elimination of continuum background emission and simple line-overlap interferences in a N_2O/C_2H_2 flame; even when spectral lines were separated by as little as 0.02 nm, avoidance of interelement interferences was possible, although in practice a separation of 0.06 nm was preferred. L'Vov and co-workers (348) have also studied the use of wave-length modulation in emission spectrometry with a N_2O/C_2H_2 flame; they found the method to be especially suitable for working in reducing flames or in regions containing unresolved molecular bands. The design and operational characteristics of a square-wave optical scanning system (1113) have been discussed by Skogerboe and co-workers.

Interest in the use of transform techniques in analytical atomic spectrometry has been quite widespread (*ARAAS*, 1974, **4**, 21). Lebedev (1079) concluded, on a theoretical basis, that grille spectrometers may provide a gain in signal-to-noise ratio for detection with a photomultiplier when compared to conventional, non-multiplexed spectrometers. Dawson *et al.* (578) concluded that when the spectrum under consideration contained only a few intense lines, Hadamard transform techniques should give improved precision, but under other conditions, reduced precision of measurement should be expected. In other words, the technique should be advantageous in AA but deleterious for emission techniques. When the technique was applied to the determination of Pb (283.3 nm) and Mg (285.2 nm) by FAAS, the detection limits obtained were significantly worse than those obtained using a single fixed slit. A detailed theoretical treatment (1324) of the artefacts induced by using a Hadamard mask with transparent slits which are systematically narrower than the opaque sections of the mask has been published. Malmstadt and Martin (728) have described a non-dispersive flame AF spectrometer for multi-element analysis which utilises frequency multiplexing and demultiplexing by Fourier transform techniques. Several HCLs, each modulated at a different frequency, were used as excitation sources and a solar-blind photomultiplier monitors the optical signal. Applications of Fourier transform techniques to multi-element AFS (977) have also been considered by Fuller. Horlick and co-workers (730) concluded that although spectral measurements in the u.v.—visible region are normally limited by signal noise (rather than detector noise, where the advantages of Fourier transform are best realised) several important practical advantages should still exist. These are: the precision of the wave-number axis, the easy control of the resolution function, and the achievement of high resolution in a relatively compact system. In a review of techniques it was concluded (681) that more work and results are necessary to demonstrate the true overall capability of Fourier transform techniques for measurements in the u.v. —visible region.

Smythe and Doolan (424) have described a dual-channel, computer-controlled AF spectrometer which monitored fluorescence-plus-scatter in one channel, and scatter alone in the second channel and thus corrected for the contribution from scatter. A patent (1212) has been granted for a similar design.

Other references of interest—

Emission mirror adaptor for an AA spectrometer: 335.
Automatic frequency control: 1209.

3.2 WAVELENGTH SELECTION

3.2.1 Dispersive Systems

Developments in diffraction grating ruling in Australia (834) have been reviewed and a ruling engine with a piezo-electric inching mechanism described which rules gratings at up to 60 strokes/minute. Westwood and Lit (24) have shown that the resolving power of metallic linear gratings deposited upon a thin-film dielectric waveguide can be increased by separating sections of the grating by blank spaces. Up to 30% higher resolving power may be obtained from such a ruled section grating, even although it has a smaller total number of lines. Pouey (23) discussed the image-forming wavefront given by ruled concave diffraction gratings, using both geometrical optics and diffraction theory to examine their focussing properties.

The design of stigmatic monochromators based upon a simple rotation of holographic gratings has also been described by Pouey (25). Bartoe and Brueckner (524) have described the design of a stigmatic, coma-free, concave grating spectrograph which employs double dispersion by two classical concave gratings in tandem. Fassel and co-workers (833, 1101) outlined the problems associated with stray light in ultra-trace determinations by optical emission spectroscopy. Various forms of stray light, originating from grating defects (ghosts, near and far scatter) and from spectrometer design defects were considered, as well as methods for their reduction or elimination. Hodges and Belcher (1368) have described a method of focussing the zero-order spectrum from a grating in a direct-reading vacuum spectrometer on the entrance slit of a second monochromator as a method of extending the spectral range to 830 nm. A double monochromator (298) incorporating two plane gratings and concave mirrors for collimating and focussing has a lens in front of the exit slit which pivots about an axis parallel to the slit in association with the pivoting movement of the gratings and thereby improves resolution and intensity. The various types of quantitative multi-element analysis systems based upon the echelle spectrometer (651) have been critically reviewed. Curry *et al.* (1376) have described an automatic filter-positioning device for emission spectroscopy which avoided the need to stop and restart the arc.

3.2.2 Non-dispersive Systems

A detailed comparison of signal-to-noise ratios from dispersive and non-dispersive flame AF spectrometers has been reported by Vickers and co-workers (90). It was concluded that non-dispersive systems could not provide any improvement in signal-to-noise ratio and could result in a poorer ratio even for atomizers of quite moderate background emission, such as the separated air/C_2H_2 flame. The development and performance of a multi-element, non-dispersive AF spectrometer (99) has been described. Computer-controlled, pulsed HCLs were used as excitation sources, with a sheathed flame or non-flame atomizer, and computer-controlled synchronous integrators and signal processing.

Neville (16) critically reviewed the uses and limitations of passive interference filters A Hg-vapour filter based upon a selective specular reflection has been described by Senitzky (152). The filter bandwidth is centred about the Hg 253.65 nm line, and can be varied between 0.01 and 0.1 nm by changing the vapour pressure.

Other reference of interest —

Effects of spectrograph resolving power: 1041.
Modular apparatus for spectroscopy research: 1090.
Echelle monochromators: 1518.
Vacuum u.v. spectrometry: 593, 621.

4 Detector Systems

There has been a significant increase in interest in the potential application of vidicon, TV-type or diode-array detectors for multi-element analysis, resulting, however, in some duplication of effort. Outside this field, very little fundamental work of possible significance in analytical atomic spectrometry has been reported, a notable exception perhaps being the suggested use of photo-ionization detectors (832). Photo-ionization detectors depend upon ionization by far-u.v. radiation of a suitable gas or vapour contained in the detector cell; the ionization current is proportional to the intensity of the incident radiation and is only obtained when the wave-length of the radiation exceeds the ionization potential of the filler gas or vapour. The detector has been evaluated in a simple non-dispersive micro-wave plasma emission analyser. Tiffany (14) has reviewed the advantages of pyro-electric detectors for the soft X-ray–vacuum u.v. region. Walsh (1046) has suggested the use of a separated flame as a resonance detector. The advantage of this system is the relative ease with which the element, to which the detector is selective, may be changed.

4.1 SOLID STATE DETECTORS

Duncan (15) has described the operating characteristics of typical silicon photodiodes. Photovoltaic operation gave the best signal-to-noise ratio at up to 350kHz, but the photo-conductive mode of operation must be used at higher frequencies (up to 10 ns response time). Neiswander and Plews (1325) reported that a substantial reduction in the noise generated by a silicon photodiode and pre-amplifier may be effected by cooling the system to 200 K. A quantum efficiency at least three times better than that of a photomultiplier was claimed. The use of an anti-reflection coating (1068) to give enhancements of up to 60% in the u.v. response of photodiodes has been described. Drift problems were minimised by removal of surface impurities. Fry (1081) has reviewed the development and uses of photodiode arrays. Horlick and co-workers (89, 485, 654, 744, 1369) have extended their studies of photodiode arrays in multi-element spectroscopy (*ARAAS,* 1974, **4**, 21), both for AAS (654, 744, 1369) and for d.c. arc emission spectometry (89, 654).

4.2 VACUUM PHOTOTUBES

Pardue (486) and co-workers (471, 652, 1116) have studied the potential of vidicon spectrometers. These workers have discussed the design and performance of a vidicon derivative spectrometer (652, 1116). They have also used a vidicon spectrometer for simultaneous flame ES analysis for Na and K in serum (471), although this must be classified as a rather extravagant use. A more generally-useful computerised multi-element vidicon flame emission spectrometer (489, 653) has been described by other workers. This spectrometer was used for the multi-element analysis of clinical and geological samples using internal standardization in the multi-channel mode, and spectral stripping of interferences caused by flame and concomitant element molecular bands and overlapping lines. Marrs (1115) has discussed the use of vidicon tubes in spectrometry and the computer acquisition and treatment of data.

The slow response of vidicon camera tubes for multi-element analysis renders them unsuitable for the study of transient signals, such as those obtained with certain electro-thermal atomizers. Aldous (488) and co-workers (1128) have reported the results of a preliminary study of the use of an image-dissector multi-channel AA spectrometer. This non-storage image detector may be used to study transient signals. Danielsson *et al.* (1573) have described a stigmatic, coma-compensated echelle spectrometer incorporating an image dissector tube. A computerised television direct-reader echelle spectrometer has been developed by Wood and co-workers (1377). The instrument operates over the range 230–860 nm, and incorporates an image intensifier for u.v.-visible conversion. The visible light is interfaced with a TV camera. Schagen (1086) and Knapp (131) have discussed the principles of intensifiers, the latter with special reference to flame spectroscopy.

Computerised TV spectrometers have been described by Van der Piepen and Classe (591) and Wood, Dargis and Nash (734). It has been shown (1055) that by placing a transmission grating over the lens of a TV camera which is focussed on a source such as flame or discharge tube, simple spectra may be demonstrated for teaching purposes.

The nature and applicability of TV-type multi-channel detectors (669, 670) in spectroscopy has been critically reviewed by Talmi.

Other references of interest —

Systematic error in microphotometry: 1494.
Photographic spectrometry detection limits: 1496.

4.3 SIGNAL PROCESSING

The increasing availability of microprocessors may well lead to their more extensive use in commercial instrumentation. They can be used, for example (747, 1125, 1175) to integrate transient absorption signals, to select absorption maxima or to straighten working curves. Witmer (1076) has described a system for the automatic analysis of photographically-recorded emission spectra. A semi-classical theory of photon counting for quasi-monochromatic thermal (Gaussian) light has been applied by Bures (22) to the calculation of the standard deviation of the photo-electron distribution generated in a detector. For Gaussian light the signal-to-noise ratio equalled the square root of the number of photo-electron events, whereas from pseudo-thermal (partially-coherent) light, the signal-to-noise ratio was approximately equal to the square root of the number of degrees of freedom. Winefordner (487) has discussed the variation in signal-to-noise ratios which may be expected for the various types of multi-element atomic fluorescence spectrometric instrumentation, including linear and slew scanning, and multi-channel and multiplexed techniques.

The theoretical principles underlying the use of phase-sensitive detection to recover signals buried in noise and the response of narrow band amplifiers to various waveforms was considered by Blair and Sydenham (1083). Faulkner (1082) has published an article concerned with the optimisation of amplifier parameters in relation to those of the signal source, with particular reference to the frequency range d.c. to a few hundred kHz. The physical and mathematical properties of noise and the concepts of noise bandwidth and system bandwidth have been discussed by Usher (1085) and a general design procedure for the fabrication of measurement systems has been outlined. Although these papers were not written with direct reference to analytical spectrometry, the approach adopted is applicable in spectroscopy, and they are therefore of interest to those involved in synchronous amplifier design or evaluation.

5 Data Processing

5.1 EMISSION SPECTROSCOPY

Further computerized systems for the spectral analysis of photographic plates have been described (82, 283, 545, 1091, 1109, 1111; see also *ARAAS*, 1973, **3**, 28). In one system (1111) a scanning microphotometer controlled by a mini-computer measured transmissions at 5 μm intervals along each spectrum. The device scanned a single spectrum in 70 s, taking over 90,000 readings which were stored in a disc file which could then be searched for up to 500 spectral lines.

An integrated intensity method (218) has been devised to improve computerised semi-quantitative analysis of geological materials using the d.c. arc. The method was illustrated by the determination of Sr in a Ca matrix. A procedure has been described (235) which used time-sharing computer facilities for accurate photographic-emulsion calibration in quantitative work. The operational range was from just above gross fog to very high line-density values. A simple spectral-stripping flame emission vidicon spectrometer has been developed (130) which would simultaneously monitor a 20 nm spectral range. The system effectively overcame practically all the spectral interferences observed in FES.

Other references of interest—

Book: 453.
Calibration: 630, 717, 718, 719, 769, 1036, 1039, 1156.
Detection limits: 225.

5.2 ABSORPTION SPECTROSCOPY

The interfacing of AA instruments to programmable calculators has been reported by a number of workers (745, 1131, 1329). This type of system allows automatic manipulation of analytical results; presentation of mean, RSD and concentration for up to 20 analyses; computation of the analytical results in any chosen units; updating of calibration graphs and computation of peak area (or height) in transient atomization systems. A solid-state computer interface (126) has been used to update older double-beam instruments to give similar performance to the latest models.

The factors affecting precision and detection limits have been evaluated (138, 425) for a number of instrumental systems. Shot and flicker noise, readout noise, flame noise and sampling reproducibility were studied. It was concluded that the optimum absorbance range to perform AA measurements depends on the factor that limits the precision (*ARAAS*, 1974, **4**, 22) and is not necessarily the same as that in conventional solution spectrophotometry.

Other references of interest—

Automation: 451.
Calibration: 884.
Computer programs: 1172.
Electrothermal atomization: 68, 943.

6 Complete Instruments

Tables A, B and C contain a comprehensive summary of instruments for analytical atomic spectroscopy. We are indebted to instrument makers for their willing collaboration in the provision of details of their instruments. The tables present information available to us at January 1976. More detailed lists of instrument distributors providing national coverage are available in most countries such as the Instrument Supplement of *"Science"*, 1975, **190** (November), for America, or the surveys by Krugers published in *"Chem. Weekblad"*, 1975, **71** (1088, 1089) for the Netherlands.

Over the last two decades atomic spectroscopy has become an established tool in the analytical laboratory. During this growth phase the determination of one element at a time has met the needs of most analysts. The demand for multi-element analyses is increasing and obviously if these can be carried out simultaneously there is a saving of time and resources. Simultaneous multi-element analysis by arc and spark techniques is old-established practice in the spectroscopic laboratory but these methods are unsuitable for the average analytical laboratory. Multi-element AAS has not proved to be amenable to simple instrumentation owing to the difficulty of providing multi-element light sources and of optimising the instrument performance for each element. Notable reviews on the problems of simultaneous multi-element analysis have been published by Kirkbright (1190) and Winefordner *et al.* (1379). As a general solution, the r.f. plasma appears to be the most promising but the successful use of such equipment will require more specialist staff and laboratory facilities than those for flame-based methods.

An aspect of analytical atomic spectroscopy which is not often presented in the scientific literature is the teaching of the subject, therefore the appearance of papers such as that by Price (1191) is particularly welcome. The needs of persons attending an instrument manufacturers' training course, and the problems of designing a course to meet those needs, are considered. The importance of post-training communication by the user with other workers in the field is emphasised.

6.1 EMISSION INSTRUMENTS

Two manufacturers appear for the first time in this volume of *ARAAS*, Kontron of West Germany and Spectroscandia AB of Finland. The former market various direct reading instruments including one using a plasma source and the latter, the IDES universal Spectrometer with hollow-cathode discharge, d.c. arc and IC sources. MBLE have introduced the Model PV 8350 integrated spectrometer system and an updated range of instruments is available from RSV. Jarrel-Ash have introduced a 50-channel, ICP source spectrometer with computer control and Shimadzu-Seisakusho announced two new direct-readers.

A portable flame photometer for the testing of metals has been described (535). The testing head produces a low-voltage arc on the surface of the object and generates an aerosol which is drawn through a hollow anode into the flame photometer for excitation and analysis. The use of a flame photometer for metal testing is unusual as other instruments *e.g.* the 'Metascope' (*ARAAS*, 1972, **2**, 42) have used electrical excitation of the spectrum.

Computers are widely used in the control and data processing of more advanced spectrometric systems. A control system for high-resolution echelle grating spectrometer and d.c. argon plasma has been described (1110). Pre-programmed interchangeable cassettes facilitate the use of the system for the sequential or simultaneous determination of 2–20 elements.

Other references of interest —

Vacuum spectrometer: 1541.

6.2 ABSORPTION INSTRUMENTS

Many manufacturers have upgraded their instruments: Jarrel-Ash have introduced Dialatom III, a single-beam instrument and the 82–850 double-beam with computer-controlled parameters and Czerny-Turner monochromator (747). The Perkin-Elmer 460, a double-beam instrument with Littrow monochromator is micro-processor controlled (1178), their latest model 603 has a Czerny-Turner monochromator and incorporates auto-nitrous oxide switching and burner-head sensor and flame pressure sensor. Instrumentation Laboratory Model 351 incorporates push-button controls, auto-calibration and safety gas controls and the Model 451 is a double-beam dual-channel instrument with two Ebert monochromators. Hitachi and Shimadzu-Seisakushno have also introduced a new range of instruments.

To obtain optimum performance of an AAS instrument, time should be spent in determining the effects of changes in gas flows to the burner, the burner height and lamp current. A simplex algorithm has been applied (1383) to factorial optimization of the instrumental parameters in the determination of Ca. This method, however, is more elaborate than is appropriate in most analytical situations. Trassy and Robin (620) have reviewed the sources of noise in AAS and describe a dual-wavelength instrument in which absorption measurements were made along the axis of a 40 MHz inductively-coupled plasma generated by a specially designed torch. By correcting for correlated noise in the light source and plasma, this system gave detection limits two orders of magnitude lower for Al, Mg and Ti when used in the two-wavelength mode rather than in the normal (single-wavelength) mode.

The analysis of biological materials has frequently prompted workers to develop their own instrumentation. For the determination of Ca and Mg in small volumes (approx. 1 nl) by AAS, Danielson (321) has revived the platinum loop technique. An air/H_2 flame with adaptor was used to measure 4 x 10^{-12} moles Mg with an RSD of 0.05. Falchuk *et al.* (263) developed a multi-channel instrument for the determination of Zn, Cu and Cd. A multi-element HCL and alumina tube, as the absorption cell, were used in conjunction with a concave-grating spectrograph with photomultipliers placed behind exit slits positioned at appropriate points on the Rowland circle.

Recommendations of safety practices for AA spectrophotometers have been made (49). These suggestions relate to the gas systems, waste vessels, the use of organic solvents and eye protection.

Other references of interest—

AA spectrometer: 1535.

6.3 FLUORESCENCE INSTRUMENTS

At the present time there does not appear to be a commercial instrument available for fluorescence analysis. One or two companies offer attachments to AA instruments for selected elements *e.g.* Hg and those forming volatile halides. A single-element AF spectrometer has been described by Arnold and Smythe (851). It is a dual-channel instrument in which one channel observes the fluorescence radiation and the second, a nearby non-fluorescence line in order that correction may be made for scatter. The system was evaluated for Zn in a matrix of $Al_2(SO_4)_3$. It is claimed that the use of scattering correction lowers the limit of detection by a factor of 10. In the continuing search for simultaneous multi-element analysis instruments, two flame-based computer-controlled systems have been described. In one (727), HCLs are used to excite the fluorescence radiation while in the other (1237), a 500 W Xe arc is used. Detection limits for the latter are claimed to be comparable with those obtained by flame AA.

Table A COMMERCIALLY AVAILABLE EMISSION SPECTROMETERS

Supplier	Model	Type	No. of channels	Reciprocal dispersion/ nm per mm	Wavelength range/nm	Focal length	Type of source	Special features	Applications
Angstrom Inc., P.O. Box 248, Belleville, Mich. 48111, U.S.A.	22-101	D.R.	60	0·278 0·397 0·556	210–430 210–518 210–634	3·0 m 3·0 m 3·0 m	HV a.c. spark, HV or LV a.c. arc, LV undirectional d.c. arc, triggered capacitor discharge	Optical interlock spectrum monitor. Various excitation stands and read-out options	All ferrous and non-ferrous applications
	V-70	D.R.	50	0·556	160–440	1·5 m	As model 22-101	As model 22-101. All wavelengths in vacuum. In-focus wavelength scanning ± 3 nm from each receiver slit	Ferrous and non-ferrous metals, oils, soils and geological specimens
	V-71	D.R.	22	0·556	178–320	0·75 m	As model 22-101	Mobile and operable in hostile environments. All wavelengths in vacuum. Various excitation stands and read-out options	As model V-70
	A-70	D.R.	68	0·556	190–770	1·5 m	As model 22-101	As model V-70	As model V-70 but including determination of C, S and P
	A-71	D.R.	22	0·556	210–350	0·75 m	As model 22-101	As model V-70	As model A-70
Applied Research Laboratories Ltd., Wingate Road, Luton, Beds., England	Quanto-meter 20	D.R.	60 (20 lines)	1·388 or 0·695 0·695 or 0·35	200–800 200–400	0·75 m	Low voltage, high voltage	Air or argon excitation stands. Typewriter and digital computer options	Particularly suited to non-ferrous, e.g., Al, Mg, Cu, Zn and white metals, slags, powders, solutions, including oils
	Quanto-vac 28	D.R.	60 (28 lines)	0·70 or 0·35	175–500	0·5 m	As Quantovac 80	As Quantovac 80 but no air-conditioner	As Quantovac 80, but limited to 28 elements
	Quanto-vac 28C	D.R.	As Quanto-vac 28	0·70 or 0·35	175–500	0·5 m	As Quantovac 28	Complete computer control, teletype or visual display output. Off-line computer links	As Quantovac 28
	Quanto-vac 80	D.R.	96 (60 lines)	0·46	170–407	1·0 m	Various, low voltage, high voltage, multi-source (HVS, LV, d.c. arc)	Typewriter, teletype and digital computer options. Single or dual stand options. Second stand can be argon or air. Built-in instrument air-conditioning	All ferrous and non-ferrous alloys, powders including slags, sinters, ores, rocks, ceramics, soils, etc. Solutions, oils, etc.
(continued)	Quanto-meter 80	D.R.	As Quanto-vac 80	0·695 or 0·35	190–610	1·0 m	As Quantovac 80	As Quantovac 80	As Quantovac 80, but excluding determination of C, S and P

*New equipment since publication of Volume 4

Table A COMMERCIALLY AVAILABLE EMISSION SPECTROMETERS—*continued*

Supplier	Model	Type	No. of channels	Reciprocal dispersion/ nm per mm	Wavelength range/nm	Focal length	Type of source	Special features	Applications
(continued)	Quanto-meter 29000B	D.R.	60 (48 lines)	0·35 or 0·175 0·46 or 0·23 0·56 or 0.28 0·695 or 0·35	190–520 190–630 190–705 190–840	1·5 m	As Quantovac 80	Typewriter, teletype and digital computer options. Argon and/or air stands available	As Quantometer 80
	Quanto-meter 33000	D.R.	(64 lines) (8 reference)	0·695 or 0·35	190–610	1·0 m	As Quantovac 80	Automated sequential analysis. Computer options also available	As Quantometer 80
	Quanto-vac 33000	D.R.	As Quanto-meter 33000	0·46	170–407	1·0 m	As Quantovac 80	As Quantometer 33000. Computer options also available	As Quantovac 80
	Quanto-meter 33000CA	D.R.	As Quanto-meter 33000	0·695 or 0·35	190–610	1·0 m	H.F. plasma	Automatic loading of up to 24 samples	Solutions
	Q.A. 137	D.R.	48	0·46	185–410	1·0 m	R.F. plasma (Inductively coupled)	P.p.b. analysis. Computer options also available. Direct solids nebuliser can be fitted	Solutions of many materials ferrous/non-ferrous, slags, clinical and pollution control applications
Baird Atomic Inc., 125 Middlesex Turnpike, Bedford, Mass. 01730, U.S.A. East Street, Braintree, Essex, England Veenkade 26, The Hague, Netherlands	SB-1	Phot.	—	1·5 or 0·75	370–740	1·5 m	Arc or spark	Built-in order sorter	General spectrographic analysis
	SH-1	Phot.	—	1·0	450–750	1·5 m	Arc or spark	Built-in order sorter	General spectrographic analysis
	GW-1	Phot.	—	0·8 or 0·4	185–2400	2·0 m	Arc or spark, modular or RE-1	Dual gratings for simultaneous photography of two spectral regions	General spectrographic analysis
	GWR-1	Phot.	—	0·8 or 0·4	185–2400	2·0 m	As GW-1	High speed (f/15·5) gratings for rapid examination of transient and/or weak sources of complex spectra. Optional echelle grating for f/12 aperture	Transient, weak or complex sources. General spectrographic analysis
	Spectro-met 1000	D.R.	30	0·6 or 0·3	210–590	1·0 m	Arc or spark, modular	Compact, low-cost direct reader with minimum air-conditioning require-ments. Manual master monitor to check slit alignment	Ferrous metals (except determination of S) using C 193·1 nm, P 214·9 nm in 2nd order. Non-ferrous metals, oils

	Spectro-vac 1000	D.R.	30	0·6 or 0·3	173–767	1·0 m	Arc or spark, modular	Compact, low-cost direct reader with minimum air-conditioning require-ments. Logarithmic read-out. Manual master monitor to check slit alignment. Dual stand option	Ferrous and non-ferrous, including C, S and P
	Spectro-met II	D.R.	60	0·294 0·59	190–432 190–863	2·0 m	As Spectromet 1000	Automatic optical servo monitor continuously maintains correct slit alignment. Logarithmic read-out. Manual master monitor to check slit alignment. Temperature-compensated fixed focal length. Dual stands for argon and air available	All direct-reader applications above 190 nm
	Spectro-vac II	D.R.	60	0·29	173–432	2·0 m	As Spectromet 1000	As Spectromet II. All photomultipliers in vacuum	All direct-reader applications including C, P and S
Jarrell-Ash Div., Fisher Scientific Co., 590 Lincoln St., Waltham, Mass. 02154, U.S.A.	78-000	Phot.	—	1·1 or 0·54	210–780	1·5 m	Various available in 'Varisource' spark, low and unit including high-voltage d.c. arcs. Also versatile 'controlled wave-excitation source'. Plasma	Wadsworth spectrograph	General spectrographic analysis
	70-310	Phot.	—	1·0 to 0·24 depending upon grating	180–3000 180–1500 180–750	3·4 m		20 inch camera	General spectrographic analysis
Fisher Scientific Co., Jarrell-Ash Division Europe, Av de Lavaux 26, 1009 Lausanne-Pully, Switzerland	75-150	Phot.	—	4·4 to 1·1 3·2 to 0·8 1·6 to 0·4	200–6000	0·75m 1·0 m or 2·0 m		Choice of 3 gratings. Nitrogen purging extends range to 175 nm. Optional accessories permit use as direct reader or scanning spectrometer	Versatile instrument particularly suitable for measuring transient spectra
	90-750	D.R.	Up to 50	0·54	168–500	0·75 m	As above, except electronic con-trolled peak current	Computer controlled	Most metallurgical analyses
	90-785	D.R.	Up to 50	0·54	168–500	0·75 m			
Fisher Scientific Co., GMBH, Heilgerweg, 67/69 46 Dortmund W Germany	1500	D.R.	Up to 30	0·56 or 0·28 0·34 or 0·17	200–800 or 190–400 200–510 or 190–250	1·5 m	As above	Choice of 2 gratings	All direct-reader applications above 190 nm
Nippon Jarrel-Ash Co. Ltd., Kyoto, Japan	70-314	D.R.	30+	As 70-310	As 70-310	3·4 m	As above	Scanning optional. Easy interchange to photo-graphic (70-310) and scanning version (70-320)	All direct-reader applications above 190 nm
(continued)	90–975*	DR	Up to 50	0·5	168–500	0·75 m	ICAP	Computer controlled 1 Variable Channel	All solutions

*New equipment since publication of Volume 4

Table A COMMERCIALLY AVAILABLE EMISSION SPECTROMETERS—*continued*

Supplier	Model	Type	No. of channels	Reciprocal dispersion/ nm per mm	Wavelength range/nm	Focal length	Type of source	Special features	Applications
(continued)	84–405	Scan	—	3·3	200–900	0·25 m	None	Various scanning spectrometers	Suitable for spectroscopic investigations rather than for analytical applications
	82–410	Scan	—	1·6 and 3·3	200–900	0·25 m	Tungsten Deuterium		
	82–415	Scan	—	Depends on grating	Depends on grating selected	0·25 m	As above		
	82–000	Scan	—	As above	As above	0·5 m	Supplied by user		
	25–100	Scan	—	As above	As above	0·75 m, 1·0 m 2·0 m	As above		
	75–150	Scan	—	As above	As above	0·75 m, 1·0 m 2·0 m	As above		
Jobin-Yvon Division d'Instruments SA, 16–18 Rue du Canal, 91160 Longjumeau, France	VARAF	Scan.	—	1·8 or 0·9	200–800	0·465 m	Flame	Czerny-Turner monochromator, adjustable bandwidth 0·02–4 nm. Nine models with various burner and read-out options	Liquids and solutions
	DELTA	Scan.	—	1·2	195–770	0·6 m	Flame	Czerny-Turner monochromator, adjustable bandwidth 0·02–4 nm. Automatic wavelength scanning device. Pneumatic nebuliser, laminar flow burner. Ultrasonic nebuliser optional	Liquids and solutions
Labtest Equipment Co., 11828 La Grange Ave., Los Angeles, Calif. 90025, U.S.A.	310	D.R.	60	0·56	190–900	1·5 m	'Transource' high voltage triggered discharge. Low voltage triggered d.c. arc	Wavelength in first order. CRT. Teletype printer or computer readout systems, dual air/inert gas and solution excitation stand	Ferrous and non-ferrous alloys
	V-25	D.R.	40	0·67	170–550	1·0 m			As above
	2100	D.R.	30	0·46	188–455	1·0 m			As above
	71	D.R.	74	0·52	170–900	2·0 m			General purpose

Kontron, GmbH, 8051 Eching b, München, Oskar-von-Miller-Str. 1, West Germany	2100*	D.R.	28			1·0 m			
	V25*	D.R.	42			1·0 m			
	310*	D.R.	60			1·5 m			
	ICP* Plasmaspec 100		40				Plasma		Steel, slag, cement, rock, soil, noble metals, plating liquors, food, sewage, blood
M.B.L.E., Rue des Deux-Gares, 80 B–1070, Brussels, Belgium	PV 8300 Vacuum	D.R.	60 (80 lines)	0·55 or 0·46	170–430	1·5 m	Triggered capacitor discharge. 'Monoalterance' discharges up to 500 Hz. d.c. arc intermittent arc	Optional dual air/argon excitation stand. Readout by printer, teletype or digital computer systems	Steels, iron, non-ferrous metals and non-conductive powders Air stand for oils, d.c. arc, etc.
Pye Unicam Ltd., York Street, Cambridge, CB1 2PX, England	PV 8350* Vacuum	D.R.	20	0·46	177–410	1 m	As for PV 300	Integrated spectrometer system including source and read-out options as for PV 8300	Steels, iron, non-ferrous metals, non-conductive powders
	PV 8210 Air	D.R.	60 (50 lines)	0·55 or 0.28	190–700	1·5 m	As for PV 8300	Wavelength range covered in 1st crder. Remote-controlled roving detector. External excitation. Rotrode and inert atmosphere facilities Readout as PV 8300	All direct reader analyses above 190 nm particularly non-ferrous metals, solutions, oils and non-conductive powders
Optica, Via Gargano, 21 20139 Milano, Italy	B5	Phot.	—	0·69–0·36	200–800	1·2 m	All conventional types available	Stigmatic instrument with rotating Ebert grating	General purpose
	B5C	D.R.	16	0·69 or 0·36	220–420	1·2 m	LV triggered arc and spark. HV spark, a.c. and d.c. arc	Double spark stand both in air and inert atmostphere. Rotrode for solutions	General purpose. Metallurgical analysis, e.g., Al, Pb, Zn, Fe, Cu alloys. Wear metals in oils etc.
	B7V	D.R.	93	0·37	165–440	1·5 m	LV triggered arc and spark. HV spark	Air-vacuum instrument with all 92 exit slits accessjble from outside for adjustment. Many analytical programmes can be arranged in parallel for easy interchange. Computer facilities available.	Complex analyses involving many spectral lines

(continued)

*New equipment since publication of Volume 4

Table A COMMERCIALLY AVAILABLE EMISSION SPECTROMETERS — *continued*

Supplier	Model	Type	No. of channels	Reciprocal dispersion/ nm per mm	Wavelength range/nm	Focal length	Type of source	Special features	Applications
(continued)	ESA1	Scan.	—	0·41	200–500	1·0 m	Controlled and non-controlled HV spark, a.c. arc	Scanning monochromator with one channel for analytical line and another channel for reference using reflected beam principle	Metallurgical work. All material excitable with same source parameter
	ESA3	D.R. + Scan	9	0·36	160–500 (40 nm as poly-chromator)	1·2 m	LV triggered arc and spark	Combined vacuum mono- and polychromator. All excitable elements accessible with scanning system	Routine analysis (including C, S and P) of iron and steel. Non-ferrous alloys
	ESA4	Scan.	—	0·41	165–500	1·0 m	LV triggered arc. HV spark, a.c. arc	Scanning vacuum mono-chromator with one channel for analytical line and another channel for reference. Facilities for analysing two elements simultaneously	Metallurgical work. Analysis of ferrous and non-ferrous alloys
Rank Hilger Westwood Industrial Estate, Margate, Kent, CT9 4JL England	E1000 Polyvac	D.R.	60	0·293–1·155	159·6–864·3	1·5 m	Various, including high repetition condensed arc	Dual spark stands. Computer-controlled instrument. Dual gratings give 7 systems	Ferrous and non-ferrous alloys. Geological samples. Wear metals in oil
	E952	D.R.	36	0·546 or 0·741	174·0–447·7	0·75 m	As E1000	Curved entrance and exit slits. Solid state electronics or computer controlled. Air or vacuum	Ferrous and non-ferrous alloys. Wear metals in oils
	E742/3 Large Quartz/ Glass	Phot.	—	—	191–800	1·57 m	D.C. arc, HV spark condensed arc	Adjustable slit. Spectral length 0·67 m, of which 0·24 m can be selected for a given exposure	Analysis of high-purity specimens having complex spectra. Determinations of trace element concentrations
	E777/8	Phot.	—	0·26–0·97	200–1200	1·5 m	Various	Czerny-Turner mono-chromator	Routine qualitative and quantitative analysis. Flash photolysis. Examination of line profiles
	E549 Medium Quartz	D.R.	12	0·5–10	200–600	0·53 m	HV spark, con-densed arc, d.c. arc, thyratron-controlled a.c. arc	Solid state electronics. Quartz plate for atmospheric pressure compensation	Non-ferrous metals, soils, additives and wear metals in oils

RSV–Prazisions-mehgerate, GmbH., 8031 Hechendorf Pilsensee, West Germany	SPN 3.5	Phot. D.R.	30	0·14–0·48	200–1000	3·5 m	Glow discharge lamp, high, medium or low voltage spark, a.c. or d.c. arc, continuous and intermittent	Paschen-Runge mounting specially designed for range below 200 nm. Direct reading attachment available	General analysis
Siemens Ltd., Great West House, Great West Road, Brentford, Middlesex, England	SPN 2.0	Phot. D.R.	30	0·24–0·84	200–1000	2·0 m	As above	As above	As above
	SPN 1.5	Phot. D.R.	15	0·37–1·1	200–1000	1·5 m	As above	As above	As above
	SPN 1.0	Phot.	—	0·56–1·7	200–1000	1·0 m	As above	As above	As above
	SPN 1.0 (vac)	Phot.	—	0·4–1·7	300–1300	1·0 m	As above	As above	As above
	Analymat I-air	D.R.	40	0·31 or 0·54	200–650	1·5 m	Glow discharge lamp (others available)	Exhibits no background; no matrix effects. Linear calibration for all elements 0–100%	As above
	Analymat II-vac	D.R.	40	0·31 or 0·54	150–490	1·5 m	As above	As above	As above
	Analymat III-vac	D.R.	40	0·42 or 0·5	110–500	1·0 m	As above	As above	As above
	Analymat* IV	D.R.	250	0·22 2×2·5 m spectrum length	2X 200–600	2·0 m	As above	As above	As above
	Analymat* V	D.R.	250	As above	2X 150–600	2·0 m	As above	As above	As above
	Analymat* VI	D.R.	250	As above	2X 120–630	2·0 m	As above	As above	As above
	Analymeter* I		Scan	0·16	200–630	2·0 m	As above	As above	As above
	Analymeter* II		Scan	0·16	150–630	2·0 m	As above	As above	As above
	Analymeter* III		Scan	0·16	110–630	2·0 m	As above	As above	As above
Shimadzu Seisakusho Ltd., 1 Nishinokyo-Kuwabaracho, Nakagyo-ku, Kyoto, Japan	GCT-100 (Czerny & Turner)	Photo & D.R.	3 (max)	0·83 (1200 G/mm)	200–850 (10″ camera)	1·0 m	Modular-source DCA, ACA LVS, LVA	High speed	General purpose
	GE-170 (Ebert)	Photo	—	0·48 (1200 G/mm)	200–1200 (10″ camera)	1·7 m	HVS SDCA		General purpose
(continued)	GEW-170* (Ebert)	Photo & D.R.	55 (max)	0·48 (1200 G/mm)	200–1200 (20″ camera)	1·7 m			

*New equipment since publication of Volume 4

Table A COMMERCIALLY AVAILABLE EMISSION SPECTROMETERS—*continued*

Supplier	Model	Type	No. of channels	Reciprocal dispersion/ nm per mm	Wavelength range/nm	Focal length	Type of source	Special features	Applications
(continued)	GQM-75*	D.R.	35 (max)	0.52 (2400 G/mm)	190–430 & 510.5 589·0, 518·3	0·75 m	HVS, LVS DCA SG-400	3 kinds of read out electronics are available 1. Computer built-in 2. Digital with linearizer 3. Pen recorder	General purpose
	GVM-100	D.R.	60 (max)	0·46	170–410	1·0 m	HVS, LVS SG-400		Solid, liquid, powder Metal
Spectrametrics Inc., 204 Andover St., Andover, Mass. 01810, U.S.A.	AE 2	Phot. D.R.	1	0·06	190–900	0·75 m	Plasma jet	Optimised AE system using a high dispersion high energy throughput echelle spectrometer and a high temperature plasma jet excitation source	Routine analysis
	D.R.10	D.R.	20 (inter-changeable cassettes)	0·06	190–900		Plasma jet		Routine quantitative multi-element analysis
Techmation Ltd., 58 Edgware Way, Edgware, Middlesex HA8 8JP, England	ES 9	Phot.	—	0·06	190–900	0.75 m	Plasma jet, flame or arc stand	Built-in computer	
	RS 1	D.R.	1 (variable wavelength)	0·06	190–900	0.75 m	Plasma jet, flame or arc stand		Qualitative and semi-quantitative analysis. Spectroscopic research
	Spectra-span III*	Phot D.R.	20 Element inter-changeable cassettes	0·06	190–900	0.75 m	d.c. argon Plasma	Optimised AE system using high-dispersion high energy throughput echelle grating spectro-meter and a high-temperature plasma jet excitation source. Built-in micro-processor. Most spectral and matrix effects are eliminated.	Routine sequential. Quantitative analysis and multi-element analysis
Spex Industries Inc.,	1870	Scan.	—	1·6	175–1280	0·5 m	—	Multi-purpose unit	Routine analysis
3880 Park Ave., Metuchen,	1702	Phot.	—	1·1	175–1500	0·75 m	—	—	Research
N.J. 08840, U.S.A.	1704	Phot.	—	0·8	175–1500	1·0 m	—	—	Research
	1802	Phot.	—	0·8	180–1500	1·0 m	—	Direct reading accessory available	Routine analysis
Glen Creston, The Red House, 37 The Broadway, Stanmore, Middlesex HA7 4DL, England									

Spectroscandia AB, SF–21660 Nagu, Finland	IDES* 2080	D.R.	100 (300 lines)	0·16 at 200 0·32 at 400 0.52 at 650 0.63 at 800	200–800	0.5 m	Hollow cathode discharge, plasma, d.c. arc	Channels not preselected, changeable at anytime. Channel minimum spacing 0.2 nm. Wavelength accuracy 0·001 nm Plane samples 0·8 to 0·1 cm. CRT Lineprinter or Teletype readouts. Digital computer as standard	Ferrous and non-ferrous metals, slags, powders, ores, geological specimens, trace elements in metal, crganic materials and solutions. High accuracy in low and high concentrations
VEB Carl Zeiss Jena, 69 Jena, Carl-Zeiss Str. 1, German Democratic Republic	PGS-2	Phot.	—	0·74 or 0·37	200–2800	2·075 m	Arc or spark	Automatic expansion of measuring range. Stigmatic depiction. Dispersion doubled by double passage of light. Pre-disperser for order sorting and isolation. Gratings interchangeable. Automatic transport of plate holder	General spectrographic analysis. Also examination of line profiles, hyperfine structure etc.
Carl Zeiss Scientific Instruments Ltd., PO Box 43, 2 Elstree Way, Boreham Wood, Herts, WD6 1NH England	Q-24	Phot.	—	0·76	210–550	0·54 m	Arc or spark	Full range of accessories available	General spectrographic analysis

*New equipment since publication of Volume 4

Table B COMMERCIALLY AVAILABLE ATOMIC ABSORPTION SPECTROMETERS

Supplier	Model	Single/ double beam	Monochromator	Grating lines per mm	Reciprocal dispersion/ nm per mm	Resolution nm	Wavelength range/nm	Read-out; scale expansion	Other features
Beckman Instruments, 2500 Harbor Boulevard, Fullerton, Calif. 92634, U.S.A.	485	Double	Littrow	1200	2·7	0·2	190–860	Meter; × 50	Single and triple pass optics; automatic filter selection
	495	Double	Littrow	1200	2·7	0·2	190–860	Digital; × 100	As model 485
Beckman Instrument GmbH, 8 Munich 45, Frankfurter Ring 115, West Germany	1233	Double	Littrow	1200	2·7	0·2	190–860	Meter; × 55	Single and triple-pass optics; % T, abs. or concn. read-out
	1236	Double	Littrow	1200	2·7	0·2	190–860	Digital; × 55	As model 1233
Beckman RIIC Ltd., Eastfield Industrial Estate, Glenrothes, Fife, KY7 4NG, Scotland	1248	Double	Littrow	1200	2·7	0·2	190–860	Meter; × 10	Auto zero and calibrate integration
	1272	Double	Littrow	1200	2·7	0·2	190–860	Digital; × 10	As model 1248 plus curvature correction
Carl Zeiss, 7082 Oberkochen, Wurttemberg, West Germany	FMD 3	Single	Ebert	600	2·5	0·05	193–300	Digital	4-lamp turret, 2 stabilized power supplies. Curve correction, auto zero, optional auto calibrate and background correction
Corning Ltd., Halstead, Essex, CO9 2DX	EEL 140	Single	0·25 m modified Ebert-Fastie	1180	3·5			Non-linear meter	Single-lamp turret
	EEL 240	Single	As EEL 140	1180	3·5			Meter	4-lamp turret; integration
Diano Corporation, 75 Forbes Boulevard, Mansfield, Mass. 02048, U.S.A.	Multi-spec	Single	Double grating 0·25 m modified Czerny-Turner	1200	1·5	0·2	190–800	Meter; × 10	3-lamp turret with 3 stabilized power supplies; 4-way gas control; % T, abs. or concn. read-out
GCA/McPherson Instrument, 530 Main St., Acton, Mass. 01720, U.S.A.	EU 703	Single		1180	2·0	0·1	180–1100	Digital	Modular AA; flame emission; various detectors and grating available; convertible to single or double beam UV spectrometer
Hitachi Ltd., Nissei Sangyo Co. Ltd., 15–12 Nishi-Shimbashi, 2-Chome, Minato-Ku, Tokyo, Japan	170–10*	Single	Littrow	1440	2·25	0·4	190–900	Meter: ×0·1–×1 ×1–×10 Digital: (Optional)	Single, N_2O–air simultaneously exchanged Concentration read-out continuously variable time constant

	170–30*	Single	Littrow	1440	2·25	0·4	190–900	As 170–10	Concentration read-out, time weighted. Averaging Signal AA/AE measurement auto zero, N,O–Air simultaneously exchange
	170–50*	Double	Littrow	1440	2·25	0·1	190–900	As 170–10	Background correction. Base line drift correction. Curve corrector time weighted. Averaging signal, auto zero
Instrumentation Laboratory Inc., 113 Hartwell Av., Lexington, Mass. 02173, USA	351*	Double	0.33 m Ebert	1200	2·5	0·03	190–900	Digital: × 50	4 lamp turret, wavelength drive, full time integration peak height or peak area. Auto-calibration, curve correction, background correction, dual-grating. Push button operation–zoom lens. Full automatic safety gas controls.
Instrumentation Laboratory (UK) Ltd., Technical Services Div., Edgeley Rd. Trading Estate, Cheadle Heath, Stockport, SK3 0XE, England	251	Double	0·33 m Ebert	1200	2·5	0.03	190–900	Digital: × 50	4 lamp turret, wavelength drive, full time integration, peak height or peak area, off line calibration, curve correction, background correction, zoom lens, autogas
	151	Single	As 251						
Jarrel-Ash Division Fisher Scientific Co., 590 Lincoln St., Waltham, Mass. 02154 USA	Dial* Atom III	Single	0·25 m Czerny-Turner	1180	3·3	0.02	193–860	Digital	Laminar flow burner, integral gas flow controls auto-zero, concentration calibration, curvature correction; 2 lamp turret
Fisher Scientific Co., Jarrel-Ash Division Europe, Av-de Lavaux 26, 1009 Lausanne-Pully, Switzerland	82–810	Double dual channel	0·4 m Ebert	1180	2.08	0.03	190–900	Digital: × 25	Laminar flow burner, curvature correction. 2 lamp turret
	82–850	Double	0·4 m Czerny-Turner	1180	2.08	0.03	190–900	Digital	Computer-controlled parameters
Fisher Scientific Co. GmbH, Heilgerweg 67/69, 46 Dortmund, W Germany									

*New equipment since publication of Volume 4

Table B COMMERCIALLY AVAILABLE ATOMIC ABSORPTION SPECTROMETERS — *continued*

Supplier	Model	Single/double beam	Monochromator	Grating lines per mm	Reciprocal dispersion/nm per mm	Resolution nm	Wavelength range/nm	Read-out; scale expansion	Other features
Jobin-Yvon, Division d'Instruments, 16–18 Rue du Canal, 91160 Longjumeau, France	VARAF	Single	0·465 m Czerny-Turner	1220	1·8			Digital	
Optica, Via Gargano 21, 20139 Milano, Italy	6000	Single	0·35 m Ebert					Digital; × 50	Auto filter insertion, auto concn., integration. Flame temp. regulation. Pre-focussed water-cooled hollow-cathode lamps available
Perkin-Elmer Corp., Norwalk, Conn. 06856, U.S.A.	103	Single	0·27 m Littrow	1800	1·6	0·2	190–860	Meter; × 50	All mirror optics; integration; auto zero and flame ignition
Perkin-Elmer Ltd., Post Office Lane, Beaconsfield, Bucks. HP9 1QA, England	107	Single	0·27 m Littrow	1800	1·6	0·2	190–860	Digital; × 50	As Model 103
	360	Double	0·27 m Littrow	1800	1·6	0·2	190–860	Meter; × 50	All mirror optics; auto zero, gain control, flame ignition and optional nitrous oxide switching; integration; curve correction; peak reader
	370	Double	0·27 m Littrow	1800	1·6	0·2	190–860	Digital; × 50 and × 0·5	As model 360
	460*	Double	0·27 m Littrow	1800	1·6	0·2	190–860	Digital: × 0·01–× 100	All mirror optics, micro-processor controlled, auto zero, auto concentration, automatic gain control, auto curve correction with up to 3 standards, peak height, peak area, integration time selectable from 0.2 to 6·0 seconds, flame ignition, auto nitrous oxide switching, optional double beam background correction
	306	Double	0·4 m Czerny-Turner	u.v. 2880 vis. 1440	0·65 1·3	0·03	180–440 400–900	Digital; × 0·1–× 100	All-mirror optics; auto zero and flame ignition; peak reader; integration; curve correction; auto concn.

	503	Double	0·4 m Czerny-Turner	u.v. 2880 vis. 1440	0·65 1·3	0·03	180–440 400–900	Digital; × 0·1–100	As Model 306 plus signal averaging; auto nitrous oxide switching; flame, pressure and burner head sensors
	603*	Double	0·4 m Czerny-Turner	u.v. 2880 vis. 1440	0·65 1·3	0·03	180–440 400–900	Digital: 0·01–× 100	All mirror optic micro-processor controlled, auto zero, peak height peak area, integration times from 0·2 to 6·0 seconds, auto concentration, auto curve correction with up to 3 standard flame ignition, auto nitrous oxide switching burner head sensor, flame and pressure sensor
Bodenseewerk Perkin-Elmer & Co. GmbH, Postfach 1120, D–7770 Uberlingen, West Germany	400S	Double	0·33 m Czerny-Turner	1800	1·3	0·2	190–860	Meter; × 50 and × 0·2	Auto zero and flame ignition; integration
	400	Double	0·33 m Czerny-Turner	1800	1·3	0·2	190–860	Digital; × 50 and × 0·2	As Model 400S plus auto concn.; curve correction
Pye-Unicam Ltd., York Street, Cambridge, CB1 2PX, England	SP 191	Single	Ebert	1200	3·3	0·2	190–850	Digital; × 25 and × 0·1	4-lamp turret; auto zero and ignition; integration; curve correction
	SP 190	Single	Ebert	1200	3·3	0·2	190–850	Digital; × 25 and × 0·1	As Model 191 without emission capability
	SP 1950	Double	Ebert	1800	2.2	0·1	190–850	Digital; × 20 and × 0·1	Auto zero and ignition; integration; curve correction
	SP 1900	Double	Ebert	1800	2.2	0·1	190–850	Digital; × 20 and × 0·1	As Model 1950 plus 6-lamp turret
Rank-Hilger, Westwood Industrial Estate, Ramsgate Road, Margate, Kent, CT9 4JL, England	ATOM-SPEK H 1170	Single	Silica prism	—	1·7 at 200 nm 44·6 at 500 nm			Meter	6-lamp turret
	ATOM-SPEK H 1550	Single	Czerny-Turner	1200	2·6	0·1	190–850	Digital	6-lamp turret; auto zero and flame ignition; curve correction; integration, background correction optional

*New equipment since publication of Volume 4

Table B COMMERCIALLY AVAILABLE ATOMIC ABSORPTION SPECTROMETERS — *continued*

Supplier	Model	Single/double beam	Monochromator	Grating lines per mm	Reciprocal dispersion/nm per mm	Resolution nm	Wavelength range/nm	Read-out; scale expansion	Other features
Seiko Instruments, Tokyo, Japan	SAS 721	Single							
	SAS 740	Double; dual channel							Microcomputer and line printer
Shandon Southern Instruments Ltd., Frimley Road, Camberley, Surrey, GU16 5ET, England	A3400	Single	0·25 m Czerny-Turner	632	6·0	0·2	190–860	Meter; × 25	4-lamp turret; auto zero and flame ignition; curve correction; integration wavelength scan
Shandon Labortechnik GmbH, Frankfurt/Main 50, West Germany	A3600	Single	0·25 m Czerny-Turner	632	6·0	0·2	190–860	Meter; × 25	Integration
Shimadzu-Seisakusho Ltd., 1 Nishinokyo-Kuwabaracho, Nakagyo-ku, Kyoto 604, Japan	AA-610S	Single	Czerny-Turner	1500	1·9		190–900	Meter: × 10	Wavelength drive, two lamp holders
	AA-620*	Single	Czerny-Turner	1500	1·9		190–900	Meter: × 10	Wavelength, drive, two lamp holders, auto ignition, flame monitor, gas pressure monitor
	AA-650*	Double	Czerny-Turner	1200	1·9		190–900	Digital: × 180	Wavelength drive, two lamp holders, auto zero integration, curvature correction, BG correction, peak detector, auto ignition, flame monitor, gas pressure monitor
Varian Techtron Pty., 679 Springvale Road, Mulgrave, Vic. 3170, Australia	1100/1200	Single	0·25 m Czerny-Turner	1276	2·8	0·2	185–900	Meter/Digital; × 0·3–× 50	4-lamp turret, auto zero, integration, curve correction, peak reader. f/8 aperture, optional automatic gas-box
Varian Associates Ltd., Russell House, Molesey Road, Walton on Thames, Surrey, England	1150/1250	Single; dual channel	0·25 m Czerny-Turner	1276	2·8	0·2	185–900	Meter/Digital; × 0·3–× 50	As Models 1100/1200 plus simultaneous background correction, optional automatic gas box on Model 1250

Varian Instrument Div., 611 Hansen Way, Palo Alto, Calif. 94303, U.S.A.	AA6	Single; dual channel	0·51 m Ebert	638	3·3	0·05	185–1000	Meter/ digital; × 0·3–× 50	Modular construction 4-lamp turret, auto curve correction, integration, peak reader, f/10 aperture, optional automatic gas-box and simultaneous background correction
VEB Carl Zeiss Jena, 69 Jena, Carl-Zeiss-Str. 1, German Democratic Republic C Z Instruments Ltd. 2 Elstree Way, Boreham Wood, Herts WD6 1NH, England	AAS 1	Single	0·5 Ebert	1300	1·5	Continuously adjustable	190–820	Meter; × 10	4-lamp turret, auto zero, single or triple pass optics, continuously adjustable slit

*New equipment since publication of Volume 4

Table C COMMERCIALLY AVAILABLE ELECTROTHERMAL ATOMIZERS

Supplier	Model	Type	Max. sample volume/μl	Control unit	Sensitivity for 1% abs. (S) Detection limit (d.l.)		Special features
					Cu	Si	
Barnes Engineering Co., 30 Commerce Road, Stamford, Conn. 06902, U.S.A.	Glomax	Tantalum strip	50	Programmable; dry, ash, atomize, burn off. Max. temp. 2400°C	d.l. 10^{-11} g (50 μl)	N.S.	Fits most AA spectrometers. Air-cooled; inert-gas and hydrogen shielding
Beckman Instruments GmbH, 8 Munich 45, Frankfurter Ring 115, West Germany	1217*	Graphite furnace	100	Programmable; dry, ash, atomize, burn off. Max. temp. 3100°C	d.l. 4×10^{-12} g (100 μl)	d.l. 10^{-11} g (100 μl)	Water-cooled; inert-gas shielding. Safety feature for failure of water or purge gas, gas stop. Fits Beckman and Pye-Unicam instruments
Instrumentation Laboratory Inc., 113 Hartwell Avenue, Lexington, Mass. 02173, U.S.A. Instrumentation Laboratory (UK) Ltd., Technical Services Div., Edgeley Rd Trading Estate, Cheadle Heath, Stockport, SK3 0XE, England	455	Graphite and tungsten furnace	100	Programmable; six stages, ramp or step Max. temp. 3500°C	S. $3{\cdot}4 \times 10^{-12}$ g d.l. $0{\cdot}9 \times 10^{-12}$ g	S. 10^{-11} g d.l. 10^{-11} g	Fits all AA spectrometers, water-cooled, safety interlock system, automatic cell door, automatic cleaning pressurisation. Solid sampling, temperature. Controlled heating via tungsten sensor. True temperature readout
Jarrel-Ash Div., Fisher, Scientific Co., 590 Lincoln Street, Waltham,	MTA-2	Tantalum strip	50	Programmable; dry, ash, atomize. Max. temp. 2400°C	d.l. 2×10^{-12} g (50 μl)	N.S.	Fits most AA spectrometers. Inert-gas and hydrogen shielding
Mass. 02154, U.S.A.	FLA 10	Graphite furnace	50	Programmable; dry, ash, atomize. Max. temp. 2800°C	d.l. 2×10^{-11} g (5 μl)	N.S.	Fits most AA spectrometers. Inert-gas shielding but an air ash possible
S. & J. Juniper & Co., 7 Potter Street Harlow, Essex, England Spectronic Services, E & J Brereton, 4 White Rose Way, Garforth, Leeds LS25 2EF England	110	Graphite furnace	50	Programmable; dry, ash, atomize, burn out. Max. temp. 3500°C	S. 3×10^{-11} g (10 μl)	N.S.	Water-cooled; inert-gas shielding. All programme stages cover full temperature range
Optica S.A.S., Via Gargano 21, 20139 Milan, Italy	CAT 6	Tantalum strip	50	Programmable; dry, ash, atomize	d.l. 10^{-11} g (50 μl)	N.S.	Water-cooled; inert-gas shielding

Bodensee Perkin-Elmer & Co. GmbH, Postfach 1120, 7770 Uberlingen, West Germany	HGA 74	Graphite furnace	100	Programmable; dry, ash (2), atomize. Max. temp. 2700°C	d.l. 2×10^{-12} g (100 μl)	d.l. 5×10^{-11} g (100 μl)	Fits Perkin-Elmer and Zeiss AA spectrometers. Water-cooled; inert-gas shielding. Permits ramp ashing, gas stop operation. Closed system. Safety feature for failure of water or purge gas
Perkin-Elmer Corp., Norwalk, Conn. 06856 U.S.A.	HGA 2100	Graphite furnace	100	Programmable; dry, ash, atomize. Max. temp. 2800°C Ramp accessory provides linear-type ramp temperature increase in all 3 cycles plus auto high temperature at end of programme	d.l. 2×10^{-12} g (100 μl)	d.l. 5×10^{-11} g (100 μl)	Water-cooled, inert-gas shielding. Gas-stop and reduced gas flow operation, and direct temp. calibration. Safety feature for failure of water or purge gas
Pye Unicam Ltd., York Street, Cambridge CB1 2PX, England	SP9-01*	Graphite furnace	50	Programmable; dry, ash, atomize, tube clean, tube blank, with cancel and delay stages. Max. temp. 3000°C	S. $4{\cdot}4 \times 10^{-11}$ g	N.S.	Water-cooled, inert-gas shielding. Safety feature for failure of water. Tube life indicator and remote recorder control for 1, 2, 3 or all phases
Rank Hilger, Westwood Industrial Estate, Ramsgate Road, Margate, Kent, CT9 4JL, England	H1975/ FA256	Graphite furnace	100	Programmable; dry, ash, wait, atomize. Max. temp. 2600°C	S. 5×10^{-11} g	N.S.	Water-cooled, inert-gas shielding. Background correction when fitted to H 1550 Atomspek
Shandon Southern Instruments Ltd., Frimley Road, Camberley, Surrey, GU16 5ET, England	A3470	Graphite rod	25	Programmable; dry, ash (2), atomize. Max. temp. 3000°C	d.l. 5×10^{-12} g (5 μl)	d.l. 6×10^{-11} g (5 μl)	Fits selected AA spectrometers. Air-cooled; inert-gas shielding. Pyrolytic graphite coating for rods available in situ. Flameless accessory can be left in instrument during flame measurements
Shimadzu-Seisakusho Ltd., 1 Nishinokyo-Kuwabaracho, Nakagyo-ku, Kyoto 604, Japan	GFA-2*	Graphite furnace	50	Programmable: current stabilised dry, ash, atomize. Max. temp. 3000°C	d.l. 5×10^{-12} g		Current stablised to obtain highly reproducible results. Graphite tube replacement is very easily carried out
Varian Techtron Pty. Ltd., 679 Springvale Road, Mulgrove, Vic. 3171, Australia	CRA 90*	Graphite furnace (graphite tube), Threaded graphite furnace, graphite cup	25	Programmable; dry, ash, atomize. Max. temp. 3000°C	4×10^{-12} g (25 μl)	8×10^{-11} g (25 μl)	Fits most AA spectrometers. Water-cooled; inert-gas shielding and hydrogen flame option. Automatic ramp-hold atomization. Pyrolytic graphite coating on cups and tubes

*New equipment since publication of volume 4

7 Ancillary Topics

7.1 THEORETICAL STUDIES

The question of the detection limit and its interpretation continues to stimulate publications. Alkemade (882) has drawn attention to the assumption that the background fluctuations of the paired sample and blank readings are statistically uncorrelated; in fact, correlation may arise due to low-frequency noise components in the background emission or in the measuring instrument. The effect on these components is minimised by reducing the time interval between the measurement of the sample and the blank, and by repeating the measurement several times. This paper is a theoretical treatment of the well-established practical approach of "sample flow modulation" used in emission and absorption flame photometry. The origin and effects of noise in flame AAS have been examined in two papers (755, 883). They confirm that at low absorbances, lamp and flame noise are dominant, while at high absorbances, atomization noise dominates.

Omenetto, Winefordner and Alkemade (1458) have derived an expression for the effect of self-absorption on fluorescence and thermal emission intensities which takes into account stimulated emission. This theoretical treatment considers the simple idealised case of a two-level atomic system in a flame homogeneous with respect to temperature and composition, and uniformly illuminated by an external quasi-continuum source. Non-dispersive flame AFS has been considered in two papers (90, 852). It is concluded that such systems generally do not show any improvement in signal to noise ratio or detection limits in comparison with dispersive systems, except where optimum energy throughput lies beyond the range of adjustment of the dispersive system, or when the background signal is relatively small. In practical analysis, the problem of background scatter is best overcome by the use of an auxiliary line source to provide a signal due only to scatter. This signal can then be used to correct the measured signal.

7.2 STANDARD REFERENCE MATERIALS

Tables D, E, F and G contain updated information on suppliers of spectrographic standards, spectrographic graphite electrodes, standard metal solutions and reagents for AAS and organo-metallic compounds. Miller (*Spex Speaker,* 1975, **20** (2) 1) has comprehensively reviewed the preparation and characterisation of high-purity inorganics and notes that as the "term 'high-purity' remains very relative, it is far better to have more information than required that not enough". A modification of the statistical method of Dean and Dickson for the testing for normal distributions and for cancelling the out-liers from the set, has been described (1304) and applied to the estimation of the homogeneity of spectrometric standard samples. The problem of the stability of organometallic compounds used as standards in spectrochemical analysis has been examined (1301). It was found that the acetylacetone complexes and the metallic salts of dithiocarbamic acids are acceptable but accuracy can be maintained only if the standard solutions are freshly prepared from solid compounds just before the analysis. Pinta has reported (1267) on inter-laboratory comparison analysis of plant leaves. The elements measured included: N, P, K, Ca, Mg, Fe, Cu, Mn, Zn, S, Cl and Na. For determination of trace impurities in metals by electrothermal AAS, new solid standards have been prepared by ion implantation and dissolution in molten urea (184). The standards were Ti in Al foil and Ti_2SO_4 in urea.

7.3 DOCUMENTATION

For workers in the field of AAS, Perkin-Elmer continue to publish their valuable list of references (8,549). Two new journals have appeared during 1975 which promise to make useful contributions to the field of analytical atomic spectroscopy. First is the specialist journal "ICP Information Newsletter" which is devoted to inductively-coupled plasma-optical analytical spectroscopy. The "Newsletter" contains articles on all aspects of the subject, including techniques, instrumentation and abstracts. The second journal is "European Spectroscopy News" of which Volume 1, No. 1 was published by Heyden in July 1975. The "News", appears bimonthly, incorporates the "British Bulletin of Spectroscopy," and presents news items from all European Spectroscopy Groups on both atomic and molecular spectroscopy. Articles of scientific interest, both reviews and instrument news contribute to the interest and value of this new publication.

Table D. SPECTROGRAPHIC STANDARDS (see following page)

Supplier	Irons	Steels	Al-base	Co-base	Cu-base	Mg-base	Ni-base	Pb-base	Pt and Pd in Cu-base	Solders	Sn-base	Ti-base	Zn-base	Refractories	Binary Alloys
Aluminium Company of America, Alcoa Laboratories, Alcoa Center, Pennsylvania 15069, USA			X												
Apex Smelting Co., 6700 Grant Avenue, Cleveland, Ohio 44105, USA			X			X							X		
BNF Metals Technology Centre, Grove Laboratories, Denchworth Road, Wantage, Berks OX12 9BJ, England					X										
Bundesanstalt fur Materialprufung (BAM), 1 Berlin 45, Unter den Eichen 87, Germany		X	X		X			X			X		X		
Bureau of Analysed Samples Ltd., Newham Hall, Newby, Middlesbrough, Cleveland TS8 9EA, England	X	X	X		X		X							X	
CKD Research Institute, Na Harfe 7, 190 02 Praha, Czechoslovakia	X	X			X										
Comité de liaison des Industries de metaux non-ferraux de la Communauté Européenne, Boulevard de Berlaimont, 1000 Brussels, Belgium									X						
G. L. Willan Ltd., Sheffield Works, Catcliffe, Rotherham, South Yorkshire, England	X	X		X			X								
Johnson Matthey Chemicals Ltd., 74 Hatton Garden, London EC1P 1AE, England		'Spectromel' powders							'Specpure' metals						
Moore Boundy Hamill Ltd. Station House, Potters Bar, Herts EN6 1AL, England	X	X	X	X	X	X	X	X		X	X	X	X	X	X
Office of Standard Reference Materials, National Bureau of Standards, Washington DC 20234, USA	X	X	X	X	X		X				X		X		
	Various other metals including high-purity metals														
Pechiney, 23 Rue Balzac, Paris 8e, France			X												
Spex Industries Inc., PO Box 798, Metuchen, NJ 08840, USA (Glen Creston, The Red House, 37 The Broadway, Stanmore, Middlesex, England)	X	X			X			X			X			X	
Zinc & Alliages, 34 Rue Collange, 92307 Lavallois-Perret, France													X		

Table E. SPECTROGRAPHIC GRAPHITE ELECTRODES

1. Baird-Atomic, Inc, 125 Middlesex Turnpike, Bedford, Mass 01730, USA
2. Carbon Products Division, Union Carbide Corp, 270 Park Avenue, New York, NY 10017, USA (ARL Ltd, Wingate Road, Luton, Beds, England)
3. General Graphites Inc, First and Monroe Street, Bay City, Mich 48706, USA
4. Johnson-Matthey Chemicals Ltd, 74 Hatton Garden, London EC1P 1AE, England
5. Le Carbone (GB) Ltd, Portslade, Sussex, England
6. Le Carbone Lorraine, 45 Rue des Acacias, 75821, Paris, France
7. Met-Bay Inc, 900 Harrison Street, Bay City, Mich 48706, USA
8. Poco Graphite, Inc, PO Box 2121, Decatur, Texas 76234, USA
9. Ringsdorff-Werke GmbH, 53 Bonn-Bad Godesberg, West Germany (Mining & Chemical Products Ltd, Alperton, Wembley, Middlesex HA0 4PE, England)
10. Spex Industries, Inc, 3880 Park Avenue, Metuchen, NJ 08840, USA (Glen Creston, The Red House, 37 The Broadway, Stanmore, Middlesex, England
11. Ultra Carbon Corp, PO Box 747, Bay City, Mich 48706, USA (Heyden & Son Ltd, Spectrum House, Alderton Crescent, London NW4, England)

Table F. STANDARD METAL SOLUTIONS (MS) AND REAGENTS FOR AAS (R)

1. Aldrich Chemical Co., Inc, 940 W. St. Paul Avenue, Milwaukee, Wis. 53233, USA (R)
2. J. T. Baker Chemical Co., 222 Red School Lane, Phillipsburg, N.J. 08865, USA (MS, R)
3. Barnes Engineering Co., 30 Commerce Road, Stamford, Conn. 06902, USA (MS)
4. B.D.H. Chemicals Ltd., Poole, Dorset BH12 4NN, England (MS, R)
5. Bio-Rad Laboratories, 32nd and Griffin Avenues, Richmond, Calif. 94804, USA (MS)
6. Carlo Erba, Divisione Chimica Industriale, Via C. Imbonati 24, 20159 Milano, Italy (MS)
7. Eastman Organic Chemicals, Eastman Kodak Co., 343 State Street, Rochester, NY 14650, USA (R)
8. Fisons Scientific Apparatus Ltd., Bishop Meadow Road, Loughborough, Leics. LE11 0RG, England (MS, R)
9. Harleco, Div. of American Hospital Supply Corp., 60th and Woodland Avenues, Philadelphia, Pa. 19143, USA (MS)
10. Hopkin & Williams Ltd., PO Box 1, Romford, Essex RM1 1HA, England (MS, R)
11. V. A. Howe & Co. Ltd., 88 Peterborough Road, London SW6, England (MS)
12. Instrumentation Laboratory Inc., 113 Hartwell Avenue, Lexington, Mass. 02173, USA (MS)
13. Johnson Matthey Chemicals Ltd., 74 Hatton Garden, London EC1P 1AE, England (R)
14. Koch-Light Laboratories Ltd., Colnbrook, Bucks., England (R) (Anderman & Co. Ltd., Battlebudge House, 87–95 Tooley Street, London SE1, England)
15. May & Baker Ltd., Dagenham, Essex RM10 7XS, England (R)
16. E. Merck, D 61 Darmstadt, West Germany (R)
17. Spex Industries Inc., 3880 Park Avenue, Metuchen, N.J. 08480, USA (MS)
18. Ventron Corp., Alfa Products, 44 Congress Street, Beverly, Mass. 01915, USA (MS)

Table G. ORGANOMETALLIC STANDARDS

1. Angstrom Inc., P.O. Box 248, Belleville, Mich. 48111, USA
2. Baird-Atomic Inc., 125 Middlesex Turnpike, Bedford, Mass. 01730, USA
3. J. T. Baker Chemical Co., 222 Red School Lane, Phillipsburg, N.J. 08865, USA
4. B.D.H. Chemicals Ltd., Poole, Dorset BH12 4NN, England
5. Messrs Burt and Harvey Ltd., Brettenham House, Lancaster Place, Strand, London WC2, England
6. Carlo Erba, Divisione Chemica Industriale, Via V. Imbonati 24, 20159 Milano, Italy
7. Conostan Div., Continental Oil Co., P.O. Drawer 1267, Ponca City, Okla. 74601, USA
8. Durham Raw Materials Ltd., 1–4 Great Tower Street, London EC3R 5AB, England
9. Eastman Organic Chemicals, Eastman Kodak Co., 343 State Street, Rochester, N.Y. 14650, USA
10. Hopkin and Williams Ltd., P.O. Box 1, Romford, Essex RM1 1HA, England
11. E. Merck, D 61 Darmstadt, West Germany
12. Moore Boundy Hamill Ltd., Station House, Potters Bar, Herts. EN6 1AL, England
13. National Spectrographic Laboratories Inc., 19500 South Miles Road, Cleveland, Ohio 44128, USA
14. Office of Standard Reference Materials, National Bureau of Standards, Washington, D.C. 20234, USA
15. Research Organic/Inorganic Chemical Corp., 11686 Sheldon Street, Sun Valley, Calif. 91352, USA
16. Ventron Corp., Alfa Products, 44 Congress Street, Beverly, Mass. 01915, USA

PART II

METHODOLOGY

Introduction

In part II, the term Methodology covers all aspects of the application of the techniques and instrumentation of AAS, AES and AFS to chemical analysis.

The format adopted for previous volumes has been retained, with the subject matter treated under the two principal headings of (1) General Information, covering new methods, inter-laboratory comparisons and referee methods, and (2) Applications, where specific methods of analysis are reviewed and tabulated.

The classification of the range of applications into nine main fields of analysis also follows the established pattern. Some duplication of entries may be found in instances where a method is relevant to more than one section.

EXPLANATION OF THE TABLES

Each of the Applications sections, 2.1 to 2.9, is accompanied by a table which summarises the principal analytical features of the references from which the corresponding section is compiled. All relevant references are included in the appropriate table, while the accompanying text discusses only the more noteworthy contributions.

These Applications Tables form a convenient source of information for analysts interested in particular elements, matrices, sample treatments or atomization systems. In many cases, sufficient detail is given for the analytical procedure to be followed; absence of such detail usually means that the information was not directly available to the compiler of the table and the original reference should be consulted. The key to the tables is given below.

ELEMENT	The elements determined are listed in alphabetical order of chemical symbol, except that, for space economy, multi-element applications (5 elements or more) are given at the end of some tables.
λ/nm	The wavelength, in nanometres, at which the analysis was performed.
MATRIX	An indication, necessarily brief, of the material analysed.
CONCENTRATION	The concentration range or level of the element in the original matrix, expressed as % or $\mu g\ g^{-1}$ for solids and $mg\ l^{-1}$ or $\mu g\ ml^{-1}$ for liquids.
TECH.	The atomic spectroscopy technique is indicated by A (absorption), E (emission) or F (fluorescence).
ANALYTE	The form of the sample, as presented to the instrument, is indicated by S (solid), L (liquid) or G (gas or vapour). 'd.l.' = detection limit in the analyte.
SAMPLE TREATMENT	A brief indication is given of the sample pre-treatment required to produce the analyte.
ATOMIZATION	The atomization process is indicated by the abbreviations A (arc), S (spark), F (flame) or P (plasma), usually with some additional descriptive detail.
REF.	The number refers to the main Reference Section, which gives the title of the paper and the name(s) of the author(s), with address.

1 General Information

1.1 NEW METHODS

1.1.1 Introduction

This section describes novel methods of analysis which are considered to be of sufficient general interest to merit discussion here rather than in a subsequent applications section. Inevitably, the number of genuinely new methods appearing in the literature in any one year is limited. However, the criterion of novelty is applied less rigorously to papers where the authors have made a detailed study of experimental parameters of widespread interest.

1.1.2 Sample Preparation and Extraction Techniques

A novel method for the introduction of samples into either a flame or microwave plasma has been proposed by Skogerboe *et al* (237). The dried sample is converted to the chloride form by reaction with gaseous HCl, and the chloride volatilised into the emission cell. The method is applicable to 30 elements and is claimed to be less prone to matrix interference than direct solution nebulisation.

A comprehensive approach to the dissolution of *siliceous materials*, prior to their AAS determination, has been given by Whiteside (385). After initial heating of the sample with HF in a pressure vessel, H_3BO_3 was added and the pressure vessel sealed and reheated. This has the effect of dissolving insoluble fluorides formed in the first step, and in addition the fluorboric acid formed acts as a releasing agent and obviates the need for La addition.

A simple means of controlling the addition of a spectroscopic buffer soution to a sample has been described using a branched capillary tube fitted to the nebulizer of a flame spectrometer (11, 12). The ratio of sample to buffer uptake rates is controlled by the tube lengths.

A detailed scheme for the determination of *ambient forms of Hg in air* (508) involves a series of packed absorption tubes for the separation of the various forms of Hg. Tube packings used were: (i) glass wool, for particulate Hg, (ii) 3% SE–30 on Chromasorb W, to remove Hg (II), (iii) Chromasorb W treated with 0.05M NaOH to remove methyl-mercury (II), (iv) silvered glass beads to remove dimethyl-mercury. Hg is released from each section separately by heating, and determined by d.c. discharge emission spectroscopy. *Paper chromatography* has been used to separate Nb and Ta prior to emission spectrography (1203). The separated paper portions are ashed with 1 + 1 H_2SO_4, mixed with graphite and determined in an a.c. arc.

In a study which shows that great care must be taken in the analysis of water samples, surfactants have been found to have a serious effect on recoveries of Fe and Cu, when these elements are extracted as APDC complexes into MIBK (427). Recoveries ranging from 40 to 150% (compared to standards) were found in the presence of 1 μg ml^{-1} surf-actant, the lowest recovery coming with linear alkylate sulphonate solutions and the highest with soap solutions. Mn and Pb were found, however, to be little affected.

The use of *amines as extractants* continues to be of interest, and two groups of workers have studied the use of long chain alkylamines for extraction of Mo, W and Re complexes. Kim *et al*, in an extensive paper (996) studied the relative extraction efficiencies of a number of primary, secondary and tertiary amines. In the proposed method the metals, as thio-cyanate complexes in acid solution, were taken into a 0.1 – 1% solution of Amberlite LA1 resin in $CHCl_3$.. The desolvated residue was then redissolved in MIBK and the elements determined by conventional AAS in a N_2O/C_2H_2 flame. Good recoveries, high sensitivity and freedom from interferences were claimed for the method. Udelevich *et al* (287), on the other hand, obtained 98–100% recoveries by a simpler procedure where Mo, W and Re were extracted with trioctylamine from 0.1N H_2SO_4 into toluene; again, high sensitivity was claimed. Micro-quantities of Re can also be extracted with 99% efficiency into APDC-MIBK at room temperature (377). A further amine extraction procedure proposed (997) uses octylaniline in toluene to selectively remove Pt group metals from 3M HCl solution.

A number of metal dithiocarbamates can be rapidly concentrated by up to 2 orders of magntude after absorption on to a column of silylated Chromasorb W at pH 4 – 5 (925). The complexes (Zn, Cu, Ni, Pb, Hg, Co, Cd or Fe) can be displaced by 1 ml MIBK for AAS determination, but the whole procedure takes up to 30 min per run. In a general study (812) a number of solvents and extractants were compared for pre-concentration of Cu, Zn, Cd, Pb, Fe, and Mn. DDC in xylene was favoured for the stability of its complexes, in spite of poor burning properties in air/C_2H_2 and air/H_2 flames.

1.1.3 Emission

Naganuma and Kato (374) have made a thorough study of the influence of sample electrode pores on the intensity ratios of spectral lines. Both absolute intensities and intensity ratios were found to be altered considerably by the distribution, numbers and shapes of the pores, with the greatest effect at high pore densities. Linear working curves only resulted from porous electrodes prepared under reproducible conditions.

1.1.4 Absorption

A potentially very useful extension to the range of elements determinable by the *Delves cup technique* has been described where the high cup and furnace temperature provided by the N_2O/C_2H_2 flame have proved useful for the determination of elements with low volatility (*e.g.* Cr, Cu, Fe, Co, Ni, Mn, Ag and Sn) (573). The most suitable cup material was considered to be a Pt–Rh alloy.

In an interesting addition to the range of indirect methods for anions, *pyrophosphate* (5×10^{-6}M to 5×10^{-5}M) has been determined indirectly, in the presence of similar concentrations of phosphate, other condensed phosphates and some other anions, by its inhibitory effect on the extraction of Cu(II) by Amberlite LA–1 resin into MIBK or ethylacetate (35). The Cu was measured by AAS.

1.2 ANALYTICAL PARAMETERS

1.2.1 Sensitivity, Precision and Detection Limits

In two papers, Ingle has assembled a comparative study of sensitivity and limit of detection in quantitative spectrometric methods (67) and presented a detailed theoretical study of factors affecting precision in AAS (138). An alternative to the methods of Kaiser and Boumans for the calculation of detection limits in emission spectrography has been proposed (225). The complex procedure is based on the Siedel values for a number of standards and samples, but it is difficult to see how this adds to existing procedures.

The premixed O_2/H_2 flame, with ultrasonic sample nebulization, has been proposed as a superior emission source to the premixed laminar N_2O/C_2H_2 or O_2/C_2H_2 flames, or turbulent O_2/H_2 flame (244). Lower flame background is claimed to improve detection limits (including 2pg ml^{-1} for Ca), but there is a lack of interference studies, and in the light of past experience (*ARAAS*, 1971, **1**, 13) with similar flames, the analytical utility of the flame is open to question.

1.2.2 Interferences

The success or failure of the r.f. ICP as a versatile analytical tool will probably be decided by its ability to handle a variety of sample matrices without the need for frequent re-calibration. In an important paper, Larson *et al* (228) have examined in detail some potential interferences in the r.f. ICP. The interferences of Al and PO_4^{3-} on Ca (well characterised in flames) were found to be negligible in the plasma. Also Na at concentrations up to 6900 $\mu g\ ml^{-1}$ exerted only slight interference on the emission intensities of three elements of widley differing ionization potential (Ca, Cr and Cd). The determination of Mo in several matrix solutions showed a maximum signal change of only -6%, compared to changes of -70% to $+50\%$ for the same solutions in the N_2O/C_2H_2 flame. However, Koirtyohann and Lichte (855) found that mineral acids (especially HNO_3) could influence the emission signal for Cd in the ICP and that HNO_3 also appeared to have a much longer clearance time from the nebulizer than metal ions, Mermet and Robin (182) found that significant ionization interferences on P were not explicable in terms of the Saha equation.

A number of studies of matrix interferences in *carbon furnace AAS* has been described (248, 813, 814, 815) and the advantage of using oxyanions, rather than halides, as analytes is said to give more reproducible atomization (814, 815).

It is often found that new spectroscopic techniques are proposed without data relating to any really thorough interference studies. As a consequence, later workers may find difficulty in applying the technique to their particular samples. *Hydride generation* techniques probably fall into this category so the interference study by Smith (399) is particularly valuable. Smith has found many interferences in the AAS determination of As, Bi, Ge, Sb, Te and Se by hydride generation and atomization in an Ar/H_2–diffusion flame, and his conclusions were: *(i)* elements of Groups I, II, III, IIIA and IVA do not interfere, *(ii)* Cu, Ag, Au, Pt, Rh, Ru, Ni and Co always interfere (except Ge with Ag) and *(iii)* As, Bi, Ge, Sn, Te and Pb all mutually interfere.

The absorbance from Pt solutions in an air/C_2H_2 flame (9) was found to be higher with more thermally stable complexes and relationships between stability and *cis-trans isomerism* were indicated. The interferences of a number of mineral acids on the AAS of refractory elements (866) have been explained in terms of lateral diffusion effects.

1.2.3 Standard Reference Materials and Calibration

A method for the production of solid standards for use with electrothermal *atomization* has been published by Gries and Norval (see *ARAAS*, 1973, **3**, 17). Inorganic and organic matrices are used and the required metal concentrations (up to 1000 $\mu g\ g^{-1}$) introduced by ion implantation for metal matrices and dissolution for a urea matrix. Both standard types are suitable for other methods of trace analysis and the urea matrix will be of particular interest because of its relative simplicity of preparation in the laboratory.

2 Applications

2.1 PETROLEUM AND PETROLEUM PRODUCTS

2.1.1 Petroleum

The new Institute of Petroleum (I.P.) Methods Book (436) gives procedures for Ba, Ca, Mg and Zn in lubricating oils by AAS after dilution with white spirit. A similar method is used for Na, Ni and V in fuel oils and crudes. Both methods involve the use of organometallic standards; it is very probable that the relatively poor precision obtainable by these methods may be improved by ashing and dissolving the residue in dilute acid. Methods for Li and Na in grease by flame emission spectrometry and for Ba, Ca, Zn and P by direct reading emission spectrometry are also included. The criticism of the I.P. method for V is confirmed (927) in that bis (1-phenyl-3, 3-butanedione) oxovanadium (IV) and vanadyl tetraphenyl-porphyrin give different absorption values. The former is widely used as an organometallic standard while the latter typifies the form of V present in naturally occurring petoleum. This effect has been pursued by other workers (1548) and extended to Ni. The accuracy of the Zn determination in lubricating oils can be improved (1232) by comparison with standards containing the blend additive and the same base stock oil.

Viscosity effects are also known to be troublesome in direct dilution procedures. In the determination of Fe (1067) and Cd (903), the crude oils were diluted with xylenes to a viscosity of 1.3 cP. Similarly corrections for viscosity could be made by measurement of uptake rate with subsequent mathematical correction (962).

The direct determination of Pb in leaded and unleaded petroleum depends on the form of Pb present. Normally Pb occurs as a mixture of methyl and ethylleads, ranging from pure tetraalkyl derivatives to dimethyldiethyl and monomethyltriethyl, etc., Pb compounds. The effects of these derivatives may be normalised by the addition of I_2 (*ARAAS* 1971, **1**, 78). Better performance is claimed if an organic liquid ion exchanger is added with the I_2 (675). Direct dilution of gasoline with iso-octane/acetone (678) and determination of Pb using a total-consumption burner is apparently not affected by the Pb compound present. Coker (230) has determined individual Pb alkyls by GLC/AAS using an isothermal separation taking 5 min/sample.

The N_2O/H_2 flame has been recommended (890) for use with solvents such as benzene or xylenes which promote carbon formation with hydrocarbon gas flames. A special burner (1540) has been designed to use the sample as its fuel. It can be used with a wide range of different samples.

2.1.2 Lubricating Oils

Ti is a difficult metal to determine in used lubricating oils because of its relative insensitivity by AAS and the low levels usually encountered. It may be pre-concentrated by extraction using aqueous HCl/HF and then determined directly in an N_2O/C_2H_2 flame (872).

Additives in lubricating oils have been determined (73) by emulsifying the oil with an addition agent. The sample was mixed with an equal volume of low-viscosity mineral oil, 5 ml of acetone and 0.2 g of the emulsifier mixture (Emulsogen M 62.5%; Emulgator HS 12.5%; Olein 12.5%; Dipropylene glycol 12.5%)). Because this mixture is then diluted with water, aqueous standards and ionisation suppressants may be used. A similar method for Ba (499) used dilution with 80 ml of 2-methylpropan-2-ol and toluene with 10 ml of 1% potassium naphthenate as ionization suppressant (342). This solvent can incorporate additions of 5 ml of water, and hence, aqueous standards can be used.

2.1.3 Electrothermal Atomization

Electrothermal techniques have great potential for the analysis of petroleum products because of their sensitivity and *in situ* ashing capability. Elements determined on carbon rod atomizers this year were Mn (139), Cr, Mn (897), Be (900), Be, Cr, Mn and Al (685, 900). Other elements which have been determined after wet digestion are: Cd, Sb, Pb (899); Ni, V (894) and Co, Mo (898). Pyrolitic graphite boats have been used (802) to ash 10 mg of crude oil in an air/C_2H_2 flame or oven before insertion into a tungsten electrothermal atomizer for V determination.

Both carbon rod and graphite tube atomizers were improved by coating the middle of the atomizer with an aqueous or organic solution containing La or Zr (see *ARAAS* 1973, 3, 17). This is claimed to prevent deterioration of the tube and to inhibit carbide formation with elements such as Al, Be, Cr and Mn (685, 900). A similar approach (702) has been used in the determination of Si, in this instance the tube was soaked in sodium tungstate solution.

2.1.4 Instrumental Developments

The widespread practice of determining wear metals in lubricating oils has prompted the search for an analytical technique which combines the speed and multi-element capability of direct-reading emission spectroscopy with the ruggedness, simplicity and precision of AAS. Multi-element AAS using a silcon target vidicon detector (80) has been proposed for Ag, Cu, Fe and Mg determinations in used lubrication oils. A multi-element hollow cathode lamp and background absorption correction were employed; read-out was by means of an oscilloscope or strip chart recorder. A contiuum source (high-pressure Hg/Xe arc, 200 W) was used (731) for the determination of several wear metals below 300 nm by AAS using an echelle grating spectrometer. Determinations above 300 nm are comparatively easy and were described in an earlier paper (*ARAAS*, 1974, 4, 18). Winefordner and co-workers (1237) determined 23 elements in lubricating oils by AFS with a separated air/C_2H_2 or N_2O/C_2H_2 flame. The determinations were virtually simultaneous in that a computer-controlled slewed-scan spectrometer was used in conjunction with a 500 W Xe arc source.

The demountable hollow cathode lamp has been suggsted as an emission source for wear metals (751). The sample, prepared in a number of ways, was placed in an ASTM PC-2 porous cup electrode which formed the cathode.

The use of an ICP for the analysis of metals in petroleum products has been described in a preliminary communication (1149).

TABLE 2.1 PETROLEUM AND PETROLEUM PRODUCTS

Element	λ/nm	Matrix	Concentration	Tech.	Analyte form	Sample treatment	Atomization	Ref.
Ag	338.1	Used oils	0.7–7.6 μg/g	A	L	Dilute (1:4) with MIBK. (Vidicon detector, simultaneous determination of Ag, Cu, Fe, Mg)	F Air/C_2H_2	80
Ag	—	Oils	10–50 μg/g	E	L	—	S Rotrode (in argon)	1527
As	—	Petroleum products	10 ng/g	A	G	Decompose with H_2SO_4/HNO_3, reduce As to AsH_3 with $KI/SnCl_2/Zn$	F —	895
Ba	553.6	Mineral oils	0.75%	A	L	Mix (1:1) with pure low-viscosity oil, add acetone + emulsifier, dilute with H_2O and shake vigorously before aspiration. Emulsifier=62.5% Emulsogen M + 12.5% Emulgator HS + 12.5% Olein + 12.5% dipropylene glycol	F N_2O/C_2H_2	93
Ba	553.5	Oils	0.02–1%	A	L	Dilute with mixed solvent (3 parts 2-methylpropan-2-ol + 2 parts toluene) add 1.0% K (as napthenate) and compare with aqueous standards similarly diluted.	F N_2O/C_2H_2	499
Be	—	Petroleum products	Trace levels	A	L	Dilute (1:1) with solvent. Calibrate by standard-addition method.	Graphite furnace (HGA-70; CRA-63)	900
Be	—	Petroleum products	1–50 ng/g	A	L	—	Graphite furnace	1249
Ca	422.7	Mineral oils	0.05%	A	L	See Ba, ref. 93	F N_2O/C_2H_2	93
Ca	422.7	Oil products	μg/g levels	E	L	Description of new burner design	F —	1540
Cd	228.8	Petroleum products	1–100 ng/g	A	L	Digest with H_2SO_4 and ash. (Interlaboratory comparison of flame and non-flame methods).	F Air/C_2H_2 Graphite furnace	687
Cd	—	Petroleum products	From 10 ng/g	A	L	Digest with H_2SO_4 and ash. Calibrate furnace method by standard additions.	F Air/C_2H_2 Graphite furnace	899
Cd	228.8	Petroleum	—	A	L	Dilute with p-xylene	F Air/H_2	903
Co	—	Crude oils	100 μg/g levels	E	L	Study of effect of solvent matrix on analytical results	S Rotrode	587

TABLE 2.1 PETROLEUM AND PETROLEUM PRODUCTS — continued

Element	λ/nm	Matrix	Concentration	Tech.	Analyte form	Sample treatment	Atomization	Ref.
Co	—	Crude oils, fuels, distillates	ng/g levels	A	L	(a) Acid digestion + flame; (b) acid digestion + furnace; (c) direct atomization in furnace (comparison of methods)	F Air/C_2H_2 Graphite furnace	898
Cr	—	Petroleum products	From 10 ng/g	A	L	Dilute with tetrahydrofuran. Calibrate by method of additions	Graphite furnace (CRA-63)	897
Cu	324.7	Used oils	0.8–27.8 μg/g	A	L	See Ag, ref. 80	F Air/C_2H_2	80
Cu	—	Crude oils	—	A	L	Ash in graphite boat in air/C_2H_2 flame and insert boat + ash into graphite cuvette	F Air/C_2H_2 + Graphite furnace	1143
Cu	—	Oils	10–50 μg/g	E	L	—	S Rotrode (in argon)	1527
Cu	327.4	Oil products	μg/g levels	E	L	See Ca, ref. 1540	F —	1540
Fe	248.3	Beach asphalts	132 μg/g	A	L	Dissolve in C_6H_6, filter, evaporate to dryness and dissolve weighed aliquot of residue in CCl_4	Graphite furnace (HGA-2000)	5
Fe	—	Propylene glycol	ng/g levels	A	L	Automated peak discrimination and integration system	Carbon rod	68
Fe	248.3	Used oils	0.5–42.5 μg/g	A	L	See Ag, ref. 80	F Air/C_2H_2	80
Fe	—	Crude oils	—	A	L	Dilute with xylene to standard viscosity (1.3 cSt)	F Air/C_2H_2	1067
Fe	—	Crude oils	—	A	L	See Cu, ref. 1143	Graphite furnace	1143
Fe	—	Oils	10–50 μg/g	E	L	—	S Rotrode (in argon)	1527
Hg	—	Petroleum	Trace levels	A	S	Separate by volatilization and collect on Au. Redissolve Hg and concentrate by electrolysis on Au cathode. Heat cathode in furnace to release Hg	Graphite furnace	447
Hg	253.7	Petroleum products	20–100 ng/g	A	L,G	(a) Decompose with acid in closed system; (b) use Wickbold C_2/H_2 combustion method	Cold vapour	686

Hg	—	Petroleum products	ng/g levels	A	G	(a) Decompose with HNO_3/H_2SO_4; (b) combust and absorb product in acid $KMnO_4$	Cold vapour	901
Mg	285.2	Used oils	0.7–5.6 μg/g	A	L	See Ag, ref. 80	F Air/C_2H_2	80
Mg	285.2	Mineral oils	0.18%	A	L	See Ba, ref. 93	F Air/C_2H_2	93
Mn	—	Petroleum	10–300 ng/g	A	L	Calibrate by standard-addition method	Graphite furnace (CRA-63)	139
Mn	—	Petroleum products	From 10 ng/g	A	L	See Cr, ref. 897	Graphite furnace	897
Mn	279.5	Oil products	μg/g levels	E	L	See Ca, ref. 1540	F —	1540
Mo	—	Crude oils· fuels, distillates	ng/g levels	A	L	Digest with acid or dry-ash and measure Mo by flame or furnace atomization. Add KCl for flame method, + $AlCl_3$ if Ca present. See also Co, ref. 898	F N_2O/C_2H_2 Graphite furnace	898
Na	589.0	Gas-turbine fuel oil	0–2 μg/g	E	L	(a) Add H_2O, shake and remove aqueous layer. Compare with NaCl standards; (b) shake ultrasonically with H_2O and aspirate directly. Compare with Na naphthenate standards	F —	1517
Ni	232.0	Beach asphalts	75 μg/g	A	L	See Fe, ref. 5	Graphite furnace (HGA-2000)	5
Ni	—	Distillate fuels	ng/g levels	A	L	Decompose by acid digestion. Calibrate by standard-addition method	Graphite furnace	894
Ni	—	Crude oils	—	A	L	See Cu, ref. 1143	Graphite furnace	1143
Ni	—	Fuel oils	—	A	L	Note on sources of error	F —	1548
Pb	283.3	Gasoline	0.2–1.0 g/l (Detection limit 0.2 mg/l)	A	L	Separate Pb alkyls on g.c. column at 130°C and pass into flame. (Method for individual or total Pb alkyls, also applicable to trace Pb in unleaded fuels. Results compared with ASTM/IP standard XRF method	F Air/C_2H_2	230
Pb	283.3	Gasoline	Up to 26 mg/l	A	L	Add excess iodine and a liquid ion-exchanger. Compare with standards of complexed $PbCl_2$ in gasoline	F N_2O/H_2 (fuel-lean)	675
Pb	283.3	Gasoline	0.01–2.5g (per U.S. gallon)	A	L	Dilute with iso-octane/acetone	F Air/H_2	678
Pb	—	Petroleum products	—	A,E	L,S	—	F,A —	899
Pb	283.3	Oil products	μg/g levels	E	L	See Ca, ref. 1540	F —	1540

TABLE 2.1 PETROLEUM AND PETROLEUM PRODUCTS — continued

Element	λ/nm	Matrix	Concentration	Tech.	Analyte form	Sample treatment	Atomization	Ref.
Sb	—	Petroleum products	From 10 ng/g	A	L	Treat with H_2SO_4, ash and redissolve in acid. Calibrate by method of standard additions.	Graphite furnace	899
Se	—	Crude oils, petroleum products	From 10 ng/g	A	G	Decompose with $HNO_3/H_2SO_4/HClO_4$. Reduce to H_2Se with nascent H	F Ar/H_2 + entrained air	896
Ti	—	Lubricating oils	—	A	L	Separate Ti from oil by treatment with HCl/HF. Centrifuge and spray	F —	892
V	318.4	Beach asphalts	200 μg/g	A	L	See Fe, ref. 5	Graphite furnace (HGA-2000)	5
V	318.2	Beach asphalts	200 μg/g	A	L	Dissolve in CCl_4. Atomize at 2700°C. (Comparison with NAA method)	Graphite furnace	575
V	—	Crude oils	—	E	L	See Co, ref. 587	S Rotrode	587
V	—	Crude oils	p.p.b. levels	A	L	Dry-ash in furnace (up to 20mg of sample)	Graphite/tungsten furnace (IL-455)	802
V	—	Distillate fuels	ng/g levels	A	L	See Ni, ref. 894	Graphite furnace	894
V	318.4	Petroleum	0–36 μg/ml	A	L	Study of form of V combination	F N_2O/C_2H_2	927
V	—	Crude oils	—	A	L	See Cu, ref. 1143	Graphite furnace	1143
V	—	Fuel oils	—	A	L	See Ni, ref. 1548	F —	1548
Zn	213.8	Mineral oils	0.03–0.15%	A	L	See Ba, ref. 93	F Air/C_2H_2	93
Zn	—	Lubricating oils, additives	—	A	L	Prepare standards from base oil and additive	F —	1232
Zn	213.8	Oil products	μg/g levels	E	L	See Ca, ref. 1540	F —	1540
Various (9)	—	Oils, greases	—	A,E	L	Standard IP methods, including Ba, Ca, Mg, Zn, P in lubricating oils, Li, Na in greases and Na, Ni, V in crude oils and fuel oils.	F,S —	436
Various	—	Petroleum	—	A	L	Improvement of detection limits for Be, Cr, Mn, Al by treatment of furnace with La or Zr	Graphite furnace	685
Various (8)	—	Petroleum products	—	A,E	L	Flame study, applied to Pb, Mn in gasoline; Ca, Zn in oils; B in transmission fluids and Cu, Fe, Ni in feedstocks	F N_2O/H_2	890

Various (12)	—	Petroleum products	Trace levels	E	—	—	—	—	902
Various	—	Hydraulic oils	Trace levels	A	L	Application of atomic absorption continuum (AAC) technique (200 W Xe/Hg source)	F	—	731
Various	—	Oils, fuels	Minor and trace levels	E	L,S	—	Hollow cathode		751
Various	—	Petroleum products	—	E	L	Study of viscosity effects	F	—	962
Various	—	Used oils	Trace levels	A,E	L	Review of wear-metal monitoring techniques	—	—	1148
Various	—	Petroleum products	—	A,E	L	Review	F,P	— Graphite furnace	1149
Various (23)	—	Used oils	Trace levels	F	L	Multi-element analysis with computer-controlled rapid-scan AFS system, using xenon arc continuum source	F	Air/C_2H_2 N_2O/C_2H_2	1237
Various (10)	—	Petroleum products	$\mu g/g$ levels	E	S	Evaporate sample on graphite/sulphur powder (buffer mixture)	A	D.C. arc (10 A)	1262
Various	—	Crude oil spills	—	A	L	Review of methods, including AAS	F	—	1519

2.2 CHEMICALS AND MISCELLANEOUS APPLICATIONS

2.2.1 Gunshot Residues

The applications described have all involved the determination of elements in gunshot residues either in victims or on the hands of the gun-person. Samples can be collected by swabs of cotton wool mounted on plastic shafts, (10, 88, 532, 540, 1219, 1220) or by adhesive transparent tape or cellulose acetate lifting film (758). The collector is then dried and leached with 2 ml of dilute acid, if blood is present a low-temperature ashing with an O_2 plasma may be used (88). Sherfinski (10) used a graphite tube atomizer for Ba determinations and found degradation in performance only after 80-120 burns. The onset of degradation was indicated by a small positive or negative peak after the Ba signal.

2.2.2 Paints

Paints have been screened for Pb and Cd by dipping a nichrome wire into the sample and then inserting the wire into an air/C_2H_2 flame (162). Alternatively, liquid or dried paint can be wrapped in filter paper held together with a paper clip and inserted into a flame (542). Pb and Cd have been determined using the Delves cup technique (496) after first ashing the samples (10 mg) in the nickel cups in a muffle at 400°C. Standard-additions procedures were used and matched cups were necessary.Liquid paints were suspended in water or mixed with (1+1) toluene/dimethylformamide before adding 10 ml to the cup. Solution creep was overcome by coating the inside of the cup with a resin. A similar approach using a sampling boat was described by Holak (162) who preferred initial ashing of the sample with 3 ml of HNO_3 in a PTFE vessel.

Electrothermal atomization with a graphite furnace (364, 566) has been used to detemine Co, Pb, Fe, Cu and Al after dilution of the paint with MIBK or ashing of dried paint films with HNO_3/H_2SO_4. Hg was determined by the cold-vapour technique using AFS. After separation As and Sb were determined by the hydride-generation technique using AAS.

An unusual oxidation agent for PTFE bombs was $H_2SO_4/KMnO_4$ used in the determination of Cr in surface coatings by conventional AAS (229). The coating was dried and charred in an Al dish on a hotplate before wet-pressure digestion.

Micro examinations of paint surfaces using laser microprobe spectrography have been reported (1330) and the technique compared with arc spectrography (1211).

2.2.3 Organic Chemicals, Polymers

Solid polyacrylamide and sodium alginate samples have been analysed directly for Fe, Cu and Cr (156) by graphite tube AAS using background correction. Detection limits were Ca 0.01 $\mu g\ g^{-1}$ with 5 mg samples but matrix effects presented serious problems and calibration by standard additions was necessary.

Au has been determined in photographic film after decomposition with a hydrolytic enzyme (0.1% Mezymforte in 0.05M NH_4Cl at 38°C for $\frac{1}{2}$–2 h) and ashing with HNO_3/H_2O_2. The Au (down to 0.7 $\mu g\ cm^{-2}$) was extracted into MIBK from an HBr solution for determination by graphite furnace AAS (108).

2.2.4 Indirect Determinations

Anthranilic acid (379) was determined by extraction of its Co(II) complex, formed in the presence of bathophenanthroline at pH 6.5, into MIBK. At least two molar excesses of Co and ten molar excesses of bathophenanthroline were necessary for complete extraction.

The Co was subsequently determined by AAS in an air/C_2H_2 flame.

1,2 Diols (122) have been cleaved with periodic acid to yield iodate as the reaction product. Excess Ag was added and $AgIO_3$ precipitated, filtered off, dissolved in ammonia and Ag determined by AAS.

Barbiturates (1291) have been precipitated as Cu pyridine barbiturate complexes from 0.08 M Na_2CO_3 by the addition of 1–3 ml of 3 mg ml^{-1} Cu solution, 1 ml pyridine and 10 ml water. Cu was determined by AAS. A methanolic solution of *P-aminobenzoic acid* was evaporated to dryness with $CuCl_2$, washed with ethylacetate and the dry resdiue extracted with an MIBK solution of bathophenanthroline (1285). The extracted Cu was determined by AAS.

Fluorine compounds have been examined by several workers. Effluents from a GLC column were passed into a silica absorption tube at 800°C containing Na vapour (20). The decrease in Na absorption was proportional to the F content. No interference was observed from Cl but Br and I interfered. Gatsche *et al* (1464) determined F by generating SiF_4 with $SiO_2 + H_2SO_4$ and determining Si by AAS. Fluoride was determined indirectly in arc-welding electrodes and cryolite (1354) by precipitation of CaF_2 with excess $CaCl_2$. After filtration, the precipitate was boiled with $CeCl_3$ solution to deposit CeF_3. The displaced Ca was determined in solution by AAS.

Tyre-cord dip solutions were doped with $Sr(NO_3)_2$ to measure the degree of pick-up. Sr was measured by AAS on a sample of the cord after passing through the dip solution (483).

2.2.5 Industrial Products, Intermediates

Traces of Pt, Pd, Rh, Ir, Ru and Au have been determined in chloride/sulphate liquors containing large excesses of Ni, Cu, Se and Fe. The solution was boiled with 10% HCl and 0.5 g of ion-exchange fibre (1162) to collect the noble metals. The fibre was separated by filtration, ashed and mixed with Na_2SO_4, $Fe(NO_3)_3$ and carbon for excitation in a 15A a.c. arc or pressed into discs with Cu powder for 4A a.c. arc excitation. A similar application of ion-exchange fibres (1157) is described in Section 2.5.2.

Radioisotope studies (1155) showed that traces of Ca, Cr, Mn, Fe and Co are completely retained during the dry ashing of graphite at 700°C for 7 hours. Sodium chloride was added to the residue for excitation in a 15A d.c. arc.

The absorption signal observed in cold-vapour AAS of Hg in selenides and tellurides is depressed by the formation of HgSe or HgTe by the presence of large excesses of these elements. When selenides or tellurides are present this can be overcome by the use of formaldehyde as the reducing agent (514, 981).

Heavy metal contaminants (Ag, Bi, Co, Cd, Cu, Fe, In, Ni, Pb, Tl, Zn) in alkali and alkaline earth salts (178) were precipitated in aqueous solution by DDC. As the deposit was not filterable it was absorbed by filtering through a paper containing 50 mg of activated carbon which was then leached with HNO_3 to give an aqueous solution of the impurities. The solution was then analysed by AAS.

Electrolysis for 5 min. with four C electrodes was used to separate Cr, Cu, Pb, Ni, Mn, Mg, Co, Al, Bi, V, Ga, Fe and Ca from 50 ml of a 20% m/v H_3BO_3 sample solution. These electrodes were then arced and using an echelle grating spectrograph (1169) detection limits of 0.001–0.03 $\mu g\ g^{-1}$ H_3BO_3 were obtained.

TABLE 2.2 CHEMICALS, MISCELLANEOUS MATERIALS

Element	λ/nm	Matrix	Concentration	Tech.	Analyte form	Sample treatment	Atomization	Ref.
Ag	328.1	Pb salts	μg/g levels	A	L	Dissolve in HNO_3, evaporate to dryness, redissolve in H_2O. Precipitate and centrifuge matrix. Evaporate, after solid–liquid extraction, and dissolve in HNO_3 + acetone	F —	438
Ag	338.3	Pb/Sn plating baths	Up to 500 μg/ml	A	L	Prepare solution of Pb/Sn alloy in HF/HNO_3. Use standard addition method for 100–500 μg/ml range of Ag, Cu, Pd. No interferences at low levels	F Air/C_2H_2	1008
Al	—	Pressed board	—	A	L	Wet-ash with HNO_3	F —	538
Al	309.2	Paint films	—	A	L	Digest with HNO_3/H_2SO_4	Graphite furnace (HGA-70)	566
Al	—	GaAs films	—	E	L	Study of sources of error	— —	1021
Al	—	Soaps, phosphate materials	1 μg/g	A	L	Dilute phosphate materials to 0.01 M H_3PO_4 and apply background correction (for Cu, Zn). Use extended ashing stage for Al	Graphite furnace (HGA-70, 72)	1268
As	193.7 189.0	General solutions	From 0.06 ng	A	L	Use standard-addition method	Graphite furnace (HGA-72)	194
As	193.7	Paint	—	A	L	Form hydride by $NaBH_4$ reduction	Heated silica tube	364
As	193.7	Metal salt solutions	ng/ml levels	A	L	Char at 600°C and volatilize at 2300°C. (Study of enhancement and interference effects).	Graphite furnace	762
As	313.3 (Mo)	Ultra-pure water	60 pg/ml	A	L	Convert to molybdophosphate or molybdoarsenate complex, extract into butyl acetate (with ethanol addition for As), decompose with NH_3 and back-extract into H_2O	Graphite furnace	1269
As	193.7	Sodium chloride	μg/g levels	A	G	React solution with Zn + HCl, trap arsine at −150°C and release in argon flow to flame	F Ar/H_2	1350
Au	242.8	Photographic film	0.7–10 ng/g	A	L	Decompose film with enzyme solution of 0.1% Mezymforte in 0.05 M NH_4Cl at 38°C. Evaporate to dryness, react with HNO_3/H_2O_2 and extract Au, as bromide, into MIBK	Graphite furnace	108

B	—	Graphite	ng/g levels	E	S	Heat to 150°C with equal weight of H_2SO_4/H_2SO_5 (9:1). Cool, reflux-extract with methanol + chlorotriphenylmethane, evaporate and mix residue with B-free carbon	A	—	441
B (isotopes)	208.9 209.0	B compounds	—	A	L	—	F	N_2O/C_2H_2	858
B	—	GaAs films	—	E	L	See Al, ref. 1021	—	—	1021
B	—	Si surface structures	—	E	L	Etch with HF/HNO_3 and evaporate solution on graphite powder + 0.4% NaF	A	—	1395
Ba	553.6	Gunshot residues	0.1–1.0 μg/ml in extract	A	L	Study of 'memory' effects and tube degradation	Graphite furnace (HGA-2100)		10
Ba	553.6	Calcium carbonate	0–0.5 μg/ml in extract (0.5–50 μg/g in $CaCO_3$)	A	L	Dissolve 2.5 g sample in 15 ml 4 M HNO_3 and dilute to 100 ml	Graphite furnace (A-3470) Purge gas: argon + CH_4		19
Ba	553.6	Gunshot residues	0–3 ng/10 μl in extract	A	L	Swab residue with 5% HNO_3 on cotton. Extract Ba, Sb by (a) leaching with I M HNO_3 or (b) LTA (plasma) ashing. Take 10 μl sample	Tantalum strip (2500°C in argon)		88
Ba	—	Paint extenders	—	A	L	—	F	N_2O/C_2H_2	364
Ba	—	Gunshot residues	Trace levels	E,A	S,L	Examine clothing by ES and hand-washings by AAS	A,F	—	532
Ba	—	Gunshot residues	From 0.1 ng in 10μl samples	A	L	Investigation of sampling techniques	Tantalum strip		758
Ba	—	Sputtered deposits	From 0.3 μg/ml (in extract)	A	L	Add K + ethanol to analysis solution to increase sensitivity	P	—	963
Ba	—	Gunshot residues	—	E	L	—	P	ICP	1107
Ba	—	Gunshot residues	—	A	L	—	Graphite furnace		1140
Ba	553.6	Gunshot residues	0.07–2 μg	A	L	Extract with 10% HCl on swab, followed by H_2O extraction	Graphite furnace		1220
C	229.7	Zirconium carbide	8–14%	E	S	Excite directly in Al electrode	A	—	1170
Ca	—	Paint extenders	—	A	L	—	F	N_2O/C_2H_2	364
Ca	—	Sputtered deposits	From 0.03 μg/ml (in extract)	A	L	See Ba, ref. 963	P	—	963

TABLE 2.2 CHEMICALS, MISCELLANEOUS MATERIALS — continued

Element	λ/nm	Matrix	Concentration	Tech.	Analyte form	Sample treatment	Atomization	Ref.
Ca	422.7	Lead oxalate	10–100 μg/g	A	L	Ignite at 500°C, dissolve in HNO_3 and add conc. HNO_3 to preciptate Pb. Filter, evaporate and extract with H_2O	F —	1561
Cd	228.8	Processing solutions	From 0.002 μg/ml	A	L	Match Standards for HCl, H_2SO_4 content of samples	F Air/C_2H_2	46
Cd	326.1	Paint	1–300 μg/g	A	S,L	(a) Suspend paint in H_2O or toluene/DMF (1:1) and take 10 μl aliquot; (b) weigh sample (0.01 g), ash at 500°C and insert in flame. Use method of standard additions	F Air/C_2H_2 + Delves cup	496
Cd	—	Pressed boards	—	A	L	See Al, ref. 538	F —	538
Cd	228.8	Miscellaneous solids	—	A	S	Fold sample in filter-paper and introduce into flame, below optical path (Rapid sorting test)	F Air/C_2H_2 (air-rich)	542
Cd	—	Paint	Trace levels	A	L	Dry-ash at 500°C and extract with HNO_3 + NH_4 acetate solution	F Air/C_2H_2	779
Cd	228.8	GaAs, acids	ng/g levels	A,F	L	Dissolve GaAs in HCl/HNO_3. Evaporate acids and redissolve in I N HCl	Graphite furnace	1317
Co	240.7	Paints and driers	—	A	L	Dilute with MIBK	Graphite furnace (HGA-70)	364
Co	—	Radioactive solutions	10–300 ng/ml	A	L	Adjust solution to pH 3 and extract with APDC/ethyl acetate	Graphite furnace (CRA-63)	1144
Cr	357.9	Polymers	Trace levels (from 0.01 μg/g)	A	L,S	Use solid samples directly, introduced to furnace by tantalum spoon or treat sample with 0.16 M HNO_3 and use 50 μl portons of extract	Graphite furnace (HGA-2000)	156
Cr	357.9	Paint	100–800 μg/g	A	L	Ash in aluminium dish at 450°C and dissolve in $H_2SO_4/KMnO_4$ in acid digestion bomb	F Air/C_2H_2	229
Cr	—	Gallium arsenide	—	A	L	Separate Ga by extraction chromatography with dichlorodiethyl ether or di-isoamyl ether	Graphite furnace	1153

Cr	283.5	Polymethyl-siloxane	—	E	L	Dissolve in toluene, evaporate 0.05 ml on graphite electrode containing NaCl + AgCl	A	11 A a·c.	1399
Cr	357.9	Industrial materials	4–25%	A	L	Dissolve in acid, add ethanol, heat, evaporate, dissolve in HNO_3 and transfer to flask containing acetic acid + $Ca(NO_3)_2$. Dilute to volume	F	Air/C_2H_2	1570
Cu	324.7	Graphite, zinc oxide	0.02%	E	S	Study with computer-coupled photodiode-array spectrometer	A	13 A d.c.	89
Cu	324.7	Polymers	Trace levels (from 0.01 μg/g)	A	L,S	See Cr, ref. 156	Graphite furnace (HGA-2000)		156
Cu	—	Gunshot residues	Trace levels	E,A	S,L	See Ba, ref. 532	A,F	—	532
Cu	324.7	Paint films	—	A	L	See Al, ref. 566	Graphite furnace (HGA-70)		566
Cu	324.7	Pb/Sn plating baths	Up to 500 μg/ml	A	L	See Ag, ref. 1008	F	Air/C_2H_2	1008
Cu	—	Soaps, phosphate materials	1 μg/ml	A	L	See Al, ref. 1268	Graphite furnace		1268
Cu	324.7	Aniline	From 0.5 μg/g	A	L	Dilute with H_2O + ethanol (1:1)	F	—	1504
F	589.0 (Na)	Organic vapours	Trace levels	A	G	React organic F-containing compound with Na vapour at 800°C. Relate F concentration to decrease in atomic Na concentration	Thermal (Reaction tube at 800°C)		20
F	422.7 (Ca)	Electrode coatings, fluorides, cryolite	% levels	A	L	Precipitate as CaF_2, boil, cool, add $CeCl_3$ to precipitate residual F, filter and dissolve CaF_2 + CeF_3 in $HClO_4$	F	—	1354
F	251.6 (Si)	Various	1–12 μg/g	A	G	Mix with excess SiO_2 + K_2SO_4, heat to produce SiF_4 and pass vapour through flame or heated cell	F Graphite furnace	N_2O/C_2H_2	1464
F	712.8	Phosphates	0·1–10%	E	S	—	Hollow cathode (He/Ne; 400 V)		1484
Fe	248.3	Disposable pipette tips	Up to 12 ng/g	A	S	Anaylse for Fe, Zn before and after washing with 1% HCl (×3) and H_2O (×1)	Graphite furnace (HGA-2000)		13
Fe	248.3	Polymers	Trace levels (from 0.01 μg/g)	A	L,S	See Cr, ref. 156	Graphite furnace (HGA-2000)		156

TABLE 2.2 CHEMICALS, MISCELLANEOUS MATERIALS — continued

Element	λ/nm	Matrix	Concentration	Tech.	Analyte form	Sample treatment	Atomization	Ref.
Fe	248.3	Paint films	—	A	L	See Al, ref. 566	Graphite furnace (HGA-70)	566
Fe	248.3	Aniline	From 1 μg/g	A	L	See Cu, ref. 1504	F —	1504
Ga	287.4	Semi-conductor materials	0–25 μg/ml (in solution)	A	L	Add 1000 μg/ml Mg as releasing agent	F Air/N_2/H_2 or Air/Ar/H_2	186
Ge	326.9	Graphite, zinc oxide	1%	E	S	See Cu, ref. 89	A 13 A d.c.	89
Hf	264.1	Hf/Zr extracts	0.1—10%	E	L	Use Zr as internal standard	S Spark + aerosol	506
Hg	253.7	Pharmaceuticals	—	A	L	—	Cold vapour	279
Hg	253.7	Paint	—	F	G	Review paper	Cold vapour	364
Hg	253.7	Selenides, tellurides	—	A	L,G	Reduce Hg with 40% formaldehyde solution at pH 12.	Flameless method	514
Hg	—	Se and Te preparations	—	A	L	Reduce Hg(II) at pH 12 with formaldehyde, to avoid interference by Se, Te	Cold vapour	981
Hg	253.7	Gases	—	F,A	G	Absorb Hg on thin gold wires and release by heating to 500°C in stream of argon	Cold vapour	1317
In	303.9	Semi-conductor materials	Up to 10 μg/g	A	L	Dissolve in HF/HNO_3. Add 1000 μg/ml MgI_2 as interference suppressant	F Air/Ar/H_2	1478
Ir	285.0	Anodic coatings	8–60 μg/ml (in extract)	A	L	Fuse coating with KOH/KNO_3, redissolve and dilute with HCl + buffer	F Air/C_2H_2	694
K	—	Pressed boards	—	A	L	See Al, ref. 538	F —	538
K	—	Drugs	—	E	L	Mix drug preparation (5g) with H_2O (50 ml) + synthetic gastric juice (50 ml). Centrifuge and analyse supernatant liquid	F —	940
K	766.5	Ag/Cr/H_2SO_4 solutions	0.5–3 μg/ml	A	L	Add H_2SO_4 to standards. Ag and Cr do not interfere	F —	1266
Mg	—	Paint extenders	—	A	L	—	F N_2O/C_2H_2	364
Mg	285.2	Aniline	From 0.08 μg/g	A	L	See Cu, ref. 1504	F —	1504

Mn	279.5	Tellurium	0.4 μg/g	A	L	Treat with H_2SO_4, evaporate to dryness and redissolve in HNO_3	Graphite furnace (HGA-70)	566
Na	589.0	Antibiotics	—	E	L	Extract sulphated ash with H_2O	F —	95
P	313.3 (Mo)	Ultra-pure water	25 pg/ml	A	L	See As, ref. 1269	Graphite furnace	1269
P	313.3 (Mo)	Semi-conductor films	ng/g levels	A	L	—	F N_2O/C_2H_2	1503
Pb	217·0 283.3	Organic colouring dyes	From 0.4 μg/g	A	L	Digest with $HNO_3/HClO_4$, evaporate to dryness, redissolve and extract with DEDC/xylene	F —	113
Pb	283.3	Paint	From 0.025%	A	L	Treat sample with HNO_3 at 150°C in PTFE vessel, under pressure	F Air/C_2H_2 (+ sample boat)	162
Pb	283.3	Paints and driers	—	A	L	See Co, ref. 364	Graphite furnace (HGA-70)	364
Pb	283.3 261.4	Paint	6–250 μg/g	A	L,S	See Cd, ref. 496	F Air/C_2H_2 + Delves cup	496
Pb	217.0	$PbCrO_4$ pigments	Trace levels (soluble Pb)	A	L	Extract with 0.25% HCl. (Comparison of several methods)	F —	512
Pb	—	Gunshot residues	Trace levels	E,A	S,L	See Ba, ref. 532	A,F —	532
Pb	—	Gunshot residues	Trace levels	A	L	—	F —	540
Pb	283.3	Miscellaneous solids	—	A	S	See Cd, ref. 542	F Air/C_2H_2 (air-rich)	542
Pb	—	Paint	Trace levels, from 0.01%	A,E	S,L	Comparison of methods	A,F —	659
Pb	—	Paint	Trace levels	A	L	See Cd, ref. 779	F Air/C_2H_2	779
Pb	—	Gunshot residues	—	E	L	—	P ICP	1107
Pb	—	Toothpaste	Up to 5 μg/g	A	L	Standard method, BS 5136: 1974	F —	1215
Pb	283.3	Gunshot residues	13–81 μg	A	L	See Ba, ref. 1220	Graphite furnace	1220
Pd	276.3	Pb/Sn plating baths	Up to 500 μg/ml	A	L	See Ag, ref. 1008	F Air/C_2H_2	1008
Pd	276.3	Catalysts	0.1–1%	A	L	—	F Air/C_2H_2	1346
Ru	349.9	H_2SO_4 solution	Up to 40 μg/ml	A	L	Add Na salt to overcome suppression	F Air/C_2H_2 (fuel-rich)	127
Ru	349.9	Organo-Ru compounds	1–10 ug/ml (in solution)	A	L	Dissolve in MIBK. Standardise with Ru tris-acetylacetonate in MIBK	F Air/C_2H_2	169

TABLE 2.2 CHEMICALS, MISCELLANEOUS MATERIALS — continued

Element	λ/nm	Matrix	Concentration	Tech.	Analyte form	Sample treatment	Atomization	Ref.
Ru	349.9	Anodic coatings	5–15 μg/ml (in extract)	A	L	See Ir, ref. 694	F Air/C_2H_2	694
Ru	—	Industrial solutions	—	A	L	Dilute with 10%HCl and add Ti as releasing agent	F —	764
S	384 (S_2)	S compounds in solution	ng levels (in analyte)	E	L	Add H_3PO_3 to overcome interference by metal ions. Buffer to pH 7 with NH_4 phosphate for SCN^- and S^{2-} analysis. Resolve anions on time-of-response basis.	F H_2 + MECA rod	872
S	384 (S_2)	S salts	—	E	L	(Interference study).	F Air/H_2/N_2 + MECA	921
S	384 (S_2)	S-containing compounds	ng/g levels	E	S,L	Add H_3PO_4 or phosphate buffer to overcome cation interference.	F N_2/H_2 + MECA cavity	1514
Sb	217.6	Gunshot residues	0–10 ng/10 μl in extract	A	L	See Ba, ref. 88	Tantalum strip (2500°C in argon)	88
Sb	217.6	Paint	—	A	L	See As, ref. 364	Heated SiO_2 tube	364
Sb	—	Gunshot residues	Trace levels	E,A	S,L	See Ba, ref. 532	A,F —	532
Sb	—	Gunshot residues	From 0.2 ng in 10μl sample	A	L	See Ba, ref. 758	Tantalum strip	758
Sb	259.8	Polyester fibres	20–400 μg/g	E	S	Mix with ZnO, polyoxyethylene glycol and $CHCl_3$, evaporate and fuse at 380–400°C. Mix residue with NaCl/graphite buffer. Use Zn 260.8nm as internal standard	A 15 A d.c.	781
Sb	—	Gunshot residues	—	E	L	—	P ICP	1107
Sb	217.6	Gunshot residues	0.4–2.5 μg	A	L	See Ba, ref. 1220	Graphite furnace	1220
Si	251.6	Streptomycin	From 5 μg/g	A	L	—	F N_2O/C_2H_2	146
Si	251.6	Organo-Si compounds	—	E	L,S	Dissolve in toluene and impregnate filter-paper with solution. Treat with 0.5% Co acetate solution before introduction to spark. Use Co 252.5nm internal standard	S —	1028
Sn	—	Industrial solutions	60–240 μg/ml	A	L	Dilute with 10% HCl	F —	764

Sn	—	Butyl rubber	—	A	L	—	Graphite furnace	1140
Sr	—	Sputtered deposits	From 0.04 μg/ml (in extract)	A	L	See Ba, ref. 963	P —	963
Te	—	Gallium arsenide	—	A	L	See Cr, ref 1153	Graphite furnace	1153
Ti	364.3	Fe–Ti sulphide crystals	—	A	L	Prepare samples and standards in HCl solution and add K_2SO_4 as interference suppressant (2000 μg/ml)	F N_2O/C_2H_2	368
Ti	365.3 320.0	Industrial solutions	μg/ml levels	A	L	Match standards for n-butanol content of samples	F —	764
Ti	—	Polypropylene	—	A	L	—	Graphite furnace	1140
V	310.2 309.3	Graphite	50-5000 μg/g	E	S	Form compacted samples from graphite powder. Use C (I) 247.9 nm as interval standard line	Laser + spark	604
W	255.1	Alumina catalysts	Major levels	A	L	Dissolve in HF/HNO_3	F Air/C_2H_2	1007
Zn	213.8	Disposable pipette tips	Up to 11 ng/g	A	S	See Fe, ref. 13	Graphite furnace (HGA-2000)	13
Zn	213.8	Processing solutions	From 0.002 μg/ml	A	L	—	F —	46
Zn	213.8	Hydrochloric acid; semi-conductor materials	ng/g levels	A	L	Dissolve solid samples (GaAs and $Ga_{0.9}Al_{0.1}As$) in HCl/HNO_3 in PTFE. Evaporate and redissolve in HCl	Graphite furnace	107
Zn	—	Pressed boards	—	A	L	See Al, ref. 538	F —	538
Zn	213.9	Tellurium	1 μg/g	A	L	See Mn, ref. 566	Graphite furnace (HGA-70)	566
Zn	—	Soaps, phosphate materials	1 μg/ml	A	L	See Al, ref. 1268	Graphite furnace (HGA-70, 72)	1268
Zn	—	Zn bacitracin preparations	—	A	L	—	F —	1357
1,2-diols (indirect)	328.1 (Ag)	Organic (1.2-diol) compounds in organic mixtures	From 0.2 μ mole/ml	A	L	Oxidise with periodic acid, precipitate iodate as Ag IO_3, filter and dissolve in NH_4OH. Measure Ag in extract	F —	122
Disodium edetate dihydrate (indirect)	232.0 (Ni)	Streptomycin	From 4 μg/g	A	L	Form Ni complex, release by pH adjustment and determine Ni	F —	148

TABLE 2.2 CHEMICALS, MISCELLANEOUS MATERIALS — continued

Element	λ/nm	Matrix	Concentration	Tech.	Analyte form	Sample treatment	Atomization	Ref.
Anthranilic acid (indirect)	240.7 (Co)	General solutions	3–22 μg/ml in extract	A	L	Buffer to pH 6.3 with phosphate and complex with Co(II) solution + bathophenanthroline in MIBK	F Air/C_2H_2	379
Rubber dips (indirect)	— (Sn)	Tyre cords	—	A	—	Add $Sn(NO_3)_2$ to tyre pre-dip or resorcinol–formaldehyde adhesive	Graphite furnace	483
Drugs (indirect)	324.7 (Cu)	Drug extracts	—	A	L	Methods based on formation of Cu chelate of Schiff base in presence of bathophenanthroline, followed by MIBK extraction	F Air/C_2H_2	886
p-Amino-benzoic acid (indirect)	324.7 (Cu)	Pharmaceuticals	14–37 μg/ml (in extract)	A	L	**Add CuCl/methanol and evaporate. Wash** residue with ethyl acetate, add bathophenanthroline and extract with MIBK	F Air/C_2H_2	1285
Barbituric acid derivatives (indirect)	324.7 (Cu)	Pharmaceuticals	μg/ml levels (in extract)	A	L	Dissolve in 0.08 M Na_2CO_3, add Cu solution + pyridine, dilute and filter. (a) Add HNO_3 to filtrate and dilute; (b) wash precipitate with pyridine and dissolve in HNO_3	F Air/C_2H_2	1291
Various (17)	—	High-purity acids, liquids, solvents	ng/g levels	E	S	Evaporate on graphite disc and place in electrode cavity containing buffer mixture of Ga_2O_3 + NaCl + NaF (2:1:1)	A D.C. arc (12 A)	43
Various (17)	—	Uranium tetrafluoride	0–250 μg/g	E	S	Mix UF_4 directly with diluent to give a final composition of sample + 20% Y_2O_3 + 6.6% $PbCl_2$ + 0.15% Co_3O_4 + 0.11% GeO_2. Take 150 mg portions for analysis	A 10 A d.c.	79
Various (12)	—	Photo-lithographic chemicals	Trace levels	A	L	Comparison with XRF and NAA methods	Graphite furnace	165
Various (8)	—	Alkali and alkaline-earth salts	Trace levels	A	L	Complex with DEDC and absorb on activated carbon. Extract with HNO_3	F —	178
Various (15)	—	Sulphur	Trace levels (from 0.005 μg/g)	E	S	Add HNO_3/H_2SO_4 + Co (0.5 μg/g) as internal standard and heat to dryness at 300°C. Dissolve residue in 6 M HCl and absorb on carrier mixture of NaCl/graphite (1:9)	A 13 A d.c.	292

Various (42)	—	Uranium hexafluoride	Trace levels	E	S	Separate U matrix by partition chromatography	A	—	440
Various (10)	—	Ammonium bifluoride	Trace levels	E	S	Study of effect of various matrix materials on impurity emission. Graphite best for B; MgO for Mn, Si; ZnO for Al, Cr, Cu, Fe, Mo, Ni, Pb, GaO_2 and Ge_2O_3 also considered.	A	D.C. arc	598
Various	—	Propellants	—	A	S	Absorption of solid propellant flames	F	—	969
Various (7)	—	Medicinal preparations	—	A	L	Survey of AAS methods for Au, Co, Cr, Zn, Li, Mn, Pd in various pharmaceuticals	F	—	972
Various	—	GaAs layers	—	A	L	Degrease and etch successive layers by treatment with methanol/B_2. Evaporate individual extracts, dissolve in 9 M HCl and separate matrix elements by column chromatography	F	—	1020
Various (12)	—	High-purity graphite	From 1 μg/g	E	S	Ignite in platinum vessel at 700°C for 6–7 hours, and mix ground residue with NaCl	A	15 A d.c.	1022
Various (20)	—	Aluminium isopropoxide	From 0.01 μg/g (Be, Ag)	E	S	Treat with HNO_3, evaporate and ignite at 700°C. Transfer portion to electrode cavity containing Al_2O_3/NaCl (3:1)	A	10 A d.c.	1027
Various (rare earths)	—	Magnetic garnets and sulphides	Major levels	A	L	Dissolve in HCl (garnets) or HNO_3/Br_2 (sulphides), evaporate and dissolve in 0.1 M HCl. Add 0.5% $LaCl_3$	F	N_2O/C_2H_2	128
Various (15)	—	De-ionised water	Trace levels	A	L	—		Graphite furnace	1065
Various (16)	—	Ammunition, fired cartridges	—	E	L	—	P	ICP	1107
Various	—	Automobile industry materials	—	A,E	L,S	Review of AAS applications to analysis of plastics, ceramics, ores, glass, brake fluids, petroleum products, metals, electronic materials, etc.	F	— Graphite furnace	1150
Various	—	Semi-conductors	Trace levels	E	—	—	A	—	1152
Various (12)	—	High-purity graphite	ng/g levels	E	S	Ash and mix with NaCl	A	15 A d.c.	1155
Various (noble metals)	—	Solutions (from mining technology)	Trace levels	E	S	Treat sample solution ith NaClO, adjust to 10% HCl strength, dilute (×2), boil and add ion-exchange fibre. Boil, filter and ash filter + noble metals. Mix residue with C + Na_2SO_4 + Fe $(NO_3)_3$, or mix with Cu powder and form briquettes	A	15 A a.c. or 4 A d.c.	

TABLE 2.2 CHEMICALS, MISCELLANEOUS MATERIALS — continued

Element	λ/nm	Matrix	Concentration	Tech.	Analyte form	Sample treatment	Atomization		Ref.
Various (12)	—	Boric acid	Trace levels	E	S	Electroplate impurities on to carbon electrode from 20% H_3BO_3 solution at 60°C (40 V; 20 minutes)	A	10 A d.c.	1169
Various (10)	—	Graphite	μg/g levels	E	S	Mix powdered sample with Tl salt, as internal standard	A	15 A d.c.	1198
Various	—	Paint	—	E	S	Review of methods, including ES, for forensic examination of paint	A	—	1211
Various (6)	—	Gunshot residues, bullets	—	A	L	Dissolve bullets in HNO_3. Wet-ash clothing samples with $HNO_3/HClO_4$. Methods cover Ag, Ba, Cu, Pb, Sb, Zn	Graphite furnace		1219
Various	—	Paints, coatings	—	E	S	Review of forensic applications	S	Laser + spark	1330
Various (rare earths)	—	Be, Mg, Ba, Ca and Sr fluorides	—	E	L	Treat with HCl and add saturated solution of H_3BO_3 in 6 N HCl. Digest on water-bath	S	—	1396
Various (22)	—	Tellurium, tellurides	Trace levels	E	L	Dissolve in HCl/HNO_3, precipitate Te with H_2SO_3, filter, evaporate to dryness, redissolve in HCl/HNO_3 (1:1) and transfer to electrode	A	9 A d.c.	1401
Various	—	Drugs	μg/g levels	A	L	Dissolve in 6 N HCl	F	Air/C_2H_2	1526
Various (9)	—	BN	μg/g levels	E	S	Mix with graphite + 2% NaF (carrier) + 1% La_2O_3 (internal standard)	A	10 A d.c.	1532
Various (26)	—	Inorganic materials	From 0.001% μg/g levels	E	S	Mix with dry $BaCl_2$ and form pellets	A	6.5 A d.c.	1569

2.3 METALS

2.3.1 Introduction

The main developments in metal analysis have been associated with increasing practical use of techniques and accessories for which most of the previous reports have described only simple applications. There has been increasing application of furnace and hydride-generation atomic absorption accessories and of plasma sources to metallurgical problems. Many of the earlier reported difficulties are now being overcome and these types of apparatus are well on the way to becoming standard equipment in the metallurgical laboratory.

2.3.2 Sample Preparation

A dissolution method for ferro-silicon alloys has been described (636) which gives complete solution of a wide range of alloys after removal of Si by volatilization with HF. The alloy is dissolved in HF/HNO_3, evaporated to fumes and re-dissolved in HCl. If the solution is cloudy it is filtered and the residue fused with Na_2CO_3. The melt and solution are combined and diluted to volume with 2% HCl.

Harrington *et al* (168) have described a method for dissolution of base metals on surfaces in the presence of precious metals. Alkali fusion was used to strip the oxides from the substrate and was followed by acid digestion. This solution was used for determination of Sn and Sb by AAS. For Ta determination an extraction with MIBK was first carried out. Iron has been determined in Zr by AAS in the range 0.01–1.0% after electrolytic dissolution of the Zr into an organic solvent (1272). The method is rapid, requiring only 30 minutes for a complete determination.

For determination of traces of B in steel using an emission spectrograph the following method (523) has been proposed. The sample (1 g) is dissolved in aqua regia (10 ml): 5% aqueous mannitol (1 ml) is added, and the mixture evaporated almost to dryness. The residue is dissolved in H_2O (10 ml): aqueous NH_3 is added, the solution is evaporated almost to dryness and the residue is ignited at 600°C. The resulting powder is used for the spectrographic analysis.

2.3.3 Flame Atomic Absorption Spectroscopy

A method for the analysis of steels has been described (634) which uses only the air/C_2H_2 flame. This includes a comprehensive study of interference effects, particularly for Cu, Ni, Mn, Mo, and Cr. The responses of Cr and Mo in steels were seriously affected by the matrix when the air/C_2H_2 flame was used. These effects were overcome by adding oxine for Cr determinations, NH_4Cl for Mo determinations, the maintenance of iron concentration at the same level in all solutions and by careful choice of observation height and fuel flow. RSDs of 0.003-0.007 were reported.

Cobb *et al* (53) have reported the determination of Al in steel at the 0.0002-0.001% level. The sample was dissolved in HCl/HNO_3 and any residue treated with HF/H_2SO_4 followed by fusion with H_3BO_3 and Na_2CO_3. The melt was dissolved in HCl and iron extracted from the acid solution with iso-butyl acetate; Al was determined in the remaining solution. Methods for soluble, insoluble and total Al were given.

The determination of some rare-earth metals in U alloys has been described (789). Direct AAS after $HClO_4$ solution of the alloy was used to determine Sc, Y, Ho, Er, Tm, Yb and Lu in the range 0.01–0.1%. It was necessary to match U concentration in calibration solutions to that in the sample solutions as it was found to enhance the absorption of the rare-earth metals, but suppress interferences from other elements.

Two reports of the application of the hydride-generation method to metals using $NaBH_4$ as a reductant are worthy of mention. The determination of trace Bi at 0.002–1.0 $\mu g\ g^{-1}$ in Cu using the technique has been reported (691). The interference effects were found to be "minimal" with quantitative recovery of Bi and a RSD of less than 0.1

obtained. Fleming *et al.* (880) have given a comprehensive report of the determination of As, Sb, Bi, Sn, Se, Te and Pb in steels using the technique. Bismuth was found to be relatively free from interference effects. For the other elements the presence of Cu was found to be most detrimental to the formation of the hydrides. Techniques that may be used to overcome these effects are described and working methods are given for As, Sb, Bi, Fe, Te and Sn in steel. Lead however, was seriously affected by relatively low levels of Ni and Ti. Results are quoted for determination of these elements in standard steels.

2.3.4 Furnace Atomic Absorption Spectroscopy

Ottaway *et al.* (820) have given a review of the state of the art of furnace AAS analysis of metals and alloys. The general procedures used were illustrated by reference to common examples. Methods of calibration and the requirement for background correction was covered, together with the factors which affect reproducibility.

A complete method for the determination of Al in steel, which allows 10 samples to be analysed in 1½ hours has been reported (394). Preliminary investigation indicated that the method could be extended to cover Pb, Bi, Zn, Mg, Mn and Sb. Frech (920) reported the determination of Pb in steel and found that, with careful choice of ashing conditions, background correction was not necessay.

By injecting a nitric acid solution of the sample into the furnace As, Sb and Sn were determined in steel without any pre-treatment (189). Interference effects were studied; Cr, Mo, Ti, Ni, and Nb caused the most problems.

2.3.5 Atomic Fluorescence Spectroscopy

Ebdon *et al.* (175) reported the determination of Sn in steel with detection limits of 0.05 $\mu g\ g^{-1}$ using an Ar-separated air/C_2H_2 flame and 0.01 $\mu g\ g^{-1}$ using Ar/O_2/H_2 flame. Iron(III) chloride was added to calibration solutions at the same concentrations as found in sample solutions to eliminate most serious interference effects; Si had to be removed. The determination of Zn in high-purity Cu, In and Cd at the 1 $\mu g\ g^{-1}$ level was achieved using the air/C_2H_2 flame (180). For the determination of Zn in Cd, the presence of Cd impurity in the Zn EDL caused a positive interference because of ionic fluorescence of Cd at 214.4 nm. This was overcome either by using a Cd-free lamp or by adding K to the test solution to suppress ionization.

2.3.6 Emission Spectroscopy

Le Tring Tam (785) described the determination of B at 5 $\mu g\ g^{-1}$ and above in carbon and low-alloy steels with a d.c. arc using the line at 246.98 nm. Because the 249.77 nm line is not used a high-dispersion spectrograph is not necessary. At 30 $\mu g\ g^{-1}$ B an RSD of 0.039 was obtained. Determination of B down to 0.5 $\mu g\ g^{-1}$ has also been achieved (523) using the preparation method described in section 2.3.2. For the analysis of mixed oxides 10% CuF_2 was added. By using an arc current of 5A, interference from Fe at 249.77 nm was prevented. No interference from Cr or Ni was observed.

The determination of Ti in stainless steels was studied by Alvarez-Arenas *et al.* (1033). Non-representative sampling of solid materials by direct spark excitation was shown micrographically to be due to discrete zones of titanium nitride, and in these cases it is necessary to use a solution of the material for the analysis. Dissolution in HCl/HNO_3/H_2O (3+1+6) was found to be most satisfactory, followed by spark excitation on a high-dispersion spectrograph. An ICP has been used for determining alloy and impurity elements in low- and high-alloy steels (243). Detection limits in the presence of 0.5% Fe showed no significant deterioration from those observed in pure solutions. Calibration graphs were linear over a wide concentration range, interelement interferences were "negligible "and refractory metals such as Nb, W and Zr, could be determined without separation.

TABLE 2.3 METALS

Element	λ/nm	Matrix	Concentration	Tech.	Analyte form	Sample treatment	Atomization	Ref.
Ag	328.1	Lead	3 μg/g	A	L	Dissolve in HNO_3, evaporate, dissolve in H_2O to form 20% solution and add 10 ml of 14 M HNO_3 to 5 ml portion, to precipitate $Pb(NO_3)_2$. Centrifuge. Repeat. Combine supernatant liquids, evaporate, re-dissolve in 2M HNO_3 and add acetone (60% by volume)	F —	438
Ag	328.1	High purity lead, zinc	Trace levels	A	L	Dissolve in HNO_3, adjust to pH 1.0–1.5 and add dithizone solution, in methyl cellulose. Stir, and separate precipitate by N_2-bubble flotation.	F Air/C_2H_2	698
Ag	—	Ni alloys	Up to 10 μg/g	A	L	Dissolve in HCl, HCl/HNO_3 or HCl/HF, add HF, evaporate, add HCl, evaporate, repeat as necessary and dissolve final residue in HF. Transfer to anion-exchange column (Bio-Rex 9), elute matrix with 1.5 N HCl and elute Ag, Pb with 8 N HCl. Elute Zn with 0.001 N HCl and Cd, Bi with 2 N H_2SO_4. Separate Mo, Ti from Ag/Pb eluate on cation-exchange column (Bio-Rex AG-50-X8). Prepare final solutions in HCl Ag/Pb) or HNO_3 (Zn and Cd/Bi)	F Air/C_2H_2	760
Ag	328.1	Iron, steel	Up to 0.02%	A	L	Dissolve with HCl + dropwise HNO_3. Match standards for Fe content	F Air/C_2H_2	1034
Ag	328.1	Ni alloys	0–1.5 μg/g	A	L	Dissolve (10 g) in HCl/HNO_3 and separate by ion-exchange	F Air/C_2H_2	1480
Al	396.1	Ferrous alloys	Trace levels	E	L	Repetitive optical scanning method	F N_2O/C_2H_2	34
Al	309.3	Mild steel	2–10 μg/g	A	L	Digest with HCl/HNO_3, filter and extract with iso-butyl acetate. Evaporate aqueous phase to dryness and redissolve in 50% (v/v) HCl	F N_2O/C_2H_2	53
Al	—	Silver	10–100 μg/g	A	L	Separate Ag matrix by reduction with formic acid and amalgamation with Hg	F —	100

TABLE 2.3 METALS — continued

Element	λ/nm	Matrix	Concentration	Tech.	Analyte form	Sample treatment	Atomization	Ref.
Al	309.3	Iron, steel	From 2 μg/g	A	L	Dissolve in 40% HNO_3/(acid-soluble Al). Fuse residue with $Na_2CO_3/Na_2B_4O_7$ (2:1) and dissolve in HNO_3/(acid-insoluble Al). Atomise at 2660°C	Graphite furnace (HGA-72)	394
Al	396.1	Iron, steel	From 4 μg/g	A	L	Aspirate as 5% solution, using modified (wide-slot) burner with N_2 auxiliary flow	F N_2O/C_2H_2 (+ added N_2)	584
Al	309.3	Ferro-silicon alloys	From 0.1%	A	L	Dissolve in HF/HNO_3, evaporate to fumes, redissolve in HCl, filter if necessary and fuse residue with Na_2CO_3. Combine extracts and dilute to volume with 2% HCl. Add $AlCl_3$ for Ti determination and $SrCl_2$ for Ca determination	F N_2O/C_2H_2	636
Al	396.1	Low-alloy steels	—	E	L	(a) Treat with cold dilute HNO_3, for Al in solid solution in sample (b) treat with dilute HCl, for Al nitride (c) treat with $K_2S_2O_7$, for Al oxide	F N_2O/C_2H_2	768
Al	—	V/Al alloys Zr/Al alloys	0.5–6% 30–45%	E	S	(V/Al alloys). Dissolve in 1:1 HNO_3 and evaporate to dryness. Ignite at 450°C and mix residue (1:1) with graphite containing 20% K_2SO_4. (Zr/Al alloys). Fuse at 1000°C and proceed as above	S —	788
Al	—	Steel	0.018%	A	L	Use standard-addition method of calibration	Graphite furnace	964
Al	—	Steel	0.005%	A	L	—	F —	1194
Al	396.1	Alloys	—	E	L	Nebulize sample solution, containing 0.5% KCl, into arc column	A 6A d.c. (gas-stabilized)	1345
Am	351.0	Curium	—	E	L	—	Hollow cathode	1397
As	193.7	Steel	From 5 μg/g	A	L	Dissolve in HNO_3 (1:4) and dilute. (1 g sample to 100 ml final volume)	Graphite furnace (Techtron 63)	189
As	193·7	Iron, steel	From 0.002%	A	L	See Al, ref. 584	F N_2O/C_2H_2	584

As	193·7	Pb, Cu and Fe alloys	Minor and trace levels	A	L	Dissolve in HNO_3 (Pb), HCl/H_2O_2 (Cu) or $HCl/HNO_3/HClO_4$ (Fe), warm and dilute. For As $<$0.01%, remove Pb as $PbSO_4$ or separate from Cu or Fe by C_6H_6 extraction from $HCl/HBr/HClO_4$ solution	F	Air/C_2H_2 or N_2O/C_2H_2	782
As	228.8	Al metal	ng/g levels	E	G	Treat Al with HCl + $SnCl_2$, in vessel connected to liquid-N_2 trap. Add Zn to complete reaction. Sweep AsH_3 from cold trap with argon into discharge tube	P	Microwave discharge	1284
Au	242.8	Metal products	Trace levels	A	S	Wet-ash with HCl/HNO_3. Apply correction for background absorbance using Sn 242.9 nm		Heated graphite rod + flame	410
Au	—	Ag beads	Trace levels	E	S	—	S	—	413
B	249.8	Steel	From 0.5 μg/g	E	S	Dissolve in HCl/ HNO_3, add 5% mannitol solution, evaporate to low bulk, redissolve in H_2O + NH_4OH, evaporate to dryness and ignite at 600°C. Add 10% CuF_2 to residue	A	5A d.c.	523
B	249.8 518.0	Steels	9–80 μg/g	A,E	L	Dissolve in $HCl/HClO_4$ (Comparison of AAS and FES methods)	F	N_2O/C_2H_2 N_2O/H_2	618
B	249.7	Steel	From 5 μg/g	E	S	—	A	D.C. arc	785
B	—	Copper	0.003–0.1%	E	S	Study of 'liquid-layer-on-solid-sample' technique	S	—	1303
Be	234.9	Al, Cu alloys	Trace levels	A	L	Extract with acetylacetone into MIBK from aqueous solution containing EDTA + NaCl, buffered to pH 5–7 with Na acetate	F	N_2O/C_2H_2	1282
Bi	306.8	Lead	1 μg/g	A	L	Dissolve in 1 M HNO_3 and chelate Bi with Zn-dibenzyl-dithiocarbamate (0.03% solution in CCl_4). Wash organic phase with 3.5 M HCl	F	Air/C_2H_2	66
Bi	400	Cu alloys	80 μg/g	E	L	Dissolve in HCl/HNO_3, dilute and add to MECA cavity packed with CaO. Use standard-addition method	F	$Air/H_2/N_2$	494
Bi	223.1	Ni alloys	0.5 μg/g	A	L,G	Dissolve in HCl/ HNO_3 and add $NaBH_4$ to generate Bi hydride, after separation by anodic stripping voltammetry		Heated quartz tube (850°C)	603

TABLE 2.3 METALS — continued

Element	λ/nm	Matrix	Concentration	Tech.	Analyte form	Sample treatment	Atomization		Ref.
Bi	223.1	Copper	Trace levels, 0.002–1 μg/g	A	L,G	Co-precipitate with $La(OH)_3$. Dissolve in HCl. Reduce solution with $NaBH_4$ and pass bismuthine vapour to flame	F	—	691
Bi	—	Ni alloys	Up to 10 μg/g	A	L	See Ag, ref. 760	F	Air/C_2H_2	760
Bi	223.1	Ni alloys	0–7.5 μg/g	A	L	See Ag, ref. 1480	F	Air/C_2H_2	1480
Bi	223.1	Steels	From 5 μg/g	A	L	Dissolve in HCl/HNO_3	Graphite furnace		1551
C	—	Steel	Up to 2.8%	E	S	Evaluation of errors due to variable Cr and Mn contents	S	—	439
C	—	Stainless steel	0.083%	E	S	Study of performance of adjustable waveform spark source	S	—	1093
Ca	422.7	Zr alloys	Trace levels	A	L	Dissolve in HF and separate matrix (e.g., Zr, Nb) by cation-exchange. Elute Zr, Nb with 1% HF and Ca with 50% HC.	F	N_2O/C_2H_2	314
Ca	422.7	Fe/Si alloys	From 0.02%	A	L	See Al, ref. 636	F	Air/C_2H_2	636
Cd	228.8	Pb and Zn concentrates	—	A	L	Treat with HCl/HNO_3, evaporate and redissolve in acetate/HCl solution	F	N_2O/C_2H_2	187
Cd	228.8	Cu alloys	0.04–0.55%	A	L	—	F	Air/C_2H_2	275
Cd	—	Ni alloys	Up to 10 μg/g	A	L	See Ag, ref. 760	F	Air/C_2H_2	760
Cd	228.8	Ni alloys	0–0.4 μg/g	A	L	See Ag, ref. 1480	F	Air/C_2H_2	1480
Ce	—	Iron, steel	—	E	S	—	S	(in argon)	998
Co	241.2	Cu alloys	0.001–2%	A	L	Separate Cu matrix if Co $<0.1\%$	F	Air/C_2H_2	275
Cr	359.3	Iron, steel	Minor and trace levels (from 0.1 μg/ml in solution)	E	L	Dissolve in 6 M HCl + small addition HNO_3 heat, evaporate to dryness and redissolve in 3M HCl. Remove large amounts of Si by treatment with HF in platinum vessel	P	Argon plasma torch (2450 MHz)	157
Cr	301.5	Fe alloys	—	E	S	—	Laser microprobe		412
Cr	357.9	Steels	0.2–20%	A	L	Dissolve in HCl/HNO_3. Determine Cu, Mn, Ni directly. Add NH_4Cl for Mo determination and quinolin-8-ol for Cr determination. Match all solutions for Fe content	F	Air/C_2H_2	634

Cr	357.9	Steels	0.1–0.5%	A	L	Dissolve in H_2SO_4, dilute, boil, add NaF + NH_4 hexanitrato-cerate solution and extract with 4-methyl-2-pentanone	F	Air/C_2H_2	699
Cr	—	Steels	All levels	A	L	See Cu, ref. 596	F	N_2O/C_2H_2	596
Cr	—	Steel	0.001–10%	E	S	Statistical study of precision from data given by high-repetition-rate source	S	500 Hz	613
Cr	—	Stainless steel	17.6%	E	S	See C, ref. 1093	S	—	1093
Cr	267.7	Cr-plated steel	10–70% (in layer)	E	S	—	S	—	1562
Cu	324.7	Pb and Zn concentrates	—	A	L	See Cd, ref. 187	F	N_2O/C_2H_2	187
Cu	324.7	Steel	—	A	L	Study of effect of acid treatments	F	Air/C_2H_2	290
Cu	—	Steels	0.01–0.23%	A	L	Add buffering solution of NH_4Cl + $MgCl_2$ +$LaCl_3$, to avoid necessity for matching standards	F	N_2O/C_2H_2	596
Cu	324.7	Steel	0.04–0.23%	A	L	See Cr, ref. 634	F	Air/C_2H_2	634
Cu	324.7	High-purity lead, zinc	Trace levels	A	L	See Ag, ref. 698	F	Air/C_2H_2	698
Cu	—	Al alloys	From 0.1% (AAS) From 3 μg/g (ICP)	E,A	S	Vaporize with high-voltage spark and carry particulate vapour to flame (AAS) or induction-coupled plasma (ICP)	Spark F P	atomization + $Ar/H_2/O_2$ or Ar (27 MHz)	841
Cu	—	Cobalt and production materials	—	A	L	Dissolve Co metal in HNO_3, evaporate, dissolve in HCl and dilute. Dissolve Co salts directly in HCl. Add 1% 0.5 M H_2SO_4 to all solutions	F	Air/propane butane	1026
Cu	324.7	Al alloys, steel	Minor and trace levels	A	L	Extract Cu(II) from aqueous CNS^- solution into propylene carbonate	F	Air/C_2H_2	1233
Cu	327.4	Alloys	—	E	L	See Al, ref. 1345	A	6 A d.c.	1345
Fe	248.3	Cu alloys	0.007–1%	A	L	Separate Cu matrix if Fe $<$0.1%	F	Air/C_2H_2	275
Fe	302.1 374.8	Fe/Mn alloys	10–20%	E	L	Crush and grind to 400-mesh size and dissolve in hot (1:1) HNO_3. Cool, filter and dilute. See also "Various," ref. 580	P	ICP. 27 MHz (argon)	580
Fe	—	Cobalt and production materials	—	A	L	See Cu, ref. 1026	F	Air/propane/ butane	1026
Fe	248.3	Zirconium and alloys	0.01–1%	A	L	Use electrolytic dissolution into organic solvent	F	O_2/C_2H_2	1272

TABLE 2.3 METALS — continued

Element	λ/nm	Matrix	Concentration	Tech.	Analyte form	Sample treatment	Atomization	Ref.
Fe	371.9	Alloys	—	E	L	See Al, ref. 1345	A 6A d.c.	1345
Ga	294.4	Cu alloys	0.05–1.8%	A	L	Separate Cu matrix if Ga $<$1%	F Air/C_2H_2	275
Ge	265.2	Cu alloys	0.05–1.7%	A	L	Separate Cu matrix if Ge $<$1%	F Air/C_2H_2	275
Hg	253.7	Bi, In, Pb, Sn and Tl alloys	0.1–50 μg/g	E	S	—	A 10–15 A d.c.	1402
K	766.5	Ti bronze	—	E	L	Fuse with excess $NaHSO_4$ (or $KHSO_4$ for Na determination) at 650°C. Dissolve melt in H_2SO_4/H_2O_2. Prepare blank solutions from $NaHSO_4$ (or $KHSO_4$) + TiO_2	F —	511
Mg	285.2	Zr alloys	Trace levels	A	L	Use standard-addition method. See also Ca, ref. 314	F N_2O/C_2H_2	314
Mg	280.3	Nodular iron	0.03% (ferritic) 0.1% (perlitic)	E	S	Use Fe 279.8 nm as internal standard	Laser microprobe + auxiliary spark	583
Mg	—	Al alloys	0.2–5%	E	S	Comparison of excitation methods	(a) Grimm lamp (b) Spark in argon	609
Mg	—	Al alloys	From 2 μg/g	E	S	See Cu, ref. 841	P —	841
Mg	—	High-purity aluminium	—	A	L	Sensitivity study	F N_2O/C_2H_2 Air/C_2H_2	1308
Mn	279.4	Iron alloys	—	E	S	—	Laser microprobe	412
Mn	279.5	Steel	Minor and trace levels	A	L	Automated system for the analysis of ferrite and carbide constituents of steels	F Air/C_2H_2	451
Mn	257.6	Al alloys	—	E	S	Use Al 306.0 nm as internal standard (Comparison of two excitation sources)	S (a) 800 V, 50 Hz (b) 12.5 KV, 300 Hz	579
Mn	257.6 403.4	Fe/Mn alloys	70–80%	E	L	See Fe, ref. 580 and "Various", ref. 580	P ICP 27 MHz	580
Mn	—	Steels	0.2–1.0%	A	L	See Cu, ref. 596	F N_2O/C_2H_2	596
Mn	279.5	Steel	0.2–1.2%	A	L	See Cr, ref. 634	F Air/C_2H_2	634

Mn	279.5	Ferro-silicon alloys	From 0.01%	A	L	See Al, ref. 636	F	Air/C_2H_2	636
Mn	—	Cobalt and production materials	—	A	L	See Cu, ref. 1026	F	Air/propane/ butane	1026
Mn	—	Stainless steel	0.8%	E	S	See C, ref. 1093	S	—	1093
Mn	293.3	Aluminium	0.03–0.7%	E	S	See B, ref. 1303	S	—	1303
Mo	379.8	Steels	Minor and trace levels	A	L	Add 4% NH_4Cl to sample solution, to overcome interference by Ni, Co, Cr	F	Air/C_2H_2	557
Mo	313.3	Steels	0.07–2.5%	A	L	See Cr, ref. 634	F	Air/C_2H_2	634
Mo	—	Steels	2–10%	A	L	—	F	—	1336
Na	589.0	Zr alloys	Trace levels	A	L	Use standard addition method. See also Ca, Mg, ref. 314	F	Air/C_2H_2	314
Na	589.0	Ti bronze	—	E	L	See K, ref. 511	F	—	511
Nb	—	Steels	0.003–0.6%	E	S	Study of interference by Cr	S	—	293
Nb	334.3	Steels	Up to 5%	A	L	Dissolve in HCl/HNO_3, followed by HF. Add NH_4Cl + KCl + acetic acid as interference and ionisation suppressant system	F	N_2O/C_2H_2	885
Nb	—	Steel	5–50% (in steel precipitate)	E	L	Dissolve in HCl. Fuse residue with $K_4P_2O_7$ and dissolve in HCl or tartaric acid	P	ICP	1193
Nb	—	Steel	Major levels	A	L	Review	F	—	1194
Ni	232.0	Fe alloys	—	E	S	—	Laser microprobe		412
Ni	341·5	Steels	0.1–10%	A	L	See Cr, ref. 634	F	Air/C_2H_2	634
Ni	—	Steels	0.05–10%	A	L	See Cu, ref. 596	F	N_2O/C_2H_2	596
Ni	—	Cobalt and production materials	—	A	L	See Cu, ref. 1026	F	Air/propane/ butane	1026
Ni	—	Stainless steel	9.5%	E	S	See C, ref. 1093	S	—	1093
P	313.3 (Mo)	Iron,steel	0·01–0.05%	A	L	Extract P into butyl acetate as phosphomolybdate complex	F	—	1353
Pb	283.3 261.3 364.0	Pb and Zn concentrates	—	A	L	See Cd, ref. 187	F	Air/C_2H_2	187
Pb	—	Ni alloys	Up to 10 μg/g	A	L	See Ag, ref. 760	F	Air/C_2H_2	760

TABLE 2.3 METALS — continued

Element	λ/nm	Matrix	Concentration	Tech.	Analyte form	Sample treatment	Atomization	Ref.
Pb	283.3	Steel	1–100 $\mu g/g$	A	L	Dissolve in HCl/HNO_3. Ash at 625°C before atomization. Calibrate by standard addition method. Match standards for any Mo present	Graphite furnace	920
Pb	217.0 283.3	Aluminium	28–48 $\mu g/g$	A	L	Dissolve in HCl. To test Al cooking ware, treat with 5% acetic acid at 85°C for 6 hours	F —	1038
Pb	217.0	Al and alloys	0.0005–0.1%	A	L	Dissolve in HCl, evaporate to low bulk, dilute, add NH_4 acetate and then 1% pyrrolidine-1-carbodithioic acid until no further precipiation. Filter, extract filtrate Pb with MIBK and precipitated Pb with HNO_3. Extract MIBK with HNO_3 and determine total Pb in HNO_3 extract	F Air/C_2H_2	1199
Pb	363.9	Cu alloy	—	E	S	(Computerized direct-reading 'television' spectrometer)	A —	1377
Pb	217.0	Non-ferrous alloys	—	A	L	(Interference study)	F N_2/H_2/air Ar/H_2/air	1378
Pb	283.3	Ni alloys	0–7.5 $\mu g/g$	A	L	See Ag, ref. 1480	F Air/C_2H_2	1480
Pb	283.3	Fe, Ni alloys	—	A	L	Dissolve in acid, add KI + ascorbic acid and extract with TOPO/MIBK	F Air/C_2H_2	1516
Pd	—	Ag beads	Trace levels	E	S	—	S —	413
Pt	—	Ag beads	Trace levels	E	S	—	S —	413
Rh	—	Ag beads	Trace levels	E	S	—	S —	413
S	384 (band)	Alloys	3–50 $\mu g/ml$ (in extract)	E	L,S	—	F Ar/H_2 + graphite cup	729
Sb	217.6	Precious metal on titanium substrate	—	A	L	Remove from substrate by alkali fusion and digest with acid. Determine Sb and Sn directly. Extract Ta as fluoride complex into MIBK	F N_2O/C_2H_2	168
Sb	217.6	Steel	From 5 $\mu g/g$	A	L	See As, ref. 189	Graphite furnace (Techtron 63)	189

Sb	—	Pb alloys	0.1–3.5%	A	L	Dissolve in HNO_3/citric acid	F	—	1214
Se	**196.0**	Crude copper	0.02–0.06%	A	L	Dissolve in $HNO_3/HClO_4$, evaporate to fuming and dissolve in 6 M HCl. Precipitate Se with $Fe(OH)_3$ collector, dissolve in HNO_3 and extract Se by complexing with 3,3′-diamino-benzidine	Heated Mo tube (2500°C)		924
Se	196.0	Copper, steels	From 0.1 μ/ml (in extract)	A	L	Form organo-Se complex with aceto-phenone in $HCl/HClO_4$ solution. Extract with $CHCl_3$, evaporate and redissolve in MIBK	F	Air/C_2H_2	951
Si	251.6	Ferro-manganese	0.05–3%	A	L	Dissolve in HCl/ HNO_3, complex Si with HF, filter and add solution of 2% H_3BO_3 + 5% $Na_2B_4O_7$.10 H_2O. Dilute to volume (1 g sample to 250 ml final volume)	F	N_2O/C_2H_2	77
Si	—	Al alloys	—	E	S	See Mg, ref. 609	(a) Grimm lamp (b) Spark in argon		609
Si	**251.6**	Tungsten production metals	—	A	L	Treat with $HF/HNO_3/H_2SO_4$. Impregnate graphite with 5% $NaWO_3$ solution to improve results	Graphite furnace (HGA-72)		702
Si	—	V alloys	0.1–1.0%	E	S	Mix powdered sample (1:1) with carbon powder containing 20% Co_2O_3. See also Al, ref. 788	A	12 A a.c.	788
Si	—	Stainless steel	0.5%	E	S	See C, ref. 1093	S	—	1093
Si	**288.1**	Steel	0.03–0.6%	E	S	See B, ref. 1303	S	—	1303
Sn	224.6	Precious metal on titanium substrate	—	A	L	See Sb, ref. 168	F	N_2O/C_2H_2	168
Sn	**303.4**	Steel	From 1 $\mu g/g$ Optimum range 0.005–0.1%	F	L	Dissolve in HCl/HNO_3 and (a) evaporate, redissolve in HCl, filter and dilute, or (b) add HF, boil, evaporate to low bulk and redissolve in HCl. Match standards for Fe concentration	F	Air/C_2H_2 (Ar-separated) or $Ar/O_2/H_2$	175
Sn	**224.6**	Steel	From 10 $\mu g/g$	A	L	See As, ref. 189	Graphite furnace (Techtron 63)		189
Sn	224.6	Cu alloys	0.007–0.3%	A	L	Separate Cu matrix if Sn $<$1%	F	Air/C_2H_2 or N_2O/C_2H_2	275

TABLE 2.3 METALS — continued

Element	λ/nm	Matrix	Concentration	Tech.	Analyte form	Sample treatment	Atomization		Ref.
Sn	286.3	Iron, steel	From 8 μg/g	A	L	See Al, ref. 584	F	N_2O/C_2H_2	584
Sn	326.2	Cu alloy	—	E	S	See Pb, ref. 1377	A	—	1377
Sn	286.3	Fe, Ni alloys	—	A	L	See Pb, ref. 1516	F	N_2O/C_2H_2	1516
Ta	271.5	Precious metal on titanium substrate	—	A	L	See Sb, ref. 168	F	N_2O/C_2H_2	168
Ti	365.3	Ferro-silicon alloys	From 0.1%	A	L	See Al, ref. 636	F	N_2O/C_2H_2	636
Ti	364.3	Iron, steel	0.05–0.5%	A	L	Digest with HCl + dropwise HNO_3, filter and fuse any insoluble residue with Na_2CO_3/H_3BO_3. Dissolve melt in HCl and combine with original extract. (Interference suppression by addition of La or Al)	F	N_2O/C_2H_2	930
Ti	334.9 336.1 337.1	Stainless steels	0·2–0.7%	E	L	Dissolve in HCl/HNO_3 (3:1). Use Fe 337.1, 339.2, 340.7 nm internal-standard lines	S	—	1033
Ti	—	Steel	5–50% (in steel precipitate)	E	L	See Nb, ref. 1193	P	ICP	1193
V	—	Steels	0.08–0.6%	A	L	See Cu, ref. 596	F	N_2O/C_2H_2	596
V	—	Steel	5–50% (in steel precipitate)	E	L	See Nb, ref. 1193	P	ICP	1193
W	294.7	Fe alloys	—	E	S	—	Laser + spark		412
Zn	213.9	High-purity Cd, Cu, In metals	From 1 μg/g	F	L	Dissolve in HNO_3 (1:1) and dilute. (1 g sample to 100 ml final volume). Add K to suppress Cd interference	F	Air/C_2H_2	180
Zn	213.9 307.6	Pb and Zn concentrates	—	A	L	See Cd, ref. 187	F	Air/C_2H_2	187
Zn	—	Ni alloys	Up to 10 μg/g	A	L	See Ag, ref. 760	F	Air/C_2H_2	760
Zn	—	Al alloys	From 150 μg/g	E	S	See Cu, ref. 841	P	—	841
Zn	—	Brass	32%	E	S	See C, ref. 1093	S	—	1093
Zn	636.2	Cu alloy	—	E	S	See Pb, ref. 1377	A	—	1377
Zn	213.9	Ni alloys	0–3.8 μg/g	A	L	See Ag, ref. 1480	F	Air/C_2H_2	1480

Zr	—	Steel	5–50% (in steel precipitate)	E	L	See Nb, ref. 1193	P	ICP	1193
Various (9)	—	Tungsten powder	Trace levels of Ag, Cd, Co, Cu, In, Ni, Pb, Tl and Zn	A	L	Dissolve in H_2O_2 and add NaDDC. Filter through activated charcoal, dry filter at 110°C and extract with HNO_3	F	Air/C_2H_2	58
Various	—	Metals, alloys	—	A	L	Review	F	—	172
Various (7)	—	High-purity aluminium	μg/g levels	A	L	Coat sample (10–25 g) with thin layer of Hg and treat with HCl, to concentrate impurities (Cd, Bi, Ga, In, Pb, Tl, Zn) in small residue	F	Air/C_2H_2	208
Various (12)	—	Steels	All levels	E	L	Dissolve in HCl/HNO_3 (6:1). (Detailed study of analytical features)	P	Induction-coupled argon plasma	243
Various (9)	—	Uranium	Trace levels	E	S	—	S	—	288
Various (6)	—	Uranium and uranium compounds	Trace levels	E	S	Convert to oxide and mix with carrier (AgCl or AgCl/SrF_2 (4:31) for trace detection of Nb, Rh, Ru, Ta, W, Zr	A	D.C. arc	289
Various	—	Iron and steel	—	A	L	Review, 28 refs.	F	—	318
Various (6)	—	Ultra micro samples	—	A	L	Electronic integrating system, for Ca, Cu, K, Li, Mg, Na in sample volumes from 1 nl	F	Air/H_2 + Pt/Ir loop	321
Various (15)	—	High-purity silver	Trace levels	A	L	Separate Ag matrix by extraction with tert-dodcyl mercaptan from 1 N HNO_3 solution. Extract Pd with dioctyl sulphoxide and Bi, Cu, In, Se, Te, Zn with dibutyl thiophosphate. Method also covers Co, Cr, Fe, Mn, Ni, Pb, Pt, Rh	F	—	352
Various (11)	—	Steels, Al alloys	—	E	S	Study of effect of atmospheres (air, O_2, N_2, Ar, Ar/O_2, He/O_2) on spectral emission intensities		Laser microprobe + auxiliary spark (Ar/O_2 atmosphere)	372
Various	—	Cast iron	—	E	S	Study of electrode temperature effects	S	—	375
Various	—	Steels, alloys	All levels	A	L	Review, including composite analysis scheme for steels	F	—	384
Various	—	Pb/Sn alloys	—	A	L	Review of acid treatments for differing Pb/Sn ratios	F	—	386

TABLE 2.3 METALS — continued

Element	λ/nm	Matrix	Concentration	Tech.	Analyte form	Sample treatment	Atomization	Ref.
Various	—	Ni alloys	All levels	A	L	Review	F —	388
Various	—	Al alloys	—	E	S	Study of spark sampling efficiency	S —	408
Various (6)	—	High-speed steels	—	E	S	Vaporize with contact spark (0.5–1.0 A, 100 V) on to Cu electrode and excite transferred metal with high-voltage spark (Co, Cr, Mn, Mo, V, W)	S Two-stage method	505
Various	—	Steels	—	E	S	Vaporize with low-voltage arc and pass aerosol to flame photometer (portable apparatus)	A,F —	535
Various (9)	—	Fe/Mn alloys	Trace levels	E	L	See Mn, Fe, ref. 580. Method covers Al, B, Co, Cr, Cu, Mo, Ni, Ti, V	P ICP 27 MHz	580
Various	—	Cu alloys	—	E	S	Investigation of low-voltage unipolar spark source	S —	592
Various	—	Alloys	All levels	E	S	Automated system for the rapid identification of alloys	S —	602
Various (12)	—	Steels	All levels	E	L,S	Applications to metals analysis of ICP system, using either solution or solid aerosol samples	P ICP	615
Various	—	Archaeological bronzes	—	E	S	Complete-burn method, using alloy chips	A A.C. arc (unipolar)	617
Various (11)	—	Ni alloys	Trace levels, from 0.1 μg/g	E	S	—	Hollow-cathode (410 V, in argon)	712
Various	—	Au platings	Trace levels	E	S	Study of surface composition and metal diffusion	Glow-discharge	713
Various (7)	—	High-temp. alloys	μg/g levels	E	S	Dissolve in HF/HNO_3, evaporate, ignite and mix with AgCl/LiF carrier	A 12 A d.c.	715
Various (10)	—	Al alloys	μg/g levels	E	S	—	Laser microprobe	720
Various	—	Steel	All levels	E	S,G	(Remote sampling aerosol generator coupled to capillary arc and optical spectrometer)	A 10 A d.c. (in argon) (excitation); A 4 A d.c. (vaporization)	767

Various (7)	—	U alloys	0.01–0.1% (Rare earths)	A	L	Determine directly in $HClO_4$ solution. (Method for Sc, Y, Ho, Er, Tm, Yb, Lu)	F	—	789
Various	—	Alloys	—	E	S	Applications of glow-discharge lamp	Grimm lamp		793
Various	—	Alloys	Trace levels, to 0.0001%	A	L	Aplications of graphite furnace AAS, e.g., Pb, Al in steels; Pb, Bi in copper alloys; Zn in cobalt; Al, Sb, Ag, Bi, Pb, Zn in nickel alloys	Graphite furnace (HGA-72)		820
Various	—	Steels	Trace levels, to 0.1 μg/g	A	S	Study of solid sampling technique	Graphite furnace		821
Various	—	Pure metals	—	A	S	Study of laser atomisation	Graphite furnace + pulsed laser		835
Various	—	Al alloys	μg/g levels	A	S	Laser excitation, e.g., for Cu, Mg, Zn in Al alloys	Laser atomization (0.6 joule)		836
Various	—	Zinc	μg/g levels	E	S	Pass laser-produced vapour to microwave cavity in argon atmosphere	Laser atomization + microwave excitation		837
Various	—	Alloys	10–100 μg/g	A	S	Applications include Al in Zn; Mg, Si, Fe in Al; Ni in Fe alloys	Cathodic sputtering, in argon		838
Various	—	Alloys	—	A	S	Applications include Pb, Al, Cr in Zn coatings and study of surface composition of Fe, Zn and Al alloys	Cathodic sputtering		839
Various	—	Steels	Minor and trace levels	A,E	S	—	Glow discharge lamp (RSV)		842
Various	—	Precious metal concentrates	—	A	L	Prepare solution in 20% HCl. Add 1% La for Pt, Pd, Rh and Au determinations, 0.5% Cu for Ru and 0.5% Cd for Ir. (Study of several releasing agents)	F	Air/C_2H_2	862
Various (7)	—	Steels	Trace levels	A	G	Generate metal hydride by reaction with $NaBH_4$ and pass directly to flame. Results obtained for As, Sb, Bi, Se, Te, Sn. Severe interference on Pb	F	$Ar/H_2/air$	880
Various	—	Metals	—	A,E	L	Review, 8 refs.	F	—	956
Various (alkalis)	—	W, Mo and compounds	—	E	L	Study of matrix effects	F	—	961
Various (8)	—	High-purity nickel	3 μg/g–3%	A	L	Dissolve in HNO_3	F	—	1023

TABLE 2.3 METALS — continued

Element	λ/nm	Matrix	Concentration	Tech.	Analyte form	Sample treatment	Atomization	Ref.
Various	—	Ni, Co alloys	—	A	L	Dissolve in HCl/HNO_3, dilute to volume, add 0.01 M H_2SO_4 (containing 1000 μg/ml Mg to mask Fe) to aliquot and dilute	F Air/propane/butane	1025
Various	—	Alloys	Major levels	E	S	Study of glow-discharge technique applied to micro-samples	Glow discharge sputtering	1074
Various	—	Al alloys	—	E	S	Time-resolved spectrometry study	S —	1095
Various	—	Complex alloys	Trace levels	A	L	Study of non-analyte corrections	F —	1130
Various	—	Al and alloys	—	E	S	Description of standard alloy system	— —	1151
Various (8)	—	High-purity nickel	3–3000 μg/g	A	L	Dissolve in HNO_3. Elements covered: Cd, Zn, Mg, Cu, Fe, Co, Mn, Pb	F —	1154
Various (6)	—	Cu/Ni alloys	From 1 ng/g	E	S	Dissolve in HCl/H_2O_2, boil with ion-exchange fibre, fuse fibre with Na_2O_2 and extract with HCl. Then, (a) evaporate portion with C powder (for Pt, Pd, Rh, Au) or (b) separate portion by cation-exchange (for Ru, Ir)	A —	1157
Various (19)	—	Non-metallic steelworks materials	—	E	S	Dilute with mixture of graphite + Co_3O_4 + LiF	A —	1208
Various	—	Metals	All levels	A	L	Review	— —	1222
Various (6)	—	Steels, alloys	Minor and trace levels	A	L	Review of methods	F —	1254
Various	—	Titanium and alloys	% levels	E	S	(Effect of microstructure)	A —	1263
Various (9)	—	Aluminium and alloys	0.001–0.1%	E	S	—	S —	1265
Various (6)	—	Steels	Up to 0.3%	E	S	Precision study	A D.C. arc	1280
Varous (7)	—	Ferro-molybdenum	Trace levels	E	S	Dissolve in HNO_3 (1:1) in platinum dish and ignite at 400–500°C. Mix residue with tetraethyl ammonium iodide	A —	1318
Various	—	Uranium, plutonium	Trace levels	A	L	—	Graphite furnace	1349
Various	—	Cu alloys	—	E	S	Comparison of CuO/globule arc spectrographic method and Cu metal spectrometric method	A 6 A d.c.	1400

Various	—	Iron, steel	% levels	E	S	—	Grimm lamp		1460
Various	—	Steels	—	E	S	Review of visual (steeloscope) methods	—	—	1482
Various (8)	—	Steel inclusions	% levels	E	S	Dilute with graphite + internal standard	A	9 A d.c.	1489
Various (20)	—	Pd powder	1–1000 μg/g	E	S	—	A	12–15 A d.c.	1500
Various (33)	—	Refractory metals (Mo, Nb, Ta, W)	μg/g levels	A	L	Comparison with XRF	F Graphite furnace	—	1509
Various (12)	—	High-purity gallium	Trace levels	A	L	Treat with 1 M HCl to extract Zn. For other elements, dissolve in HCl/HNO_3 until small Ga bead remains. Transfer to flask and dissolve in HCl/HNO_3. Evaporate to dryness and redissolve in HNO_3	F	—	1515
Various	—	Ferro-alloys	—	E	S	Study of errors	A	—	1542
Various	—	Ferro-alloys	—	E	S	Study of sample preparations	S	—	1553
Various	—	Steel surfaces	—	A	L	Form contact with NH_4Cl-impregnated filter-paper and strip electrolytically	Graphite furnace		1555
Various (8)	—	Fe alloys	—	E	S	Sample preparation study	S	—	1557
Various (10)	—	High-purity aluminium	μg/g levels	A	L	Dissolve in HCl (1:1) + few drops H_2PtCl_6 + H_2O_2. Evaporate to low bulk bulk and dissolve in H_2O	F	Air/C_2H_2 N_2O/C_2H_2	1560
Various	—	High-purity indium and tellurium	Trace levels	E	S	Dissolve in HNO_3, evaporate and ignite. Mix In_2O_3 with graphite (2:1) + 5% NaCl. Mix TeO_2 with graphite (4:1) + 3% KBr	A	A.C. arc (Te) D.C. arc (In)	1563
Various (9)	—	High-purity aluminium	Trace levels	A	L	(a) Dissolve in NaOH, add thioacetamide and co-precipitate impurities with $La(OH)_3$ (b) dissolve in HCl and extract with NaDDC/MIBK at pH 3	F	Air/C_2H_2	1568
Various	—	Rail steel inclusions	% levels	E	S	—	Laser + spark		1572

2.4 REFRACTORIES, METAL OXIDES, CERAMICS, SLAGS AND CEMENTS

2.4.1 Introduction

The main developments in the analysis of oxide materials during the year have been in methods of dissolution and consequently a section on sample preparation has again been written. The sections on individual techniques given in previous years have been replaced by one dealing with the analysis of materials where interesting developments have occurred.

2.4.2 Sample Preparation

Further development of both fusion and acid treatment methods of sample dissolution have been reported. Campbell and Passmore (571) have used a mixture of $BaCO_3$ and B_2O_3 as a flux for dissolving high-alumina, zirconia and magnesia refractories. The flux is claimed to give very low blanks for alkali and for alkaline earths other than Ba. When used with flame spectrometry it gives a low flame background and reduced matrix effects. High-alumina materials have also been dissolved by fusion with KHF_2 followed by addition of H_3BO_4 to complex the excess fluoride (422).

The problem of keeping Si in solution when a fusion is dissolved in acid has been examined by several workers (6, 763, 1141). Burdo and Wise (1141) used an acidic molybdate to prevent polymerisation. They claim that the solution is very stable and can be used for the determination of Si by AAS with an accuracy of 0.2 to 0.3%. Sellers *et al* (763) used lithium tetraborate as a flux and worked in graphite crucibles to prevent wetting. It should be noted that the use of this and similar fluxes has been studied in the context of X-ray fluorescence spectrometry by workers on British Ceramic Reasearch Association Panels (*e.g.* Ambrose, B.C.R.A. Special Publication No. 86). Trouble with loss of certain elements into graphite crucibles has also been reported by these workers.

Acid dissolution of oxide materials such as chrome-magnesite, slags, iron ore, chrome ore, silica brick, magnesite and fluorspar has been achieved by Hafton and Baines (1030) by using phosphoric acid. They determined Cr, Ca, Al, Mg and Si though it appears that the last element is not always completely dissolved. The solutions are said to be very stable and give very little background in AAS. Acid dissolution with HF/H_2SO_4 followed by slow evaporation to dryness has been used for high-purity lead glasses (550). The trace-metal impurities can then be leached from the $PbSO_4$ residue with HCl.

An extraction method has been used for the determination of Sb in glass (1287). The element is converted to iodide and extracted with a high molecular weight amine for AAS determination.

2.4.3 Individual Materials

Kanda (381) has analysed *titanium dioxide* for Pb at the $\mu g\ g^{-1}$ level by mixing it with graphite and placing the mixture directly into a graphite tube furnace. The method is calibrated by introducing aliquots of Pb solution into the furnace. The same material has been examined spectrographically using GaF_3 as a carrier (125).

The determination of *rare-earth* elements using a carbon arc in air suffers from interference by CN bands. Early work by Wiggins (*Analyst* 1949, **74**, 101) on this problem used a steam atmosphere to displace air. The use of oxygen for this purpose is now being proposed (504, 664). The use of oxygen may result in rapid combustion of the electrodes while a mixture of Ar + 30% O_2 gives a reasonable rate of combustion and a similar voltage drop across the arc to that obtained in air or oxygen. Higher Ar contents give a lower arc voltage and hence lower wattage dissipation.

The analysis of *cement* for calcium by AAS (974) and FES (865) has been investigated by Voinovitch. The possibility of using Sr, Mg or La as an internal standard was examined and it was concluded that their use did not improve the precision of the determinations. The accuracy of the FES determination was improved. A spectrographic method for Mg, Si, Fe, Al, Ti and Cr in this and similar materials by the rotating-disc technique has been described (1491). The author examined the effects of excitation parameters on inter-element effects rather than on precision, as done by earlier workers (*Appl. Spectrosc.*, 1965, **19**, 69 and 1969, **23**, 111). The precision obtained in the new work is poorer than that of the earlier investigations.

Uranium oxide has been examined by hollow-cathode lamp for isotope ratio (1295) and for F and Cl content (626); these are the tasks for which the lamp with its narrow line width and high excitation energy is well suited.

TABLE 2.4 REFRACTORIES AND METAL OXIDES, CERAMICS, SLAGS, CEMENTS

Element	λnm	Matrix	Concentration	Tech.	Analyte form	Sample treatment	Atomization	Ref.
Al	—	Slags (Cr/Mo/V steel)	—	E	S	Grind and dilute (×15) with buffer mixture of CuO + $SrCO_3$ + carbon powder (6:13:31)	A 15 A a.c.	503
Al	—	Cement	—	A	L	Study of operating conditions	F —	936
Al	—	Slags, ores, refractories	—	A	L	Dry, grind to 200-mesh size and digest with H_3PO_4 in Pt crucible. Heat to 300°C, cool, extract with H_2O, filter and dilute	F —	1030
Al	309.2	Concretes, cements	1–10%	E	S,L	Add Co internal standard	S Rotrode (in argon)	1491
Ba	—	Alkaline earth niobates	—	A	L	Fuse at 950°C with K_2CO_3, dissolve melt in HCl/H_2O_2 and dilute with HCl (1:1)	F N_2O/C_2H_2	1024
Ca	422.7	Cement	Major levels	A	L	Add La + excess HCl, to reduce sensitivity	F Air/C_2H_2 (rotated burner)	556
Ca	422.7	Refractory oxides	—	E	L	Grind and mix with flux of H_3BO_3 + $BaCO_3$ (2+1) in ratio 10 parts flux + 1 part sample. Fuse and dissolve	F Air/C_2H_2 or O_2/H_2	571
Ca	—	Cement	Major levels	E	L	Add La + Sr to sample solution and use Sr as internal standard	F Air/C_2H_2	865
Ca	—	Cement	—	A	L	See Al, ref. 936	F —	936
Ca	422.7	Cement	—	A	L	(Study of Sr or Mg as internal standard for Ca)	F —	974
Ca	—	Slags, ores, refractories	—	A	L	See Al, ref. 1030	F —	1030
Cl	725.6	Uranium oxide	From 10 μg/g	E	S	—	Hollow cathode	626
Co	240.7	Lead silicate glasses	0.02–1 μg/g	A	L	(a) Treat with HF, evaporate, add HNO_3, evaporate and dissolve residue in HNO_3. Add buffer solution and extract with batho-phenanthroline into MIBK (Fe, Cu, Ni, Co); (b) treat with HF/H_2SO_4, evaporate slowly to dryness, shake residue with HCl and extract impurities from HCl with NaDDC into MIBK (Fe, Cu, Ni, Co, Mn, Cr)	Graphite furnace (HGA-70)	550

Co	240.7	Uranium oxides	From 2·5 μg/g	A	L	Convert metal samples to oxide by treatment with HNO_3. Evaporate and redissolve in HNO_3	Graphite furnace (HGA-70)		561
Cr	357.9	Corundum materials	—	A	L	Fuse with KHF_2, followed by H_3BO_3	F	Air/C_2H_2 (fuel-rich)	422
Cr	357.9	Lead silicate glasses	0.02–1 μg/g	A	L	See Co, ref. 550 (Cr recoveries low by method (b))	Graphite furnace (HGA-70)		550
Cr	357.9	Uranium oxides	From 0.5 μg/g	E	L	See Co, ref. 561	Graphite furnace (HGA-70)		561
Cr	—	Slags, ores, refractories	—	A	L	See Al, ref. 1030	F	—	1030
Cr	267.7	Concretes, cements	1–10%	E	L	See Al, ref. 1491	S	Rotrode (in argon)	1491
Cu	324.7	Lead glasses	0.02–1 μg/g	A	L	See Co, ref. 550	Graphite furnace		550
Dy	407.8	High-purity Eu_2O_3	0.005–0.1%	E	S	Mix (1:1) with graphite powder	A	15 A d.c. (in O_2)	504
Dy	353.2	Gd_2O_3	Up to 0.5%	E	S	Mix (1:1) with graphite powder	A	10 A d.c. (in O_2)	664
Eu	381.9	Gd_2O_3	Up to 0.25%	E	S	See Dy, ref. 664	A	10 A d.c.	664
F	703.8	Uranium oxide	From 3 μg/g	E	S	—	Hollow-cathode		626
Fe	248.3	Lead silicate glasses	0.04–10 μg/g	A	L	See Co, ref. 550	Graphite furnace (HGA-70)		550
Fe	248.3	Uranium oxides	From 1.25 μg/g	A	L	See Co, ref. 561	Graphite furnace (HGA-70)		561
Fe	—	Cement	—	A	L	See Al, ref. 936	F	—	936
Fe	259.8	Concretes, cements	1–10%	E	L	See Al, ref. 1491	S	Rotrode	1491
Gd	364.6	High-purity Eu_2O_3	0.005–0.1%	E	S	See Dy, ref. 504	A	15 A d.c.	504
K	—	High-calcium glasses	—	E	L	Decompose with HF/H_2SO_4, heat at 900°C, dissolve in H_2O and precipitate. Ca and Al with $(COO.NH_4)_2$. Filter	F	—	129
K	—	Refractory oxides	—	E	L	See Ca, ref. 571	F	Air/C_2H_2 or O_2/H_2	571
Li	—	Refractory oxides	—	E	L	See Ca, ref. 571	F	Air/C_2H_2 or O_2/H_2	571
Mg	—	Refractory oxides	—	A	L	See Ca, ref. 571	F	Air/C_2H_2	571

TABLE 2.4 REFRACTORIES AND METAL OXIDES, CERAMICS, SLAGS, CEMENTS — continued

Element	λnm	Matrix	Concentration	Tech.	Analyte form	Sample treatment	Atomization	Ref.
Mg	—	Cement	—	A	L	See Al, ref. 936	F —	936
Mg	—	Slags, ores, refractories	—	A	L	See Al, ref. 1030	F —	1030
Mg	279.5	Concretes, cements	1–10%	E	L	See Al, ref. 1491	S Rotrode	1491
Mn	—	Slags (from Cr/Mo/V steel)	—	E	S	See Al, ref. 503	A 15 A a.c.	503
Mn	279.5	Lead glasses	0.02–1 µg/g	A	L	See Co, ref. 550	Graphite furnace	550
Mn	279.5	Uranium oxide	From 0.1 µg/g	A	L	See Co, ref. 561	Graphite furnace	561
Mo	—	Slags (from Cr/Mo/V steel)	—	E	S	See Al, ref. 503	A 15 A a.c.	503
Na	—	High-calcium glasses	—	E	L	See K, ref. 129	F —	129
Na	—	Refractory oxides	—	E	L	See Ca, ref. 571	F Air/C_2H_2 or O_2/H_2	571
Nd	406.1	High-purity Eu_2O_3	0.005–0.1%	E	S	See Dy, ref. 504	A 15 A d.c. (in O_2)	504
Ni	232.0	Lead silicate glasses	0.02–1 µg/g	A	L	See Ca, ref. 550	Graphite furnace (HGA-70)	550
Pb	283.3	Titanium dioxide	0.05–5 µg/g	A	S	Mix, by grinding, with 5- to 9-fold excess of graphite powder. Take 1–5 mg for analysis. Calibrate with aqueous Pb standards (0.025–0·3 µ/ml)	Graphite furnace (HGA-2000)	381
Rb	—	Refractory oxides	—	E	L	See Ca, ref. 571	F Air/C_2H_2 or O_2/H_2	571
Sb	217.6	Glass	—	A	L	Adjust aqueous solution to 0.8 M HCl + 0.25 M KI and extract Sb complex with Amberlite LA-1 in ethyl acetate	F Air/C_2H_2	1287
Si	—	Ceramics	50–70% (as SiO_2)	A	L	—	F N_2O/C_2H_2	254
Si	288.1	Garnets, oxides	Up to 5% (garnets) Up to 100 µg/g (oxides)	E	S	—	A 6 A a.c.	442

Si	—	Slags (Cr/Mo/V steel)	—	E	S	See Al, ref. 503	A	15 A a.c.	503
Si	—	Tungsten oxides	0.5–100 ng (as SiO_2)	A	—	Study of tube impregnation methods	Graphite furnace		702
Si	—	Cement	—	A	L	See Al, ref. 936	F	—	936
Si	—	Slags, ores, refractories	—	A	L	See Al, ref. 1030	F	—	1030
Si	—	Glasses, minerals	14–94% (as SiO_2)	A	L	Fuse with $Na_2CO_3/Na_2B_4O_7$ and dissolve melt in acidic molybdate solution	F	—	1141
Si	288.1	Concretes, cements	1–10%	E	L	See Al, ref. 1491	S	Rotrode	1491
Sm	363.4	High-purity Eu_2O_3	0.005–0.1%	E	S	See Dy, ref. 504	A	15 A d.c.	504
Sm	360.9	Gd_2O_3	Up to 0.05%	E	S	See Dy, ref. 664	A	10 A d.c.	664
Sr	—	Refractory oxides	—	E	L	See Ca, ref. 571	F	Air/C_2H_2 or O_2/H_2	571
Sr	—	Alkaline earth niobates	—	A	L	See Ba, ref. 1024	F	N_2O/C_2H_2	1024
Tb	367.6	High-purity Eu_2O_3	0.005–0.1%	E	S	See Dy, ref. 504	A	15 A d.c.	504
Tb	367.6	Gd_2O_3	Up to 0.25%	E	S	See Dy, ref. 664	A	10 A d.c.	664
Ti	319.9	Concretes, cements	1–10%	E	L	See Al, ref. 1491	S	Rotrode	1491
U (isotopes)	502.7	Uranium oxide	0.6–20% (^{235}U)	E	S	—	Hollow-cathode (N_2 + Ar)		1295
V	—	Slags (from Cr/Mo/V steel)	—	E	S	See Al, ref. 503	A	15 A a.c.	503
W	294.7	Slags from Fe/W plant)	—	E	S	Study of sources of error	A	D.C. arc	40
W	255.1	Alumina catalysts	Major levels	A	L	Dissolve in HF/HNO_3	F	Air/C_2H_2	1007
Y	398.3	Gd_2O_3	Up to 0.5%	E	S	See Dy, ref. 664	A	10 A d.c.	664
Various (Rare-earths)	—	Ba and Pb titanates, zirconates	Up to 4 μg/ml in extract. (La, Pr, Nd, Eu and Sm)	E	L	Comparison of porous cup and aerosol methods of sample excitation	S	—	45
Various (11)	—	High-purity titania	Trace levels. (From 0.5 μg/g Pb, Sn, Si, Mg, Fe, Al, Cu, Co; 1 μg/g Mn, Ni and 5 μg/g V)	E	S	Mix powdered sample with 5% GaF_3. Use Ti 311.25 nm internal standard	A	10 A d·c.	125

TABLE 2.4 REFRACTORIES AND METAL OXIDES, CERAMICS, SLAGS, CEMENTS — continued

Element	λnm	Matrix	Concentration	Tech.	Analyte form	Sample treatment	Atomization	Ref.
Various (10)	—	Potassium fluoro-tantalate	μg/g levels	E	S	Grind (<200 mesh) and adhere powder to graphite electrode. Use Ta 298·7 nm as internal standard	A A.C. arc	291
Various (12)	—	Tantalum oxide	Minor and trace levels	E	S	Mix sample with ZnO and graphite powder (1:1:1). Use Ta 297.5 nm as internal standard line	A 15 A d.c.	444
Various (11)	—	Tungstic oxide	1–1000 μg/g	E	S	Mix with 20% graphite powder	A 9–18 A d.c. Magnetic field	696
Various	—	Refractories, U_3O_8	Trace levels	A	S	—	Graphite furnace	823
Various (9)	—	Slags, ores, sinters and refractories	All levels	A	L	Fuse with Li tetraborate in graphite crucible, cool and dissolve in 10% HCl. For SiO_2 refractories, fuse with Li metaborate and treat with $HClO_4$	F N_2O/C_2H_2	763
Various	—	High-purity glass	Trace levels	A	L	—	Graphite furnace	1058
Various (16)	—	Uranium oxide	Trace levels	E	S	(a) For B, Cd, V, Fe, Cr, Pb mix with 5% NaF (b) for Ni, Co, Zn, Mn, Mo mix with 5% AgCl/NaF (4:1) (Results also given for Cu, Mg, Eu, Sm, Gd)	A 8 A d.c.	1200
Various	—	Glasses, refractories, ceramics	Major levels	A,E	L	Review	— —	1224
Various (14)	—	Quartz	1–1000 μg/g	E	S	—	A —	1315
Various (11)	—	Vanadium oxide	μg/g levels	E	S	Convert to V_2O_3 and mix (20:2:1) with AgCl + carbon. Introduce NaCl from counter-electrode.	A 7 A a.c.	1398
Various (19)	—	Steel works material		E	S	Dry sample, mix with C + LiF + Co_3O_4. Pack into electrode	— —	1208

2.5 MINERALS

2.5.1 Introduction

The interest of mineralogists in flame AES continues, this is particularly noticeable in the use of the nitrous oxide-acetylene flame. More recently, the ICP, which has great potential for trace element analysis, has attracted attention.

Direct atomisation from the solid sample is being exploited by using different types of graphite furnace but this is obviously dependent on good sampling techniques and most workers are still content to prepare samples by dissolution.

2.5.2 Sample Preparation

Zimmermann (6) has discussed several preparation techniques which reduced the time required for the determination of trace metals in minerals by AAS. These were based on Na_2O_2 and borate fusion followed by dissolution in cold HF to retain silica in solution. Very stable silicate solutions have also been obtained by the dissolution of a carbonate-borate fusion melt in an acid molybdate solution (1141). The use of lithium fluoroborate (450) as a flux for rock and mineral samples has also been investigated. Na_2O_2 fusion has been used in the determination of Sb in sulphide ore (791). For this procedure zirconium crucibles were used and are recommended for their long life.

Dissolution of silicate rocks in a sealed PTFE vessel has been employed by several workers. Warren and Carter (786) dissolved up to 2 g in an hour at 100°C using $HF/HNO_3/HClO_4$. A similar system (1062) has been applied to the determination of 21 elements by flame AAS. Interelement effects are said to be low and good agreement was obtained for standard rocks.

Various separation techniques have been described and three are of special interest. Trace amounts of Sb in rocks and soils have been determined after separation by sublimation as the iodide followed by recovery into MIBK with TOPO (303). The co-precipitation of Te from rocks with As by hypophosphorus acid was recommended for its determination at the ppb level. The Te was recovered as bromide by extraction into MIBK (668). Lead was separated from Pb ores, prior to determination of its isotopic composition, by sequential anodic and cathodic deposition onto platinum foil (567).

The use of graphite powder as a collector before arc AES has been advocated by several Russian workers. In the determination of Au and Pt (1157) the fibre 'Mtilon T' was used to separate the metals from boiling solutions of Cu and Ni metals obtained by smelting Cu-Ni ores. The fibre was then fused, dissolved and evaporated to dryness with a graphite collector, which was then blown through a horizontal arc. This procedure was said to take 3–5 days. In the analysis of Cu-Ni sulphide ores (1158) for the Pt metals they were precipitated with thiourea. The precipitate was dissolved in aqua regia and the solution evaporated in the presence of 150 mg carbon powder. The residue was excited with a 25A arc. The use of an ion-exchange fibre (1162) is also mentioned in section 2.2.5 for the collection of noble metals in chloride/sulphate liquors. Improved sensitivity and reproducibility were claimed (1160) in the determination of noble metals in concentrates after treatment with ion-exchange resins to remove foreign ions. The concentrate was evaporated with carbon powder and the residue mixed with Na_2SO_4. Excitation with 15A d.c. arc gave detection limits of 0.6 $\mu g\ g^{-1}$ Rh, 0.3 $\mu g\ g^{-1}$ Pd and 1 $\mu g\ g^{-1}$ Pt, Au.

As an alternative to dissolution one worker has crushed samples of rocks, soils and sediments to pass 325 mesh and suspended the material in water before its direct nebulization into the flame (1239).

2.5.3 Atomic Absorption Spectroscopy

Two developments of importance in work on minerals have been reported. The determination of Ga has a better sensitivity and imposes less stringent requirements on flame conditions when N_2O/C_2H_2 rather than air/C_2H_2 flames are used. Line overlap of Pr and Nd has been reported at 492.45 nm; as a result the determination of Nd in the presence of Pr by AAS requires the use of 463.42 nm line (1363).

The determinations of Ag and Au have received considerable attention. Background correction (555) was required when Ag was determined in calcareous samples. The scatter by Ca was about 100 000 times less than the absorption by Ag but was much more sensitive to flame conditions and observation height than the Ag absorption. Investigations into the determination of Au (102) showed that, in Cu-bearing sulphide ores, As present in the ore might aid the volatilization of the Au, roasting must therefore be performed by heating the sample slowly at low temperatures. It is reported that aqua regia is effective in dissolving all Au present after roasting only if the sample is heated with 50% HCl before its addition. A combined fire-assay/flame AAS method (421) is suggested for low Au and Ag levels in mineral raw materials by some Russian authors whereas others (410) prefer the direct determination of Au with a combined electrothermal/flame atomizer. A graphite rod with a 95 mm groove is held in a flame and resistance heated to 2500 K. For non-specific absorption the Sn 242.94 nm line is measured simultaneously with the Au 242.8 nm line at the same slit and wavelength settings. Two HCLs are powered in opposite phase and if the absorption signal is integrated the values are independent of sample composition. A comparison of fire-assay and FAAS techniques for Au (914) indicated that the scatter predicted by sampling statistics was significantly greater than that found experimentally. A method (280) for concentrating Au by fluorination with ClF_3 has been developed; this is convenient for lean ores containing high amounts of Se, As, S and SiO_2 to prevent Au loss into the slag. The use of nickel matte as a collector (306) in fire-assay concentration resulted in an enrichment factor of 1000.

The majority of workers have employed *electrothermal atomization* with solution samples. Two papers have described direct analysis of the solid sample. Langmyhr (822) used a high-frequency graphite furnace to determine eleven elements directly in silicate minerals, sulphide ores and concentrates. Lower limits of detection were claimed compared with flame methods and the risk of contamination was reduced. In the second paper (1164) the atomic vapour from the solid rock sample was diffused through the bottom of a graphite crucible and the vapour condensed on an auxillary electrode. The latter was then analysed by pulse thermal atomization for the elements Cd, Tl and Ag (D.Ls 0.02, 0.2 and 0.05 $\mu g\ g^{-1}$ respectively).

The determination of Ag in ores (311) was made more sensitive by using the carbon rod atomizer rather than a flame, the improved limit of detection was 0.002 $\mu g\ ml^{-1}$ after a liquid-liquid extraction. Similarly Pb and Cd in phosphate rock (657), Pb in carbonate rocks (995), and trace metals in coal (430) gave better sensitivity and provided accurate and precise results. Beryllium could be measured in fly ash (551) at the ng g^{-1} level with a graphite furnace.

2.5.4 Atomic Fluorescence Spectroscopy

Only two specialised application papers concerned with this technique have appeared. In the first of these (980) 5 mg samples of lunar regolith were analysed for Hg with a detection limit of 1 ng g^{-1} after selective vaporisation into an Ar atmosphere at 800–900°C. In the other Epstein *et al* (773) have examined several standard reference materials, including fly ash and coal, for Cd and Zn after a mixed acid solution procedure. Automatic scatter correction was used.

2.5.5 Atomic Emission Spectroscopy

The determination of alkaline earth elements in geological samples (340), Li in silicate rocks (1032) and the rare-earth elements in naturally occurring materials (864) are examples of the application of *flame emission techniques.* Flame emission scanning (145) in geochemical studies of alkali metals has been demonstrated with emphasis on the determination of Li. Sensitivities were Na 0.05, K5, Li 0.1, Rb 25 and Cs 200 μ gl^{-1}.

The advantages of the ICP source are beginning to be appreciated but applications in the minerals field are still rather restricted and of an exploratory nature. Michaelson (863) was enthusiastic about the performance of the plasma for the study of noble metals in mineral samples and stated that it gave versatility unattainable by other analytical methods. However, Silvester (737) in a study of 25 elements in rocks, soils and sediments has focussed attention on the effect of varying sample matrix on the slope and intercept of the analytical curves. For most of the 25 elements the soil content caused only minor changes in the slope but for certain elements such as Pb strong interference was observed. Here a regular signal increase with solution ionic strength occurred, as much as 800% enhancement in the emission signal for a 2% NaCl solution was noted relative to a 0.5M HCl reference solution.

A review (608) of factors contributing to uncertainty of analytical results in simultaneous multi-element trace analysis in geological materials, recommended the use of fusion procedures to enhance accuracy in *d.c. arc emission* spectrography. Golightly and co-workers (716) showed that improved accuracy could be obtained by using computer corrections for matrix effects. The entry of metal vapours into a d.c. arc exerts a powerful influence on both temperature (T) and electron pressure (Pe) in the arc plasma. Only recently it has been possible to attempt to routine measurement and to use the effective valves of T and Pe for correcting spectral line radiances (Decker and Eve, *Appl. Spectrosc.*, 1969, **23**, 497).

A new type of sheathed d.c. arc (174) with good stability and negligible CN emission has been evaluated on 60 rock and soil samples for the determination of a wide range of elements. Separate argon streams sheath the anode and lower part of the cathode while the plasma was supported by an Ar/O_2 mixture. A similar system based on the 'double jet' arrangement of Boumans and Maessen (*Spectrochim. Acta* B 1969, **24**, 585) stabilised the arc and reduced CN background, in direct-reading emission spectroscopy (1029).

The systematic examination of the performance of the *laser micro-analyser* (see *ARAAS* 1973, **3**, 147) has been extended to include other minerals and steel (1043, 641). Working curves for Mn, Cr and Cu were provided and RSDs of 0.1 for steel and 0.2–0.3 with minerals have been reported.

TABLE 2.5 MINERALS

Element	λnm	Matrix	Concentration	Tech.	Analyte form	Sample treatment	Atomization	Ref.
Ag	328.1	Ores	From 2 ng/g	A	L	Separate Ag by liquid–liquid extraction (**ARAAS**, 1973, **3**, ref. 1209) and use 20% HCl/10% H_3PO_4 solution for analysis	Carbon rod	311
Ag	328.1	Ores	—	A	L	Collect noble metals in Pb bead by fire-assay. Dissolve in HNO_3 (1:4) for Ag and treat undissolved residue with HCl/HNO_3 for Au	F Air/propane	421
Ag	328.1	Rocks, soils, sediments	0–10 μg/g	A	L	Study of Ca interferences	F Air/C_2H_2	555
Ag	328.1	Rocks	Trace levels	A	S	Heat sample in graphite crucible and condense vapour (Ag, Tl, Cd) on electrode for thermal atomisation	Graphite furnace	1164
Al	—	Scheelite	—	E	S	Mix (1:3) with carbon powder and add Ta_2O_5 as internal standard	A —	294
Al	396.1	Sulphide and silicate minerals	0.02–5 μg/g (E) or from 5 μg/g (A)	E,A	L	Fuse with Na_2O_2 or digest with $HNO_3/HCl/HF/HClO_4$. Add Na as ionisation buffer	F N_2O/C_2H_2	931
Al	396.1	Silicate minerals	% levels	E	S	—	Ion-beam sputtering	1459
As	234.9	Silicate rocks	From 1 μg/g	E	S	Mix with $Na_2S_2O_3$/graphite (2:3) buffer	A 9 A d.c.	1031
Au	242.8	Sulphide ores and products	0.1–40 μg/g	A	L	Roast for 2 hours at 600°C, boil with HCl, add HNO_3 and evaporate to dryness. Redissolve in 20% HCl, extract with MIBK and wash with HCl to remove any Fe before aspiration	F Air/C_2H_2	102
Au	—	Pyrite and arseno-pyrite	—	A	L	—	F —	115
Au	—	Ores	0.6–2.5 g/ton in ores and 25–33 g/ton in concentrates	A,E	L,S	Concentrate Au by treatment with ClF_3, for 20 hours at 290–370°C. Treat residue ith HCl/HNO_3 and determine Au in solution by AAS and in dry residue by ES	F,A —	280
Au	242.8	Ores	—	A	L	See Ag, ref. 421	F Air/propane	421

Au	242.8	Ores and tailings	0.6–12 μg/g	A	S	Wet ash with aqua regia. Use Sn 242.9 nm for background correction	Graphite furnace + flame	410
Au	—	Geological materials	Trace and minor levels	A	L	(Precision study)	F —	914
Au	—	Geological materials	From 0.01 μg/ml (in extract)	A	L	Pre-treat to dissolve and remove SiO_2	Graphite furnace	1264
Au	—	Ores	% levels	E	S	Prepare Ag bead by fire-assay	A 5 A d.c.	1539
Ba	—	Calcareous rocks	300 μg/g	A	L	Add $LaCl_3$ interference suppressant	F N_2O/C_2H_2	250
Ba	553.6	Silicate rock	From 3 μg/g	A	L	Treat with $HNO_3/HClO_4/HF$ in PTFE bomb. Cool, heat extract to fumes of $HClO_4$, cool, add $HCl/H_2O/H_3BO_3$, boil, add K and dilute to volume	F Air/C_2H_2 or N_2O/C_2H_2	786
Be	234.8	Fly ash, coal, environmental materials	0–12 μg/g	A	L	Treat with $HF/H_2SO_4/HNO_3$, heat to fuming, add $HClO_4$ and dilute.	Graphite furnace (HGA-2000)	551
Bi	306.8	Silicate rocks	From 0.1 μg/g	E	S	See As, ref. 1031	A 9 A d.c.	1031
Bi	—	Silicate rocks, soils	μg/g levels	A	L	Treat with $HF/HClO_4$, evaporate to dryness with HF, repeat with HCl and redissolve in HCl + few drops HNO_3. Adjust aliquot to pH 1.0 and extract with hexahydroazepinium-hexahydroazepine-1-carbodithioate into butyl acetate	F Air/C_2H_2	1523
Ca	—	Lignite fly-ashes	—	A	L	Extract with H_2O	F —	32
Ca	422.7	Silicate minerals	% levels	E	S	—	Ion-beam sputtering	1459
Cd	—	Phosphate rock	—	A	—	—	Graphite furnace	657
Cd	228.8	Coal, fly ash	μg/g levels	F	L	Digest with $HNO_3/HF/HClO_4$ in Teflon, evaporate to fuming and re-dissolve in H_2O	F Air/C_2H_2	773
Cd	—	Pb and Zn concentrates	—	A	L	Comparison with polarographic method	F —	187
Cd	228.8	Rocks	Trace levels	A	S	See Ag, ref. 1164	Graphite furnace	1164
Cd	228.8	Limestone, fossil shell	Trace levels	A	L	Decompose with HF/HNO_3, treat with HCl/HNO_3, add H_3PO_4 and concentrate to clear solution. Dilute and extract with KI/MIBK	F Air/C_2H_2	1279
Co	240.7	Rocks	—	A	L	Decompose with HF, evaporate and dissolve in H_2O. Extract with coniferron into CCl_4	Mo tube furnace	695

TABLE 2.5 MINERALS — continued

Element	λnm	Matrix	Concentration	Tech.	Analyte form	Sample treatment	Atomization	Ref.
Co	232.0	Silicate rock	From 5 μg/g	A	L	See Ba, ref. 786	F —	786
Co	—	Rocks, minerals	2–50 μg/g	E	S	Pd internal standard	A —	599
Co	—	Geological materials	From 0.01 μg/ml (in extract)	A	L	See Au, ref. 1264	Graphite furnace	1264
Cr	301.4	Rocks	0.01–0.1%	E	S	—	Laser + spark	641
Cr	357.9	Silicate rock	From 2 μg/g	A	L	See Ba, ref. 786	F —	786
Cr	301.4	Rocks	0.01–0.1%	E	S	—	Laser + spark	1043
Cr	—	Ilmenite	From 200 μg/g	A	L	Treat with HCl/HF (1:1) in Teflon. (Comparison with fusion methods)	F Air/C_2H_2 N_2O/C_2H_2	1404
Cs	—	Silicate rocks	μg/g levels	A	S	Mix powdered sample with ionisation buffer (NaCl for Li, Rb and Na_2CO_3 for Cs). Deposit on threaded Fe rod and insert in flame	F —	1261
Cu	—	Rocks	1–10 μg/g	A	L	—	Electro-contact atomizer (120 A)	338
Cu	327.4	Rocks	0.001–0.1%	E	S	—	Laser excitation	641
Cu	324.8	Rocks	—	A	L	See Co, ref. 695. Extract Cu with DDC into CCl_4	Mo tube atomizer	695
Cu	324.7	Silicate rock	From 2 μg/g	A	L	See Ba, ref. 786	F —	786
Cu	—	Pb and Zn concentrates	—	A	L	See Cd, ref. 187	F —	187
Cu	327.4	Rocks	0.001–0.1%	E	S	—	Laser + spark	1043
Cu	324.7	Limestone, fossil shell	Trace levels	A	L	See Cd, ref. 1279	F Air/C_2H_2	1279
Cu	324.7	Ilmenite	From 0.5 μg/g	A	L	Dissolve in HCl/HF (1:1) in Teflon. Dry in furnace at 100°C, ash at 750°C, atomize at 2450°C	Graphite furnace	1405
Fe	—	Lignite fly-ashes	—	A	L	See Ca, ref. 32	F —	32
Fe	258.5	Chromites	3–15%	E	S	Briquette sample (0.1 g) with buffer (0·15 g) consisting of Cr_2O_3 + Al_2O_3 + carbon (4:1:1), containing 0.15% B, and 1.25 g Cu powder	S —	36

Fe	259.9 275.5 275.3 241.3	Sphalerites, pyrites	0.15–15%	E	S	(Differentiation of isomorphous and heterogeneous Fe)	S,P	Laser + auxiliary spark	1390
Ga	287.4	Ores	1–3 μg/g	A	L	Grind, roast and extract with hot HCl. (Comparison of flame systems)	F	Air/C_2H_2 or N_2O/C_2H_2	176
Ga	294.3	Coal, coke	1–15 μg/g	E	S	Grind, heat to 600°C for 4–5 hours, mix ash with equal weight of $(NH_4)_2SO_4$ + graphite + SbO_2 (50:449:1) and use Sb 261.2 nm internal standard	A	D.C. arc	513
Ge	265.1	Coal, coke	1–15 μg/g	E	S	See Ga, ref. 513	A	D.C. arc	513
Hf	—	Mining concentrates	—	E	L	Separate Hf/Zr concentrates by extraction with TBP. Form aerosol in stream of argon and pass to carbon electrode system	S	—	506
Hg	—	Geological materials	—	A	—	Review, 83 refs.	Flameless methods		978
Hg	253.7	Rocks, lunar regolith	From 80 ng/g	F	—	Heat at 800–900°C in argon	—	—	980
Hg	253.7	Sphalerite	—	A	—	Dissolve in $HNO_3/HClO_4$ (4:1) at 150°C. Reduce with $SnCl_2$. Mix with CaO (1:1) to avoid interferences	—	—	1015
In	—	Silicate rocks and soils	μg/g levels	A	L	See Bi, ref. 1523	F	Air/C_2H_2	1523
Li	670.7	Rocks	1.1–22 μg/g	E	L	Flame emission scanning technique for Li, also applicable to K, Na, Rb, Cs	F	Air/C_2H_2	145
Li	670.7	Sr minerals	5 μg/ml in extract	A	L	Study of Sr interference. Use method of standard additions, with background correction at 667.8 nm	F	—	337
Li	—	Silicate rocks and minerals	95–3000 μg/g	E	L	Digest with acid and add excess K as buffer	F	—	1032
Li	—	Silicate rocks	μg/g levels	A	S	See Cs, ref. 1261	F	—	1261
Mg	—	Lignite fly-ashes	—	A	L	See Ca, ref. 32	F	—	32
Mn	293.3	Rocks	0.1–1.0%	E	S	—	Laser excitation		641
Mn	293.3	Rocks	0.1–2%	E	S	—	Laser + spark		1043
Mn	279.5	Ilmenite	From 50 μg/g	A	L	See Cr, ref. 1404	F	Air/C_2H_2	1404
Na	—	Lignite fly-ashes	—	A	L	See Ca, ref. 32	F	—	32

TABLE 2.5 MINERALS — continued

Element	λnm	Matrix	Concentration	Tech.	Analyte form	Sample treatment	Atomization	Ref.
Na	589·0	Cryolite	—	E,A	L	Treat with H_2SO_4, evaporate and redissolve in HCl	F O_2/H_2 (E) Air/C_2H_2 (A)	101
Na	589.0	Silicate minerals	% levels	E	S	See Al, ref. 1459	Ion-beam sputtering	1459
Nd	463.4	Minerals, ores	—	A	L	Spectral interference study	F —	1363
Ni	232.0	Silicate rock	From 7 μg/g	A	L	See Ba, ref. 786	F —	786
Os	290.9 305.8	Cu/Ni ores	μg/g levels	E	S	Add NH_4Cl $(NH_4)_2CO_3$ and charcoal to ore before roasting, to prevent Os loss. Fuse any residual alloy and dissolve in HCl/H_2O_2. Oxidise with HNO_3, distil Os as OsO_4 and absorb in 20% HCl + few drops C_2H_5OH. Evaporate with graphite powder	A —	1161
Pb (isotopes)	405.8	Ores and minerals	12–52%	E	S	Separate by sequential anodic and cathodic electrodeposition. Deposit Pb on Pt foil and use as cathode in lamp	Hollow-cathode (He-filled)	567
Pb	—	Phosphate rock	—	A	—	—	Graphite furnace	657
Pb	283.3	Carbonate rocks	1–100 μg/g	A	L	Dissolve in HNO_3 and filter to remove SiO_2 or, for complete dissolution, treat with HF	Graphite furnace (HGA-70)	995
Pb	—	Pb and Zn concentrates	—	A	L	See Cd, ref. 187	F —	187
Pb	—	Geological materials	From 0.02 μg/ml (in extract)	A	L	See Au, ref. 1264	Graphite furnace	1264
Pb	283.3	Limestone, fossil shell	Trace levels	A	L	See Cd, ref. 1279	F Air/C_2H_2	1279
Pb	283.3	Rocks	Trace levels	E	S	Mix with graphite powder containing 20% NaF (1:1)	A 10 A d.c.	1394
Pb	—	Silicate rocks, soils	μg/g levels	A	L	See Bi, ref. 1523	F Air/C_2H_2	1523
Pd	—	Ores, sweeps, slimes	—	A	L	Treat with HCl/HNO_3 and evaporate almost to dryness. Add HCl saturated with MIBK, transfer to flask and extract with NaDDC into MIBK	F —	759
Pd	—	Ores	% levels	E	S	See Au, ref. 1539	A 5 A d.c.	1539
Pt	—	Ores	% levels	E	S	See Au, ref. 1539	A 5 A d.c.	1539

Rb	—	Silicate rocks	μg/g levels	A	S	See Cs, ref. 1261	F	—	1261
Rh	—	Ores	% levels	E	S	See Au, ref. 1539	A	5 A d.c.	1539
Ru	366.1	Silicate minerals	0.1–5%	E	S	Grind and adhere (5 mg) to electrode. Prepare standards from Ru sponge + SiO_2 + Fe_2O_3, with Ca added if present in samples. Add Pd as internal standard.	A	A.C. arc	94
S	—	Iron ore, pitch	—	E	L	—	F	Air/H_2/N_2 (MECA)	434
Sb	217.6	Geological samples	1–40 μg/g	A	L	Mix with NH_4I (1:4), dry and heat to 350°C. Cool, treat with 10% HCl, add ascorbic acid and extract with TOPO into MIBK	F	Air/C_2H_2	303
Sb	217.6	Ores and concentrates	0.05–5%	A	L	Fuse with Na_2O_2 in Zr crucible, extract with HCl and dilute. Add HNO_3 if high Pb present	F	Air/C_2H_2	791
Sb	259.8	Silicate rocks	From 0.3 μg/g	E	S	See As, ref. 1031	A	9 A d.c.	1031
Se	196.1	Rocks	0.05–0.1 μg/g	A	L	Dissolve in HF/HNO_3, evaporate, add HCl and extract with 1% solution of phenol in C_6H_6. Evaporate and redissolve in H_2O	F	Air/C_2H_2 + Ta boat	700
Si	256.6	Chromites	5–40%	E	S	See Fe, ref. 36	S	—	36
Si	—	Scheelite	—	E	S	See Al, ref. 294	A	—	294
Si	288.1	Silicate minerals	% levels	E	S	—	Ion-beam sputtering		1459
Sn	—	Scheelite	—	E	S	See Al, ref. 294	A	—	294
Sn	286.3	Cassiterite	—	A	L	Extract with NH_4I, collect and dissolve in HCl (1:4)	Graphite furnace		1010
Sn	317.5	Rocks	Trace levels	E	S	See Pb, ref. 1394	A	10 A d.c.	1394
Sr	338.1	Geological samples	0.01–0.3%	E	S	Computerised integrated-intensity technique (18 standard rocks)	A	D.C. arc	218
Sr	407.7 460.7	Rocks, minerals	—	E	S	—	A	—	599
Sr	—	Silicate rocks	200–1000 μg/g	A	L	Digest with HF/HNO_3/$HClO_4$ in sealed PTFE vessel at 100°C. Evaporate to fumes, redissolve in HCl + H_3BO_3, add KCl and dilute, with addition of La	F	N_2O/C_2H_2	1044

TABLE 2.5 MINERALS — continued

Element	λnm	Matrix	Concentration	Tech.	Analyte form	Sample treatment	Atomization	Ref.
Te	—	Rocks	From 5 ng/g	A	L	Digest with HNO_3, evaporate to dryness with formic acid, extract Te into HBr and co-precipitate with As. Dissolve precipitate in HBr/Br_2 and extract Te into MIBK	F —	668
Ti	337.3 336.7	Rocks, minerals	—	E	S	—	A —	599
Tl	276.8	Rocks	Trace levels	A	S	See Ag, ref. 1164	Graphite furnace	1164
V	318.4	Silicate rock	From 10 μg/g	A	L	See Ba, ref· 786	F —	786
V	—	Geological materials	From 0.02 μg/ml (in extract)	A	L	See Au, ref. 1264	Graphite furnace	1264
Zn	213.8	Coal, fly ash	μg/g levels	F	L	See Cd, ref. 773	F Air/C_2H_2	773
Zn	—	Pb and Zn concentrates	—	A	L	See Cd, ref. 187	F —	187
Zn	334.5	Rocks	Trace levels	E	S	See Pb, ref. 1394	A 10 A d.c.	1394
Various	—	Minerals	—	A	L	(a) Fuse with Na_2O_2, dissolve in H_2O and bubble gas to avoid filtration of any precipitate e.g. Fe, Ti (method for Al, Cr, Si, V) (b) fuse with $Na_2B_4O_7$, dissolve in H_2O and add HF (c) fuse, extract with acid and add HF (method for Si)	F —	6
Various (8)	—	Coal, fly-ash	Trace levels	A,E	L,S	Analysis of coal fly-ash by AAS, AES, MS and XRF methods, including As, Be, Cd, Ni, Pb, Tl by AAS and Be, Cr, Mn, Ni, Pb by AES	F Air/C_2H_2 N_2O/C_2H_2 A D.C. arc	33
Various (17)	—	Sulphide minerals	Minor and trace levels	A	L	Decompose sample with HCl/HNO_3 (13 elements) or HNO_3 alone (4 elements). Use long-path absorption tube for improved detection limits for Ag, Bi, Cd, Pb, Sb, Sn and Tl	F Air/C_2H_2 (or Air/H_2 with long-path tube)	42

Various (12)	—	Rocks and minerals	Minor and trace levels	E	S	Dilute (1:5) with graphite powder. Prepare standards in pure SiO_2, add 6% Na and 6% K (as carbonates) and dilute (1:5) with graphite. (Elements: Ba, Co, Cu, Cr, Ga, Mn, Ni, Pb, Sr, Ti, V and Zr)	A	11–14 A d.c.	92
Various (Rare-earths)	—	Magnetic garnets and sulphides	Major levels	A	L	Dissolve garnets in HCl and sulphides in HNO_3/Br_2, followed by evaporation with HCl. Adjust solutions to 0.1 M HCl strength and add $LaCl_3$ (0.5%). Method covers Er, Eu, Dy, Gd, Ho and Y	F	N_2O/C_2H_2	128
Various (19)	—	Rocks, soils	Minor and trace levels (From 0.3 $\mu g/g$)	E	S	—	A	D.C. arc (20 A., gas-sheathed)	174
Various (6)	—	Geochemical reference samples	All levels	A	L	Comparison of AAS and XRF methods for Mg, Na, Al, Mn, Ca and K	F	—	276
Various (68)	—	Geological materials	All levels	E	S	Description of computerised system	A	D.C. arc (argon/O_2)	283
Various (Noble metals)	—	Ores and ore products	$\mu g/g$ levels	A,E	L,S	Concentrate noble metals by fire-assay, with Ni matte as collector	F,A	—	306
Various	—	Minerals	—	A	L	Review	—	—	336
Various (24)	—	Geological microsamples	Trace levels	E,A	L,S	—	Graphite furnace F	Air/C_2H_2	340
Various (5)	—	Coal	—	A	L	Ash and dissolve (Cd, Co, Cr, Cu, Pb)	Graphite furnace		430
Various (8)	—	Asbestos	Trace levels	A	L	Treat with HF at 130°C in PTFE, volatilize SiF_4 and dissolve residue in 6N HCl	F	—	443
Various (14)	—	Rocks, minerals	Trace levels	A	L	Fuse and re-dissolve. Add $AlCl_3$ for Ti determination and add NaI for Ba, Sr determinations	F	N_2O/C_2H_2	450
Various (12)	—	Iron ores	—	A	L	Digest with $HCl/HNO_3/HF$. Heat in autoclave, cool, add H_3BO_3 and dilute	F	Air/C_2H_2 or N_2O/C_2H_2 Graphite furnace (HGA-74)	547
Various (11)	—	Iron ores and agglomerates	—	A	L	Dissolve in HCl/HNO_3, add HF, heat at 140°C. Add saturated H_3BO_3, filter and dilute	F	Air/C_2H_2 or N_2O/C_2H_2	774

TABLE 2.5 MINERALS — continued

Element	λnm	Matrix	Concentration	Tech.	Analyte form	Sample treatment	Atomization	Ref.
Various	—	Rocks, ores, minerals	—	A	S	Comparison of calibration methods	Graphite furnace	822
Various (Noble metals)	—	Minerals	—	E	L	Concentrate metals by electrolysis and assay-chemical method (Anisimov)	P Argon ICP	863
Various (Rare-earths)	—	Minerals	—	E	L	—	F N_2O/C_2H_2	864
Various	—	Geological materials	—	A	S/L	Spray suspension of solid sample ($<$ 325 mesh) directly into flame	F —	887
Various	—	Rocks, minerals	—	E	S	(Study of matrix effects)	A 15 A d.c. (in Ar/O_2)	716
Various (9)	—	Sediments	Minor and trace levels	A	L	Digest with HF/HNO_3, evaporate to dryness and redissolve in HNO_3/H_3BO_3. (Comparison with NAA method)	F —	721
Various (25)	—	Rocks, soils, sediments	—	E	L	—	P ICP	737
Various	—	Minerals	—	E	S	—	A Plasma arc 10 A	581
Various (13)	—	Geological materials	Trace levels	E	S	(a) Fuse samples and synthetic standards with Li metaborate, or, (b) mix with Li_2CO_3, without fusion (Statistical study of errors)	A 10 A d.c. (argon/O_2)	608
Various	—	Geochemical survey materials	—	E,A	L,S	Review paper, covering AAS, ES and XRF methods	A,F —	628
Various (Rare-earths)	—	Rocks and minerals	—	E,A	L,S	Add 1% La or 0.1% Na as buffer. If pre-concentration is necessary, precipitate with Ca oxalate or fluoride, redissolve and collect on $Fe(OH)_3$	F —	913
Various (11)	—	Geological materials	Trace levels	E	S	Mix (1:3) with buffer mixture of graphite + LiF (4:1) containing Sc and Ge as internal standards	A 12 A d.c.	1029
Various (21)	—	Silicate rocks	All levels	A	L	Digest with $HF/HCl/HNO_3$ in sealed PTFE vessel at 90–100°C and dilute with addition of H_3BO_3	F Air/C_2H_2 N_2O/C_2H_2	1062

Various (6)	—	Cu/Ni ores	Trace levels	E	S	Absorb Au and Pt metals on "Mtilon-T" ion-exchange fibre, from hot solution. Fuse fibre + metals, redissolve and evaporate in presence of graphite powder. (Method covers Pt, Pd, Rh, Ir, Ru, Au).	A	—	1157
Various (6)	—	Cu/Ni ores	μg/g levels	E	S	Roast at 630°C for 35 minutes, to remove S and C, dissolve residue in $HCl/HNO_3/H_2SO_4$ and precipitate Pt, Pd, Ir, Ru, Rh, Au with thiourea. Dissolve precipitate in HCl/HNO_3 and evaporate in presence of graphite powder.	A	25 A d.c.	1158
Various (7)	—	Ores	From 1 μg/g to 0.15%	A	L	Treat with HCl/H_2O_2 and fuse any insoluble matter with Na_2O_2, dissolving melt in HCl. Extract Au, Ag, Pt from HCl solution with di-antipyryl-propylmethane into dichloroethane	F	Air/propane/ butane	1159
Various (6)	—	Concentrates	μg/g levels	E	S	Absorb metals (Pt, Pd, Au, Rh, Ir, Ru) on ion-exchange fibre, ash, mix with Cu powder + Na_2SO_4 + $Fe(NO_3)_3$ and excite by blowing powdered mix into arc	A	—	1162
Various	—	Geochemical materials	—	A	L,S	Study of interference effects	Graphite furnace		1205
Various (40)	—	Galena	Trace levels	E	S	Prepare standards in pure PbS	A	—	1226
Various (6)	—	Geological materials	—	A	S,L	Spray suspension of powdered sample into flame	F	—	1239
Various (13)	—	Rocks	Minor (%) and trace levels	E	S	Fuse with $Li_2B_4O_7$, cool melt in form of thin sheets and stir with aqueous suspension of cation-exchange resin. Dry resin + trace metals for ES analysis by tape-method	A	—	1256
Various	—	Rocks, minerals	—	A	L	Discussion of sample treatments. (9 refs.).	F	—	1358
Various (6)	—	Rocks	—	E	L	—	P	D.C. arc plasma	1372
Various (6)	—	Concentrates	μg/g levels	E	S	Evaporate concentrate solution, containing Rh, Pd, Pt, Ir, Au, Ag, with graphite powder and mix with Na_2SO_4 (buffer) + Co or Be (internal standards)	A	15 A d.c.	1160

TABLE 2.5 MINERALS — continued

Element	λnm	Matrix	Concentration	Tech.	Analyte form	Sample treatment	Atomization	Ref.
Various	—	Geochemical samples	—	E	S	—	A —	1375
Various (8)	—	Coal ash	—	A	L	Fuse with $Li_2B_4O_7/Li_2CO_3/RbI$ (4:1:trace) in Pt at 950°C and extract with 2% $HClO_4$	F —	1440
Various	—	Rocks, minerals	—	E	S	—	A —	1487
Various (9)	—	Rocks	% levels	E	S	Mix (1:1) with buffer, e.g. Cu, CuO. BaF_2 or AgCl	Laser + spark	1488
Various (7)	—	Carbonate formations	—	E	S	—	Laser + spark	1495
Various (13)	—	Minerals	—	E	S	—	Laser + spark	1566

2.6 SOILS, PLANTS AND FERTILIZERS

2.6.1 Introduction

Two recent reviews provide valuable surveys of the use of all analytical techniques in these areas. Dinnin (*Anal. Chem.*, 1975, **47**, (5), 97R) discussed the analysis of inorganic and geological materials covering methods of sample preparation as well as techniques for measurement. This extensively-referenced article can be recommended to anyone seeking an overall view of the field for the manner in which it puts into perspective the battery of analyical techniques now available for tackling a given problem. Similarly, Gehrke and Rexroad (*Anal. Chem.*, 1975, **47**, (5), 42R) have reviewed work on fertilizer analysis in 1973 and 1974, paying particular attention to the determination of N, P and K but also dealing with minor and trace nutrients. For on-line quality control purposes, automated molecular absorption methods are normally preferred for the major components while atomic spectroscopy finds most use in the determination of other elements, *e.g.* Ca, Co, Cu, Fe, Mg, Mn, Mo and Zn. Fertilizer analysis has also been reviewed elsewhere (516), as has the determination of many elements in soil and plant material (517, 518, 519, 520, 521).

In general, analytical interest appears to be concentrated on extraction methods with AAS normally regarded as a standard method for the final determination. In many instances, of course, analysis serves simply as a means of obtaining information of agronomic, horticultural or environmental interest (see, for example, 1407, 1409, 1411, 1415–1426, 1428–1431, 1433–1439, 1442–1446). The only use of AFS noted (273) had as its main purpose to demonstrate the applicability of the technique to a real sample, *viz.* orchard leaves.

If the beginning of a trend can be discerned in these areas it may be towards the adoption of plasma emission spectroscopy as a means of completing analyses (1454). This technique has the obvious advantage over AAS of simultaneous multi-element capability with wide dynamic ranges but, in view of the convenience and widespread acceptance of AAS and arc/spark AES, one cannot envisage an overnight change in the practice of most analysts. Commercial equipment is now available but the capital outlay required is likely to restrict its purchase to users with a high sample throughput, involving determination of several elements in each sample, who can use its capabilities fully. As development work on plasmas continues, however, one can foresee the technique becoming relatively more attractive and chemists are well advised to keep in touch with progress in this field.

2.6.2 Sample Preparation

Numerous methods are in everyday use for preparation of samples for determination of total or "available" content of metals in materials of agricultural interest. Since the variety of questions to be answered is large, there is no likelihood of universal standardisation of methods although valuable work on inter-comparison of techniques is reported each year. This provides an individual with a store of information and experience on which to base his choice of method for solution of his particular problems.

Wimberley (569) and Carel (934) have each studied the determination of total Hg in soils, releasing the metal by heating either an induction furnace or a thermogravimetric balance assembly and collecting the Hg vapour by amalgamation on gold. A fusion technique employing borax and NaOH has been recommended (537) for preparation of soil samples for AAS determination of total Ca, Mg and Mn. Extraction with $CaCl_2$ solution has been used in the determination of plant-available Mg (533) and Al and Mn (359) while other workers (266) have preferred an H_2SO_4 extraction for Mn. For trace elements, *e.g.* Co, Cu, Ni, Pb and Zn, extraction with APDC into MIBK has been used prior to AAS determination (114, 354). Sinha and Banerjee (1525), however, have found that a digestion procedure involving evaporation with HF at 200°C, eliminated the need for lengthy extraction procedures in this type of determination. A noteworthy paper (1428) describes the extraction of stable organic complexes of Al, Ca, Cu, Fe and Mg from volcanic soils. Finally, Woodrill *et al* (1147) have reported a simple method in which the soil water is separated from the soil itself by centrifugation with CCl_4. The upper, water layer is then available for determination of micro-nutrients by AAS in an electro-thermal atomizer.

For analysis of organic materials, such as plants, wet- or dry-ashing techniques can be used (391) and Isaac and Johnson (1181) have compared their efficiency concluding that both can be satisfactory for determination of Ca, Cu, Fe, K, Mg, Mn and Zn. Other workers, however, have found (783) that dry-ashing at 450–500°C caused volatilization losses in determination of Cd in tobacco and low results in comparison with those obtained after wet- or low-temperature plasma-ashing. Ganje and Page (2) also preferred wet-ashing with $HNO_3/HClO_4$ and used an electrothermal atomizer to obtain a lower limit of detection (0.01 μg g^{-1}) for determination of Cd in plant tissue but others have found (983) dry-ashing at 450°C to be satisfactory. Further examples of the successful use of wet-ashing in this area have included determination of As, Cr, Cu and Zn in preserved wood (281), Pb in roadside plants (56), Cr in Plants after extraction as the 2,4-pentanedione complex into $CHCl_3$ (1180) while Faithfull (1059) has described an automated system for detection of Fe in herbage *via* Kjeldahl digestion. Koirtyohann *et al* (*Anal. Chem.*, 1975, **47**, 1475) described the wet ashing of various organic samples in a microwave oven. Morris (142) has shown that Si can be solubilized rapidly from tobacco leaf by treatment with HF or a fluoboric-boric acid system, while several reports (227, 878, 879, 1217) have appeared describing hydride generation as a preliminary to the AAS determination of As and/or Se. Finaly, a combined liquid chromatography—AAS method has been used (889) for fractionation and determination of free and chelated Zn in plant tissue extracts.

Relatively few papers on analysis of fertilizers have been noted. Methods for determination of B by measurement of the BO_2 band emission in an air/H_2 flame after extraction by the common 2,ethyl-1,3-hexanediol procedure (1183) and Hg by cold vapour AAS (320) have been presented. An interesting method (870) for determination of $Ca(NO_3)_2$ in $CaNH_4(NO_3)_3$ which depends upon the solubility of the former in methanol has also been described.

2.6.3 Atomic Absorption Spectroscopy

As noted above, most work has been concentrated on methods of sample preparation rather than on analytical measurement techniques. Most analyses are still performed by "conventional" AAS (1258), and Faithfull's automation of this technique (1059) has already been noted. Several applications of electrothermal atomizers (2, 236, 1147, 1218, 1341) demonstrate that this approach may have advantages in some circumstances, *e.g.* if sample size is limited, concentration level is low, *etc.* Development of the "matrix modification" technique (232, 706 and Part I, Section 2.6) in which large excess of a selected compound is added to obtain the metal of interest in a convenient matrix before atomization, is likely to increase the acceptance of electrothermal atomization as an everyday technique.

2.6.4 Atomic Emission Spectroscopy

The comment was made in 2.6.1 that interest is beginning in the evaluation of plasma emission spectroscopy as a method for routine analysis of many elements in large numbers of samples, Scott *et al* (181, 304) have used an inductively-coupled plasma (1kW power) for determination of several elements, *e.g.* Al, B, Co, Cu, Fe, Mn, Ni, Pb, Zn in soils and/or plant materials and showed that for some of these elements, this method was more accurate than flame AAS. The applicability to B is a particular advantage of this technique over flame methods for which an inconvenient extraction procedure is normally needed and this has been exploited (1453) for determining water-soluble B in fertilizers. Jones (1182) has reported a valuable collaborative study of the elemental analysis of plant material by the rotating-disc spark technique which shows RSDs between 0.059 (for Mg) and 0.89 (for Na). Despite this poor precision—18 out of 70 RSDs were above 0.3— the method has been accepted as "Official First Action" by the AOAC. Other interesting applications of emission spectroscopy in this area have included the use of a rotating briquetted disc for determination of elements including B in plant material (1466), use of a glow discharge lamp for determining macrocomponents (Si, Fe, Al, P, Mg and Ca) in soil (1512) and a comparison of different analytical techniques for analysis of various elements in soil extracts (1490). The automated flame AES determination of K in fertilizers has been described (1184) but, although this technique may be very satisfactory in a laboratory, one could have reservations, on safety grounds, over its use in preference to a molecular absorption technique in a fertilizer manufacturing plant.

TABLE 2.6 SOILS, PLANTS, FERTILIZERS

Element	λ/nm	Matrix	Concentration	Tech.	Analyte form	Sample treatment	Atomization		Ref.
Ag	328.1	Soils	Trace levels	A	L	Digest with 4 M HNO_3 + 1 M HCl. For Cu and Zn, buffer to pH 4.8 with acetate and extract with APDC/MIBK	F	Air/C_2H_2	354
Al	—	Tree foliage	—	A	L	(a) Ash and extract with HCl (b) Extract directly with NH_4EDTA. (Comparison of treatments. La (0.1%) added as releasing agent. Method (a) preferred.)	F	N_2O/C_2H_2	190
Al	308.2	Plants	80–240 μg/g	E	L	Dry, ash at 500°C, extract with HNO_3 (50%), evaporate to dryness, re-ash at 500°C and dissolve in HNO_3	P	Induction-coupled plasma (27 MHz, 4 KW)	304
Al	—	Soils, plants	—	A	L	Extract soils by shaking for 5 minutes with 0.01 M $CaCl_2$ (1:2). Allow to settle and spray supernatant liquid. Extract plant ashes with HCl. (Methods for plant-available Al and Mn).	F	—	359
Al	—	Soils, plants	—	—	—	Bibliography, 36 refs	—	—	521
Al	—	Soils	—	E	L	Equilibrate with $H_2O/HCl/Ca(OH)_2$, filter and acidify with HCl	F	N_2O/C_2H_2	1430
As	193.7	Tobacco	From 50 ng/g	A	L	Digest with $HNO_3/HClO_4$. Convert As(V) to arsine and freeze out in liquid-N_2 u-trap. Re-heat and pass to flame	F	—	227
As	—	Preserved wood	—	A	L	Extract with H_2SO_4/HNO_3 and digest with H_2O_2	F	—	281
As	—	Herbage	Trace levels (As III and As V)	A	G	Generate arsine by Na BH_4 reduction (a) from 1 M HCl (total As) and (b) from solution at pH 3–5 (As III)	F	—	878
As	—	Soils, waters, aquatic organisms	—	A	G	Digest with $HNO_3/HClO_4/H_2SO_4$ (20:1:4), add $NaBH_4$ to 20% HCl solution and collect AsH_3H_2 in balloon, before transfer to flame with Ar	F	Ar/H_2	1409
B	249.8	Plants	25 μg/g	E	L	See Al, ref. 304	P	—	304
B	—	Soils, plants	—	—	—	Bibliography, 24 refs	—	—	520

B	518 (band)	Fertilizers	0.1–10%	E	L	Complex acid-soluble B with 2-ethyl-1,3-hexanediol and extract into MIBK	F	Air/H_2	1183
B	249.8	Fertilizers	From 0.001%	E	S	Extract with H_2O	P	ICP	1453
B	540 (band)	Plants	23 μg/g	A	L	Digest with HNO_3/HCO_4 (1:1). Treat aliquot with H_2SO_4/acetic acid + curcumin/acetone. Add C_2H_5OH, centrifuge to remove Ca and measure at 540 nm, using Ne hollow-cathode as radiation source	—	—	1522
Ca	422.7	Tree foliage	—	A	L	See Al, ref. 190	F	Air/C_2H_2	190
Ca	—	Plant ash	—	E	L	Extract with HCl and add $LaCl_3$	F	—	323
Ca	—	Soils, plants	—	—	—	Bibliography, 36 refs.	—	—	518
Ca	—	Soils	1%	A	L	Add La buffer	F	N_2O/C_2H_2	537
Ca	—	Ca/NH_4NO_3 fertilizers	—	A	L	Stir with CH_3OH, filter, wash and dilute with H_2O	F	—	870
Ca	—	Plant roots	—	A	L	Ash at 570°C for 2 hours, dissolve in HCl, dilute and add Sr (for Ca, Mg)	F	—	1422
Ca	—	Soils	—	A	L	Extract with $CaCl_2$/KCl and centrifuge	F	—	1427
Ca	—	Soils, earthworms	0.005–2.2%	A	L	Shake dried material with H_2O, stand, centrifuge, concentrate liquor by evaporation, treat residue with 2.5% acetic acid, centrifuge and digest residue with HNO_3. Evaporate extracts to dryness and dissolve in 0.1 N HNO_3 + 6500 μg/ml $LaCl_3$ (for Ca)	F	—	1439
Cd	228.8	Plant tissue	0.01–50 μg/g	A	L	Dry for 48 hours at 70°C, grind and treat (100 mg) with $HNO_3/HClO_4$. Dilute to volume (50 ml) and take 25 μl sample	Graphite furnace (HGA-2000)		2
Cd	—	Orchard leaves, environmental samples	—	F	L	Study of scatter-correction system	F	Air/C_2H_2 or Ar/H_2	773
Cd	—	Tobacco	—	A	L	(a) Dry-ash at $<$500°C (b) digest with $HNO_3/H_2SO_4/H_2O_2$ (c) Low-temp. plasma ash. Comparison of treatments. Use (c), with extraction of Cd by dithizone/MIBK	F	—	783

TABLE 2.6 SOILS, PLANTS, FERTILIZERS — continued

Element	λ/nm	Matrix	Concentration	Tech.	Analyte form	Sample treatment	Atomization	Ref.
Cd	228.8	Wheat	0–150 ng/g	A	L	Ash at 450°C and dissolve in 1 M HNO_3	Graphite furnace (HGA-70)	983
Cd	—	Soils	—	A	L	Extract with H_2O and concentrate Cd, Zn by dithizone/CCl_4 extraction (pH 4–10)	F —	1446
Co	240.7	Soils	—	A	L	Buffer soil extract with citrate (pH 7.8–8.3) to mask Fe, and extract Co, Ni, Pb with APDC/MIBK	F —	114
Co	345.3	Soils	4–40 μg/g	E	L	Digest with HNO_3/HCO_4 at 180°C for 1 hour. Cool, dilute to volume and decant supernatant liquid, for ultrasonic nebulisation. Results compared with those of flame AAS	P Induction-coupled argon plasma (1 KW)	181
Co	—	Plants	From 10ng/g	A	L	—	Graphite furnace	1243
Co	240.7	Soils	Trace levels	A	L	Evaporate with HF at 200°C	F —	1525
Cr	—	Preserved wood	—	A	L	See As, ref. 281	F —	281
Cr	357.9	Soils and sediments	0–4 μg/g	A	L	Extract with HCl. Add K_2SO_4 to samples and standards to remove interference by Fe, Ti, V, Ni, Co and add Al to remove interference by Mg, Ba	F Air/C_2H_2	307
Cr	357.9	Plants	From 5 ng/g	A	L	Wet-ash with $HNO_3/H_2SO_4/HClO_4$, extract Cr with 2.4-pentanedione into $CHCl_3$. Evaporate and redissolve in MIBK	F Air/C_2H_2	1180
Cu	324.7	Soils	5–2000 μg/g	E	L	See Co, ref. 181	P —	181
Cu	324.7 327.4	Plants	2.2–6.4 μg/g	A	L	Digest ground sample with $HNO_3/HClO_4$. Heat in stages to 200°C, dissolve residue in 3% $HClO_4$ and extract Cu with APDC/MIBK. Method for small samples ($<$0.05 g)	Graphite furnace (HGA-72)	236
Cu	—	Preserved wood	—	A	L	See As, ref. 281	F —	281
Cu	324.7	Plants	4–18 μg/g	E	L	See Al, ref. 304	P —	304
Cu	324.7	Soils	Trace levels	A	L	See Ag, ref. 354	F Air/C_2H_2	354
Cu	—	Soils, plants	—	—	—	Bibliography; 64 refs.	— —	519

Cu	—	Plant leaves	10–500 μg/g	A	L	Dry, homogenize, ash and redissolve. (Development of series of analysed plant standards)	F	—	1267
Cu	—	Soils, plants, manures, waters	—	A	L	Extracts soils with DTPA. Digest plants, evaporated waters and manures with $HNO_3/HClO_4$	F	—	1435
Cu	—	Barley soils	—	A	L	Digest plants with $HNO_3/HClO_4$ (2:1). Extract soils (various methods)	F	—	1436
Cu	—	Plants, tissues	—	A	L	Digest with HNO_3/HCO_4 (4:1), evaporate to dryness and dissolve in 3% HCl	F	—	1445
Cu	324.7	Soils	Trace levels	A	L	See Co, ref. 1525	F	—	1525
Fe	372.0	Plants	90–180 μg/g	E	L	See Al, ref. 304	P	—	304
Fe	—	Soils, plants	—	—	—	Bibliography; 72 refs.	—	—	517
Fe	248.3	Herbage extracts	70–200 μg/l	A	L	Dilute Kjeldahl extract with H_2SO_4 (30% w/v)	F	Air/C_2H_2	1059
Fe	—	Plant leaves	10–500 μg/g	A	L	See Cu, ref. 1267	F	—	1267
Fe	—	Plants	—	A	L	Digest with $HNO_3/HClO_4/H_2SO_4$	F	—	1416
Fe	—	Plants	—	A	L	Digest with $HNO_3/HClO_4/H_2SO_4$	F	—	1420
Fe	—	Soils	—	A	L	Shake with 0.0125 M $CaCl_2$, add metal chelate (EDTA or EDDHA), centrifuge or leach soil in column with 0.025 N $Ca(NO_3)_2$ at pH 7.1–7.2. Add 0.05 N $Ca(NO_3)_2$ + metal chelate and collect effluent	F	—	1431
Fe	—	Soils	—	A	L	Extract with acid NH_4 oxalate, neutral Na dithionate and HCl (1:1)	F	—	1434
Fe	—	Soils, plants, manures, waters	—	A	L	See Cu, ref. 1435	F	—	1435
Fe	—	Barley soils	—	A	L	See Cu, ref. 1436	F	—	1436
Fe	—	Soils, plants	—	A	L	Extract soil solution in "Baroid" press and acidify with HCl. Digest plants with $HNO_3/H_2SO_4/HClO_4$ (10:1:4), filter and dilute. Prepare plant extracts from treatment with buffer solution (pH 6.1) of 0.2 N K_2HPO_4/0.2 N KH_2PO_4. Grind, centrifuge, filter and acidify with HCl	F	—	1437

TABLE 2.6 SOILS, PLANTS, FERTILIZERS — continued

Element	λ/nm	Matrix	Concentration	Tech.	Analyte form	Sample treatment	Atomization	Ref.
Fe	—	Plants, tissues	—	A	L	See Cu, ref. 1445	F —	1445
Hg	253.7	Fertilizers	Up to 2.5 μg/g	A	G	—	Cold vapour	320
Hg	253.7	Soils, ores, organic materials	ng/g levels	A	S/G	Heat to 1000°C in induction furnace, pass vapour through $K_2CO_3/Mg(ClO_4)_2$ trap and collect Hg on Au-plated wire. Heat to release Hg for absorption measurement	Cold vapour	569
Hg	253.7	Soils	0–50 ng/g	A	G	Pass N_2-stream over heated soil, and collect Hg on Au screen, after passage through $K_2CO_3/Mg(ClO_4)_2$ scrubber. Heat screen to volatize Hg	Cold vapour	934
K	—	Plant ash	—	E	L	See Ca, ref. 323	F —	323
K	766.8	Fertilizers	0–60 μg/g (in extract)	E	L	Add La to suppress P interference and add Li as internal standard	F —	1184
K	—	Plant roots	—	E	L	See Ca, ref. 1422	F —	1422
K	—	Soils	—	E	L	See Ca, ref. 1427	F —	1427
K	—	Micaceous colloids	—	E	L	Digest with $HF/HClO_4$. (Study of As retention by hydroxy-Al)	F —	1432
K	—	Plants, soils	—	E	L	Digest plants with H_2SO_4, dilute and add Li internal standard. Extract soils with 1 N NH_4 acetate (pH 7)	F —	1433
Mg	—	Plant ash	—	E	L	See Ca, ref. 323	F —	323
Mg	—	Soils, plants	—	—	—	Bibliography; 36 refs.	— —	518
Mg	—	Soils	—	A	L	Extract with 0.025 N $CaCl_2$	F Air/C_2H_2	533
Mg	—		1%	A	L	See Ca, ref. 537	F N_2O/C_2H_2	537
Mg	—	Plant roots	—	A	L	See Ca, ref. 1422	F —	1422
Mg	—	Plants	—	A	L	Ash at 500–550°C and dissolve in HCl	F —	1423
Mg	—	Soils	—	A	L	See Ca, ref. 1427	F —	1427
Mn	—	Soils	—	A	L	Extract with 0.1 N H_2SO_4 for 3 minutes	F —	266
Mn	403.3	Plants	40–180 μg/g	E	L	See Al, ref. 304	P —	304
Mn	—	Soils, plants	—	A	L	See Al, ref. 359	F —	359
Mn	—	Soils, plants	—	—	—	Bibliography; 72 refs.	— —	517

Mn	—	Soils	0.05–0.1%	A	L	See Ca, ref. 537	F	N_2O/C_2H_2	553
Mn	—	Plant leaves	10–500 μg/g	A	L	See Cu, ref. 1267	F	—	1267
Mn	—	Plants, soils	—	A	L	Ash at 450°C, extract with 6 N HCl (plants) or extract with 1 N NH_4 acetate (soils)	F	—	1418
Mn	—	Plants	—	A	L	See Fe, ref. 1420	F	—	1420
Mn	—	Plants	—	A	L	See Mg, ref. 1423	F	—	1423
Mn	—	Soils, plants, manures, waters	—	A	L	See Cu, ref. 1435	F	—	1435
Mn	—	Barley soils	—	A	L	See Cu, ref. 1436	F	—	1436
Mn	—	Soils, plants	—	A	L	See Fe, ref. 1437	F	—	1437
Mn	—	Oats, fertilizers	—	A	L	Dry-ash plants at 470°C and dissolve in HCl. Extract fertilizers with H_2O	F	—	1438
Na	—	Fertilizers	—	E	L	—	F	—	277
Na	—	Plant roots	—	E	L	See Ca, ref. 1422	F	—	1422
Na	—	Soils	—	E	L	See K, ref. 1433	F	—	1433
Ni	232.0	Soils	—	A	L	See Co, ref. 114	F	—	114
Ni	351.5	Soils	1–1000 μg/g	E	L	See Co, ref. 181	P*	—	181
Ni	—	Plants	0–50 μg/g	A	L	Add H_2SO_4, ash at 450°C for 24 hours, dissolve in HNO_3, evaporate, re-ash at 500°C and dissolve in HNO_3	F	—	1411
Pb	—	Soils, plants	—	A	L	Extract soil samples with 0.2% acetic acid to determine available Pb. For total Pb, digest with HNO_3. Wet-ash plant samples with $HNO_3/HClO_4$	F	—	56
Pb	217.0	Soils	—	A	L	See Co, ref. 114	F	—	114
Pb	405.8	Soils	—	E	L	See Co, ref. 181	P	—	181
Pb	283.3	Soils	Trace levels	A	L	See Ag, ref. 354	F	Air/C_2H_2	354
Pb	—	Plant and soil extracts	—	A	L	Extract with HCl/NH_4F. (Study of extractable Pb in relation to soybean)	F	—	1365
Pb	—	Plants	—	A	L	Heat slowly to 490°C, ash for 4 hours, extract with 3 N HCl and dilute	F	—	1366
Pb	—	Soil, earthworms	3 μg/g–1.2%	A	L	See Ca, ref. 1439	F	—	1439

TABLE 2.6 SOILS, PLANTS, FERTILIZERS — continued

Element	λ/nm	Matrix	Concentration	Tech.	Analyte form	Sample treatment	Atomization		Ref.
Pb	217.0	Soils	10 μg/g–1.5% (dry weight)	A	L	Extract with 0.6% (w/v) HCl or conc. HNO_3. Evaporate to reduce bulk	F	—	1442
Pb	—	Soils, plants	0.5–500 μg/g	A	L	Digest soils with HF/HNO_3 (1:1). Dry-ash plants at 450°C and dissolve in 2 M HCl	F	—	1443
S (SO_4)	870 (Ba)	Superphosphate fertilizers	—	E	L	Add excess $BaCl_2$ and determine residual Ba ion.	F	—	419
Se	196.0	Plants	10–400 ng/g	A	L,G	Digest with $HNO_3/HClO_4$, dilute with 50% HCl and add $NaBH_4$. Sweep Se hydride to flame in stream of N_2	F	H_2/Air or H_2/N_2	879
Si	—	Tobacco leaf	—	A	L	Treat cured leaf with 48% HF. Add H_3BO_3 to form interference-free matrix	F	N_2O/C_2H_2	142
Si	—	Tree foliage	—	A	L	See Al, ref. 190	F	N_2O/C_2H_2	190
Zn	213.8	Soils	1–1000 μg/g	E	L	See Co, ref. 181	P	—	181
Zn	—	Preserved wood	—	A	L	See As, ref. 281	F	—	281
Zn	213.9	Plants	13–23 μg/g	E	L	See Al, ref. 304. For Zn, match standards for Ca, K, Mg content	P	—	304
Zn	213.8	Soils	Trace levels	A	L	See Ag, ref. 354	F	Air/C_2H_2	354
Zn	—	Soils, plants	—	—	—	Bibliography; 64 refs.	—	—	519
Zn	—	Orchard leaves, environmental samples	—	F	L	See Cd, ref. 773	F	Air/C_2H_2 or Ar/H_2	773
Zn	—	Plant tissues	—	A	L	Separate metal ion or chelate by liquid chromatography and pass eluate directly to nebulizer for continuous monitoring ('LCAAS' method)	F	—	889
Zn	213.9	Wheat	0–50 μg/g	A	L	See Cd, ref. 983	F	Air/C_2H_2	983
Zn	—	Plant leaves	10–500 μg/g	A	L	See Cu, ref. 1267	F	—	1267
Zn	—	Plants	—	A	L	See Fe, ref. 1420	F	—	1420
Zn	—	Soils	—	A	L	See Fe, ref. 1431	F	—	1431
Zn	—	Soils, plants, manures, waters	—	A	L	See Cu, ref. 1435	F	—	1435
Zn	—	Barley soils	—	A	L	See Cu, ref. 1436	F	—	1436

Zn	—	Soils, plants	—	A	L	See Fe, ref. 1437	F	—	1437
Zn	—	Soil, earthworms	0.2 μg/g–0.13%	A	L	See Ca, ref. 1439	F	—	1439
Zn	—	Plants, tissues	—	A	L	See Cu, ref. 1445	F	—	1445
Zn	—	Soils	—	A	L	See Cd, ref. 1446	F	—	1446
Zn	213.9	Soils	Trace levels	A	L	See Co, ref. 1525	F	—	1525
Various (6)	—	Agricultural products	Trace levels	A	L	Review of methods for "total" and "available" toxic elements (Cd, Cu, Pb, Hg, Ni, Zn) in plant products	F	—	391
Various	—	Soils, rocks	—	E	S	—	A	Gas-sheathed D.C. arc	174
Various	—	Fertilizers	—	—	—	Bibliography, 1966–1974	—	—	516
Various (11)	—	Orchard leaves	—	A	L	Digest with $HNO_3/HClO_4$	F	Air/C_2H_2	553
Various (10)	—	Soils, plants	ng/g levels (Pb, Cd)	A	L	Prepare soil extracts or plant digests and analyse directly for all elements except Pb, Cd. For these, extract with APDC or dithizone, respectively, evaporate with HNO_3 and redissolve in HCl	F	—	877
Various (7)	—	Soils	—	A	L	Complex and concentrate by biuret extraction. (Comparison with spectro-photometric method)	F	—	722
Various	—	Fertilizers, feeds	—	A	L	(Modification of official AAS method)	F	—	954
Various	—	Soils	Trace levels	A	L	Treat soil with CCl4, centrifuge and collect aqueous soil extract layer	Graphite furnace		1147
Various (7)	—	Leaf tissue	—	A	L	Comparison of (a) Dry-ashing at 500°C and (b) wet-ashing with $HNO_3/HClO_4$ (for Ca, Cu, Fe, Mg, Mn, K, Zn)	F	—	1181
Various (14)	—	Leaf tissue	—	E	L	Dry-ash and dissolve, with addition of Li buffer (Inter-laboratory study)	S	Rotrode	1182
Various (6)	—	Orchard leaves	ng/g levels	A	L,G	Dry-ash at 500°C with $Mg(NO_3)_2$. Dissolve and add $NaBH_4$ to generate volatile hydrides of As, Sb, Bi, Ge, Se, Te	F	H_2/N_2/air	1217
Various (6)	—	Orchard leaves	μg/g levels	A	L	Digest with $HNO_3/HClO_4$ (Cd, Fe, Cu, Mn, Cr) or dry-ash at 400°C and extract with HNO_3 (Pb)	Graphite furnace (CRA-63)		1218

TABLE 2.6 SOILS, PLANTS, FERTILIZERS — continued

Element	λ/nm	Matrix	Concentration	Tech.	Analyte form	Sample treatment	Atomization	Ref.
Various (9)	—	Plants, soils, waters	All levels	E,A	L	—	F Air/propane Air/C_2H_2	1258
Various (10)	—	Soil extracts	Trace levels	A	L	Use standard addition method	Graphite furnace	1341
Various (7)	—	Corn, soils, sludges	Trace levels	A	L,G	Digest with HNO_3/$HClO_4$/HF/HCl. For Hg, reduce with $SnCl_2$	F — Cold vapour (Hg)	1407
Various (6)	—	Broccoli plants	—	A,E	L	Ash at 525°C and dissolve in HCl (1:1) (AAS—Ca, Cu, Mg, Mn; FES—Na, K)	F —	1415
Various	—	Sweet orange cultures and foliage	—	A,E	L	Digest with HNO_3/$HClO_4$	F —	1417
Various (12)	—	Spruce needles	—	A	L	Digest with HNO_3/$HClO_4$ (4:1), evaporate to low bulk, dilute and add $LaCl_3$ (for Ca)	F —	1419
Various (6)	—	Plants	—	A,E	L	Dry-ash at 400–500°C and extract (×2) with 6 N HCl (AAS—Fe, Mg, Mn; FES—Ca, K, Na)	F —	1421
Various (8)	—	Plants, soils	—	A	L	(a) Wet-digest for Cu, Zn, Mn, Mg, Fe (b) dry-ash for Ca, Mg, K	F —	1424
Various (9)	—	Pine trees	—	E,A	L	Dry-ash at 480°C, dissolve in HNO_3/HCl and dilute	F —	1425
Various (6)	—	Soils	—	A	L	Heat, mix with H_2O, allow to stand, leach with H_2O + 1 N NH_4 acetate	F —	1426
Various (5)	—	Soils	—	A	L	Extract with 0.5 N NaOH, centrifuge, acidify to pH 2 with 6 N HCl, centrifuge, lyophilize, dry, shake with methanol, filter, concentrate, lyophilize, dissolve in H_2O, dialyze against H_2O, dry, ash at 750°C, fuse with Na_2CO_3 in Pt, dissolve in H_2O and acidify	F —	1428
Various (6)	—	Soils	—	A,E	L	Extract with H_2O or 1 N NH_4 acetate at pH 7 (1:25 soil/extractant ratio) or extract with 1 N HCl (1:50 ratio)	F —	1429
Various (7)	—	White clover, fertilizers	—	E	S	Dry-ash in Pt at 500°C. Mix ash with graphite powder + Ag internal standard	A D.C. arc	1444
Various	—	Plant tissue		E	S	—	P ICP	1454

Various (8)	—	Plants	Various levels	E	S	Mix ash with graphite and form into rotrode disc. Add K_2SO_4 + Na_2CO_3 (buffers) and Li + Cr (internal standards)	S	Rotrode	1466
Various (7)	—	Soils	Trace levels	E	L	Treat with 6 M HCl after removal of Si by HF digestion	S,A	—	1490
Various (10)	—	Soils and soil extracts	Trace levels	E	S	Grind and mix with $CaCO_3$ or $BaCO_3$ + Cu powder. Form into briquettes	A Grimm lamp	—	1512
Various	—	Foliage	—	E	S	Dry at 80°C, ash at 450°C, dilute with graphite + cellulose + KCl (buffer) + GeO_2 (internal standard). Form into pellets	S	—	1524

2.7 WATER, AIR AND ATMOSPHERE PARTICULATES

The demands of environmental analysis once again dominate this section of the review. Electrothermal atomizers are now well established in this field but many people still prefer flame atomization coupled with a concentration procedure if required.

WATER ANALYSIS

2.7.1 Sample Pretreatment

Solvent extraction procedures continue to find extensive application in water analysis to separate and/or concentrate trace metals. For instance both inorganic and organic forms of As have been determined in water samples by electrothermal atomization after extraction by diethylammonium diethyldithiocarbamate (29). These authors used photo-oxidation to decompose organo-As compounds to As(V), followed by reduction with acidic $NaHSO_3$-$Na_2S_2O_3$ to As(III), before extraction into chloroform.

Korkisch and his co-workers have published a number of papers on the application of ion-exchange methods to determine trace elements in natural waters (191, 192, 193, 572, 992, 1362). They used Dowex 1X8 resin either in the bromide, thiocyanate, or chloride form. Elements that have been determined are Cu, Cd, and Pb (572), Zn (192) and Mo (992); in a representative example (572) samples are acidified with HBr, reduced with ascorbic acid, and passed through the anion exchange column where the bromo-complexes of Cd(III), Cu(I) and Pb(II) are absorbed, and then later eluted with HNO_3 and determined in an air/C_2H_2 flame.

Musha and Takahashi (1278, 1288, 1289, 1296) have published a series of papers describing the determination of a range of elements in waters by concentrating them on to soybean protein coagulated by the addition of δ-gluconic lactone.

Topping *et al* (103) using reversed-phase column chromatography have determined 0.1 μg/l of Cd by concentrating as much as 3 l of acidified sample on a column of tri-n-octylamine/cyclohexane coated on an inert resin, and eluting with approx. 100 ml of 1M$(NH_4)_2SO_4$/0.1M EDTA at pH 5.5.

A number of papers describing the determination of trace metals in river or lake sediments have been published. In one of these, Agemian *et al* (1497) have determined 20 elements by a pressure-vessel dissolution method where organic matter and silica are effectively destroyed by digestion at 140°C with a mixture of HNO_3/$HClO_4$/HF.

2.7.2 Determination of Mercury

Methods for the determination of trace quantities of Hg continue to be published; although the cold-vapour procedure is the most frequently used, a considerable number of variations and modifications of this, have been proposed. Ure (153) has written a 442 reference review and this is an extremely useful paper to read if one is contemplating this determination.

There are many problems with this analysis, the major one being contamination. Bothner *et al* (249), for example, found that sea water at pH 1.5 in polythene containers increased in Hg content, whereas identical samples in Pyrex flasks did not. The authors postulated that contamination may arise from leaching effects or the passage of Hg through container walls. This is not the only paper that has been written on this topic in recent years and it appears that there is still no well defined and accepted procedure for storing samples for Hg analysis. In another paper, Hawley *et al* (239) described modifications of the cold-vapour procedure to improve detection limits and reduce analysis time. This was effected by reducing the dead volume of the apparatus, increasing the efficiency of diffusion of elemental

Hg into the carrier gas and preventing Hg losses from the samples by utilising an oxidising agent and acid as a preservative.

Mercury is often determined in sediments and Agemian *et al* (397) compared various acids for its extraction. Identical results were obtained with HCl/HNO_3 and H_2SO_4/HNO_3 while an HF/aqua region pressure vessel extraction gave slightly higher results and poorer reproducibility; either of the HF-free solvents was recommended.

In some analyses it is necessary to identify and quantify both inorganic and organic forms of Hg and Baltisberger and Knudson (155) achieved this by first determining Hg(II), in the presence of alkylmercury, by conventional Sn(II) reduction followed by total Hg determination by prior oxidation with $H_2O_2/HClO_4$ and subsequent Sn(II) reduction. Bisogni and Lawrence (30) chose to determine methylmercury by separation from the inorganic form with a benzene extraction followed by back extraction into the aqueous phase with cysteine and the conventional Hatch and Ott procedure. Floyd and Sommers (1270), however, determined organic Hg compounds in lake sediments by separating into dialkyl and monoalkylmercury fractions by steam distillation.

A number of concentration procedures have been described to determine ultra-trace levels of Hg, and a procedure has been detailed in which Hg is extracted as its pyrrolidine dithiocarbamate complex into propylene carbonate and reduced by Sn(II) in this solvent (1275). Hg has been concentrated from large volumes of solution using a micro-column of Cu (570). This was later heated by passing current through a surrounding spiral filament to expel Hg vapour into an absorption cell where it was determined by AAS. Heinrichs (447) concentrated Hg from water samples by electrolysis, in the presence of an oxidising agent, onto a gold cathode and subsequently heated this in a graphite furnace to expel Hg. In a further paper Watling (183) determined pg l^{-1} levels of Hg in sea water by amalgamating onto a silver wire after reduction with Sn(II). The wire was subsequently heated to liberate Hg for determination by a microwave-induced argon plasma emission system. A plasma emission system was also used by Kaiser *et al* (1531) who gave extensive experimental details of their system.

2.7.3 Flame Methods

Developments continue to be reported utilizing flame atomization. Ward and Biechler (12) have devised a system to overcome ionization interference in the determination of Ca in the N_2O/C_2H_2 flame by concurrently nebulizing a strong solution of Na into the flame and thereby alleviating the need for any chemical pre-treatment. Their system is similar in principle to that described by Rubeska *et al* (11).

A number of interesting papers have appeared from the New York State Department of Health. A series of elements have been determined in a variety of matrices using a molybdenum micro-sampling cup and the N_2O/C_2H_2 flame (766). Interferences were minimal, and analytical signals independent of matrix—thereby allowing aqueous standards to be used. In a further paper (672) two multi-element hollow cathode lamps and a vidicon detection system were used to determine seven common trace metals in water, simultaneously after complexation with APDC and extraction into MIBK. The detection limits were poorer than with conventional AAS but were claimed to be adequate for most waters. Samples of high salt content (ca. 3% by weight) were analysed by Isaaq and Morgenthaler (1238) using an ultrasonic nebulizer connected to a temperature-controlled heater in series with the burner head; no memory effects were found.

Anionic detergents at $\mu g\ l^{-1}$ levels expressed as sodium dioctylsulfosuccinate were analysed by formation of an association complex with Cu-phenanthroline, extraction of this into MIBK and determination of the Cu by AAS (258).

2.7.4 Electrothermal Atomization Methods

Many papers have appeared utilising electrothermal atomization; this is rapidly becoming an established technique in the water-analysis laboratory. Welz and Grobenski (1322) have described the automation of sample introduction to the graphite furnace and have obtained improvements in the precision of results compared with manual operation. Segar (792) has reviewed the application of electrothermal atomizers to environmental analysis and has described the determination of Fe, Mn, Cu, Ni and Cd in sea water without sample pre-processing (807).

Many interesting refinements are being reported for electrothermal atomization methods, Henn (232) has improved the limit of detection of Se determination to 1 $\mu g\ l^{-1}$ by the addition of Mo. This was said to react with Se to form a refractory selenide or a hetero-polymolybdate anion in which a central Se atom is surrounded by molybdate ions. A higher dry-ashing temperature can thus be used to volatilize inorganic acids and anions selectively, but nevertheless it was still necessary to carry out ion exchange pre-treatment to remove cations. It is interesting to compare this paper with that of Martin *et al* (1061) who added Ni to overcome volatility problems by forming a refractory nickel selenide.

Chapman and Dale (428) determined As, Se, Sb, Bi, Te, Sn and Ge by hydride generation but, instead of using flame atomization, transferred the gases to an electrically-heated absorption tube which they claimed had the advantages of simplicity, transparency at low wavelengths and low noise.

Van Loon and Silvester (830) have described a T-shaped furnace to determine different forms of metals in environmental samples. The furnace consists of an atomization section, purged with hydrogen, and heated with resistance wire. The separator section acts either as a gas chromatographic column or merely as a means to selectively volatilize samples into the atomizer by controlled heating. These authors have used this device for both qualitative and quantitative studies of the forms of Se, As, Cd, Hg and Pb present in environmental samples.

2.7.5 Emission Methods

Prager *et al* (223) constructed a filter photometer assembly for the determination of P in water (or air) samples, utilising ultrasonic nebulization and a specially-designed burner. The POH band emission was measured in the air/H_2 flame after first passing the water through an ion-exchange column to remove cationic interference; an impressive limit of detection of 0.9 $\mu g\ l^{-1}$ was obtained.

Talmi and his co-workers have used gas chromatography with a microwave emission spectrometric detection system (GC–MES) to determine pg levels of Se (221) and As and Sb (1227) in environmental samples. In the first paper, Se(IV) was complexed with 5-nitro-o-phenylenediamine, extracted into toluene, separated by GC and then determined by monitoring the emission intensity of the 204 nm Se line with the microwave emission detector. In the second paper, As(III) and Sb(III) were co-crystallized with thionalid, the precipitate reacted with phenyl magnesium bromide and the triphenylarsine and stibine formed extracted into ether, separated by GC and determined by MES. This system was also used to determine alkylarsnic acids in environmental samples (1251).

Fassel (906) has described the application of an ICP optical emission spectrometer to the simultaneous analysis of trace metals in water. Skogerboe *et al* (237) have demonstrated a technique of injecting volatile chlorides into the plasma for determining Pb and Cd in natural water. A new detection system for the analysis of total organic carbon in water samples has been developed (733, 1135) utilising an argon plasma at slightly less than 1 atmosphere pressure and operated at circa 100 W and 2450 MHz. Carbon emission from the plasma is monitored at 193.0 nm, and total organic carbon was determined with a detection limit of 5$\mu g\ ml^{-1}$.

AIR AND ATMOSPHERIC PARTICULATE ANALYSIS

2.7.6 Atomic-Absorption Methods

There is an increasing tendency to use electrothermal atomization for the analysis of air and atmospheric particulates by AAS methods.

Rathje (347), used hopcalite to absorb Hg vapours with virtual 100% efficiency from air samples, and removed it with successive washings with HNO_3 for subsequent analysis.

Parker *et al* (811) determined traces of Pb, Cd, Cu and Zn by collecting atmospheric particulates on millipore filters contained in graphite cups. They emphasised the need to be aware of the possible presence of contaminants in the filter materials; the authors found that consistent blanks were obtained for Pb, Cd and Cu, but the Zn content was too high and alternative filter materials had to be investigated. Risby *et al* (674) determined a number of elements by collection on porous polymer filters with subsequent electrothermal atomization.

Carbon disulphide has been determined in air samples by absorption onto activated charcoal, desorption with iso-amyl acetate and reaction with pyrrolidine to form the dithiocarbarmate. This is then complexed with Cu and extracted into isoamylacetate where the Cu is determined in the air/C_2H_2 flame (546).

Chan *et al* have combined GC separation with AAS detection for the determination of volatile alkyl Pb and Se compounds in the atmosphere (831). The GC outlet was interfaced via a stainless steel tube to the centre of an open-ended silica tube furnace, heated to ca. 980°C. Hydrogen was added to the gas from the column outlet to improve atomization. The methylation products of Pb and Se can be studied with this system at levels as low as 0.1 ng.

2.7.7 Determination of Lead

In atmospheric analysis, Pb is the most commonly determined metal because of its emission from motor-vehicle exhausts. Ellis (390) has described the operation of a battery-equipped atomic-absorption spectrometer in a motor vehicle for continuously monitoring Pb emission from the exhaust while the vehicle was in motion.

It is important to be able to determine all forms of atmospheric Pb, and GC columns at low temperature have been used to collect organic Pb compounds (660). Nitrogen elution of the heated columns volatilized the Pb for subsequent AAS. Both particulate and "molecular Pb" have been determined in air by Robinson *et al* (141, 993). A micropore carbon disc was used to collect particulate Pb while molecular Pb is absorbed on a carbon bed 50 mm deep containing 6–7 mm spectroscopic carbon pellets. These filters are subsequently analysed by electrothermal atomization.

2.7.8 Emission Methods

Sugimae has continued his work with emission spectroscopy for the determination of large numbers of trace metals by collection onto glass fibres (928, 1241).

Authors from Japan (366), have used a laser microprobe to examine various trace metals absorbed on Mylar film. A particle, suitable for analysis, is evaporated by an Nd glass laser and excited by a high-voltage spark discharge.

Braman and Johnson (508) have developed a system to analyse elemental Hg, particulate Hg, Hg(II), methylmercury and dimethylmercury (See Section 1.1.2).

Natusch and Thorpe (1105) have described a gas evolution system to specify and quantify trace metallic compounds in atmospheric particulates by subjecting the samples to a uniform temperature ramp and passing the vapours to an argon or helium plasma where there atomic spectra were excited. They claim that speciation is achieved both in terms of characteristic gas volution temperatures and in terms of elements appearing simultaneously in the plasma.

TABLE 2.7 AIR AND PARTICULATES, WATER, SEWAGE, EFFLUENTS

Element	λ/nm	Matrix	Concentration	Tech.	Analyte form	Sample treatment	Atomization	Ref.
Ag	—	Sea and river water particulates	Trace levels	A,F	L	Collect sample on membrane filter	Flameless method (electro-contact atomiser)	334
Ag	—	Sea water	Trace levels	A	L	See Cd, ref. 808	Graphite furnace (Ar/H_2-shielded)	808
Ag	328.1	Water	Trace levels	A	L	—	Tantalum strip	1554
Al	309.3	Airborne particulates	—	E	S	Collect sample on Mylar film (9 l/min for 1 hour). Dry, select individual particles under microscope and vaporize with laser discharge	Laser + spark	366
As	193.7	Water	3 ng/ml	A	L	Extract As with DDDC into CCl_4. To determine total As, pre-treat by U.V. photo-oxidation (Hg lamp, 500 W), to decompose organo-As compounds	Carbon rod	29
As	193.7	Water	0.01–0.2 μg/ml	A	L	Study of effect of oxidation states	Graphite furnace (FLA-1)	380
As	—	Environmental samples	Trace levels	A	S	Volatilize at 800°C into electrically-heated atomizer "T" tube	Tube furnace (quartz, H_2 atmosphere)	830
As	—	Waters	From 5 μg/l	A	L,G	Acidify and reduce with $NaBH_4$	F Air/Ar/H_2	1207
As	—	Environmental samples	Trace levels	E	G	Crystallise As, Sb with thionalid and react with phenyl-Mg-bromide. Extract hydrides into ether, separate on G.C. column and pass to microwave emission spectrometer	P —	1227
As	228.8	Environmental samples, pesticides	—	E	G	Reduce with $NaBH_4$ and extract arsine with ether or collect in toluene at −5°C. Separate on G.C. column. (Method for alkylarsenic acid mixtures)	P —	1251
As	193.7	Water	—	A	L		Graphite furnace	1549
Au	242.8	Waters	0.9–11.3 μg/ml 0.01–1 ng/ml	A E	L S	Add soybean milk to sample and adjust solution to pH 4.4–5.0 with δ-gluconic lactone. Analyse coagulate by AAS or ES	F Air/C_2H_2 A D.C. arc	1278

B	249.6	Water	$\mu g/ml$ levels	E	L	—	A	—	777
Ba	553.6	Mineral waters	From 5 ng/ml	A	L	Separate Ba by co-precipitation with $PbCrO_4$. Dissolve in 1,2-diaminocyclohexane-NNN'N'-tetra-acetic acid and isolate Ba by ion-exchange using Dowex 50W-X8	F	Air/C_2H_2	448
Be	234.9	Air	0–50 μg (on filter)	A	L	Extract glass fibre filter with HF/HNO_3, add H_2SO_4, heat to fuming, add HNO_3 dropwise, re-heat to fuming, cool and dilute	Graphite furnace (HGA-2000)		1064
Be	—	Air	From 0.1 μg (on filter)	A	L	Collect on paper filter, ash with oxidising acid and dissolve in H_2SO_4.	F	—	1338
C (organic)	193.0	Water	From 5 $\mu g/g$ (50 μl sample)	E	L,G	Dry sample in Ni cup and heat in tube furnace at 850°C, in argon stream. Pass vapour over hot CuO and transfer to plasma source	P	ICP	733
C (organic)	—	Water	—	E	G	Pyrolyse, oxidise to CO_2 and pass to plasma source	P	—	1135
Ca	422.7	Natural waters	0–200 $\mu g/ml$	A	L	Nebulize through T-piece capillary simultaneously with 2000 $\mu g/ml$ Na solution. Rotate burner 90°	F	N_2O/C_2H_2	12
Ca	393.3	Airborne particulates	—	E	S	See Al, ref. 366	P,S	—	366
Cd	228.8	Water	0.1–0.4 ng/ml	A	L	Acidify and extract Cd on column of XAD-2 resin coated with tri-n-octylamine/cyclohexane. Elute Cd at pH 5.5 with 1 M $(NH_4)_2SO_4$/0.1 M EDTA	F	Air/C_2H_2	103
Cd	—	Sea and river water particulates	Trace levels	A,F	L	See Ag, ref. 334	Flameless method		334
Cd	228.8	Sea and river waters	Up to 0.4 ng/ml	A	L	Extract with dithizone/CCl_4	Graphite furnace (FLA-1)		365
Cd	228.8	River waters	0–7 $\mu g/l$	A	L	Use standard addition method. Char at 230°C and atomize at 2000°C	Graphite furnace (HGA-70)		376
Cd	—	Atmospheric particulates	—	A	S	—	Graphite furnace		433
Cd	228.8	Sea water, marine organisms	Trace levels	A	L	Freeze-dry organisms, homogenize, dry at 60°C, digest with H_2SO_4/HNO_3 and dilute. Use $HClO_4/HNO_3$ in presence of high Ca levels	Graphite furnace (HGA-72)		559

TABLE 2.7 AIR AND PARTICULATES, WATER, SEWAGE, EFFLUENTS — continued

Element	λ/nm	Matrix	Concentration	Tech.	Analyte form	Sample treatment	Atomization	Ref.
Cd	228.8	Waters	Trace levels	A	L	Acidify with HBr, filter, add ascorbic acid and pass through Dowex 1-X8 ion-exchange resin. Elute Cd, Cu, Pb with 1 N HNO_3, evaporate and dissolve in methanolic HBr solution	F Air/C_2H_2	572
Cd	—	Effluents	Trace levels	A	L	Concentrate by chelation/solvent extraction	F —	805
Cd	—	Sea water	Trace levels	A	L	—	Graphite furnace	807
Cd	—	Sea water	Trace levels	A	L	Extract, at natural, pH level, with dithizone/$CHCl_3$	Graphite furnace (Ar/H_2 shielded)	808
Cd	—	Atmospheric particulates	Trace levels	A	S	Collect on filter and analyse directly, after addition of H_3PO_4	Graphite furnace (Model-63)	811
Cd	—	Environmental samples	Trace levels	A	S	See As, ref. 830	Tube furnace	830
Cd	—	Waters, sludges, sediments	Trace levels	A	L	Waters — neutralize if necessary Sludges — disperse ultrasonically in H_2O Sediments — digest with acid	F N_2O/C_2H_2 + Mo cup	766
Cd	—	Natural waters	From 0.5 ng/ml	A,E	L	Convert to volatile chloride by heating to 850–900°C in air/HCl or argon/HCl	F Air/C_2H_2 P —	237
Cd	—	Environmental samples	ng/g levels	E	G	Heat and pass evolved vapour to argon or helium plasma	P —	1105
Cd	228.8	Dust	Trace levels	A	L	Treat with H_2O, filter, ash residue at 450°C and evaporate H_2O solution to dryness. Treat combined residues with HNO_3/H_2O_2. Add NH_4 citrate + $(NH_4)_2SO_4$, adjust to pH 9.5 and extract with NaDDC/MIBK	F Air/C_2H_2	1201
Cd	228.8	Sea water	0.015 μg/ml	A	L	None	Graphite furnace (CRA-63)	1221
Cd	228.8	Water	0.1–5 ng/ml	A	L	Add soybean milk + DDTC solution to sample and adjust to pH 5.5–5.8 with δ-gluconic lactone See also, Au ref. 1278 and Hg, ref. 1288	F Air/C_2H_2	1289
Cd	228.8	Water	0.01–0.12 ng/ml	A	L	Pass through resin column (Chelex-100) at >pH 5.0. Elute Cd, Cu, Pb from column with 2 N HNO_3	Graphite furnace (HGA-70)	1293

Cd	228.8	Airborne particulates	From 0.0025 $\mu g/m^3$	A	L	Collect (2000 m^3 sample) on glass filter and extract with acid	F	—	1328
Cd	228.8	Water	Trace levels	A	L	—	Tantalum strip		1554
Cl	360 (InCl)	Drinking water	From 0.5 $\mu g/ml$	E	L	Add In solution and measure InCl emission by molecular emission cavity analysis (MECA). Use method of standard additions	F	N_2/H_2	363
Co	—	Drinking water	Trace levels	A	L	Extract with APDC/MIBK	Graphite furnace		1337
Cr	—	Water sediments	—	A	L	(a) Treat with H_2SO_4/HNO_3 (best) (b) treat with HNO_3/H_2O_2 (c) treat with HCl/HNO_3 (Comparison of above treatment against alkali-fusion method)	F	—	968
Cu	—	Tobacco smoke	—	A	L	Collect smoke by filtration or liquid N_2 cold trap, digest with HNO_3, cool, neutralize with 50% NaOH and buffer to pH 4.0 with KH phthalate. Filter and extract with APDC/MIVK	F	—	272
Cu	324.7	Natural waters	10–39 $\mu g/ml$	A	L	See Pb, ref. 191	F	Air/C_2H_2	191
Cu	324.7	Sea and river waters	0–8 ng/ml	A	L	See Cd, ref. 365	Graphite furnace (FLA-1)		365
Cu	324.7	Airborne particulates	—	E	S	See Al, ref. 366	S,P	—	366
Cu	—	Sea water	—	A	L	Sample in polythene ware, add HCl, co-precipitate Cu with CoPDC complex and redissolve in HNO_3/MIBK	F	—	401
Cu	324.7	Waters	Trace levels	A	L	See Cd, ref. 572	F	Air/C_2H_2	572
Cu	—	Effluents	Trace levels	A	L	See Cd, ref. 805	F	—	805
Cu	—	Sea water	Trace levels	A	L	—	Graphite furnace		807
Cu	—	Sea water	Trace levels	A	L	See Cd, ref. 808	Graphite furnace (Ar/H_2 shielded)		808
Cu	—	Atmospheric particulates	Trace levels	A	S	See Cd, ref. 811	Graphite furnace		811
Cu	—	Waters, sludges, sediments	Trace levels	A	L	See Cd, ref. 766	F	N_2O/C_2H_2 + Mo cup	766
Cu	324.7	Natural waters	—	A	L	Filter (0.45 μ Millipore) and adjust pH to selected value, using HNO_3 (Study of soluble/particulate Cu distribution)	Graphite furnace		917

TABLE 2.7 AIR AND PARTICULATES, WATER, SEWAGE, EFFLUENTS — continued

Element	λ/nm	Matrix	Concentration	Tech.	Analyte form	Sample treatment	Atomization	Ref.
Cu	—	Natural waters	—	A	L	Add 1% HNO_3 (by volume), to preserve sample (Inter-laboratory study)	F —	948
Cu	324.8	Effluent waters	0.5–10 μg/l	A	L	Treat H_2O (200 ml) with $HClO_4$ (2 ml) + 2% 1:10-phenanthroline (2 ml) + 14% NH_2OH. HCl (1 ml) + 80% NH_4 acetate (5 ml). Shake, add 2 ml nitrobenzene, shake and separate	F Air/C_2H_2	1189
Cu	—	Sea water	—	A	L	Study of ultrasonic nebulization of high-solids solutions	F —	1238
Cu	324.7	Water	1.4–6.3 ng/ml	A	L	See Cd, ref. 1293	Graphite furnace	1293
Cu	324.7	Sea water	Trace levels	A	L	Adjust to pH 2.0 with HCl and filter. Extract with DDC/MIBK	Graphite furnace	1299
F	529.1	Water	μg/ml levels	E	L		A —	777
Fe	—	Tobacco smoke	—	A	L	See Cu, ref. 272	F —	272
Fe	259.9	Airborne particulates	—	E	S	See Al, ref. 366	P,S —	366
Fe	—	Sea water	Trace levels	A	L	—	Graphite furnace	807
Fe	—	Natural waters	—	A	L	See Cu, ref. 948	F —	948
Fe	248.3	Sea water	0.2 μg/ml	A	L	None	Graphite furnace (CRA-63)	1221
Hg	253.7	Aquatic samples	—	A	G	Extract with C_6H_6 (Method for monomethyl Hg)	Cold vapour	30
Hg	—	Plant effluents	20 ng/g–20 μg/g	A	G	Inter-laboratory study; 37 participants	Cold vapour	151
Hg	253.7	River waters	1–25 ng/ml	A	G	(a) Oxidise with $HClO_4/H_2O_2$, in presence of Sn (II). Reduce with $SnSO_4$ solution (Total Hg) (b) Reduce directly with $SnSO_4$. (Organic Hg)	Cold vapour	155
Hg	253.7	Waters	From 3×10^{-17} g (in vapour)	E	G	Reduce with $SnCl_2$ and amalgamate Hg on Ag wool	P Microwave-induced Ar plasma (200 W)	183

Hg	253.7	Water sediments	10 ng/ml	A	G	Digest with H_2SO_4/HNO_3 at 0°C, then at 50–60°C for 2 hours. Cool, add $KMnO_4$ at 0°C, followed by $K_2S_2O_8$. Stand overnight. Reduce with $SnSO_4$	Cold vapour	185
Hg	—	Sea water	—	—	—	Study of sample contamination and storage factors	— —	249
Hg	253.7	Air	0.02–10 μg/m³	A	G	Absorb on Ag wool, release by heating at 400°C	Cold vapour	253
Hg	—	Environmental samples	—	A,F	L,G	Review (442 refs)	Flameless methods	153
Hg	253.7	Air	0.04–2.6 mg/m³	A	G	Absorb Hg vapour on Hopcalite granules and remove by washing (×4) with HNO_3	Cold vapour	347
Hg	253.7	Water sediments	100–2000 ng/g	A	G	(a) Treat dry sediment with H_2SO_4/HNO_3 (2:1) for 2 hours at 50–60°C. Cool, add $KMnO_4$, then $K_2S_2O_8$ and leave overnight, (b) as for (a), but digest with HCl/HNO_3 (1:9) (c) treat in PTFE bomb with $HF/HCl/HNO_3$ for 2 hours at 110°C. Add H_3BO_3 and proceed with $KMnO_4$ etc, as in (a)	Cold vapour	397
Hg	253.7	Waters	—	A	G	Volatilize Hg from sample and collect as Au amalgam. Extract with HNO_3, add $KMnO_4$ and electroylse to deposit Hg on Au cathode. Transfer to graphite atomizer to release Hg	Graphite furnace	447
Hg	253.7	Waters	2–50 ng/ml	A	L/G	Amalgamate on to Cu tube and release Hg by electrothermal heating	Cold vapour	570
Hg	—	Environmental samples	Trace levels	A	S	See As, ref. 830	Tube furnace	830
Hg	—	Waters	From 0.06 μg/ml	A	G	Treat with $KMnO_4$ at room temperature, then $K_2S_2O_8$ at 95°C	Cold vapour	905
Hg	—	Waters	ng/ml levels	A	G	Study of experimental parameters	Cold vapour	239
Hg	—	Waters	0–100 ng/ml	F	L	Reduce 1 ml of sample with 0.1 ml of 1% $SnCl_2$ solution	Cold vapour	923
Hg	—	Environmental samples	ng/g levels	E	G	See Cd, ref. 1105	P —	1105
Hg	—	Environmental samples	—	A	G	—	Cold vapour	1142

TABLE 2.7 AIR AND PARTICULATES. WATER, SEWAGE, EFFLUENTS — continued

Element	λ/nm	Matrix	Concentration	Tech.	Analyte form	Sample treatment	Atomization	Ref.
Hg	—	Lake sediments	—	A	—	Separate mono- and di-alkyl Hg fractions by steam distillation	— —	1270
Hg	253.7	Fresh and sea waters	5–500 ng/l	A	L	Extract Hg-APDC complex into propylene carbonate. Reduce with $SnCl_2$ and entrain Hg in N_2-flow	F —	1275
Hg	253.7	Water	—	A	S	See Au, ref. 1278	Graphite furnace + quartz cup	1288
Hg	253.7	Waste water	2–10 ng/ml	E	G	Reduce with $SnCl_2$ and sweep Hg to plasma in stream of argon	P Ar plasma arc	1290
Hg	253.7	Water	ng/ml levels	A	L	Acidify to 0.4 N H_2SO_4, stand 3 weeks, add $SnCl_2$ solution and collect Hg on Ag metal particles	Graphite furnace (500°C)	1298
Hg	253.7	Water	From 0.5 ng/ml	A	L	Digest with $KMnO_4/K_2S_2O_8/H_2SO_4$, or H_2SO_4/H_2O_2 or prepare in NaOH medium, and reduce with $SnCl_2/CdCl_2$ (Comparison of treatments)	Cold vapour	1319
Hg	—	Environmental samples	ng/g levels	A	—	—	— —	1331
Hg	253.7	Waters	0.02–2.6 μg/l	A	G	Add H_2SO_4, cool, add $NH_2OH/NaCl$, reduce with $SnCl_2$ and aerate with N_2. Dissolve any filterable residue with HNO_3/H_2SO_4 before final analysis	Cold vapour	1441
Hg	253.7	Water, air	Trace levels	E	L	—	P —	1531
Li	670.7	Water	0.001–1 μg/ml	E,A	L	General study of determination of Li, as tracer element added to water flows	F Air/C_2H_2 N_2O/C_2H_2	1259
Mg	279.5	Airborne particulates	—	E	S	See Al, ref. 366	P,S —	366
Mn	—	Tobacco smoke	—	A	L	See Cu, ref. 272	F —	272
Mn	279.5	Waters	20–120 ng/ml	A	L	Adjust to pH 8 with NH_4OH and extract with DDC/DIBK	Graphite furnace	568
Mn	—	Sea water	Trace levels	A	L	—	Graphite furnace	807
Mn	—	Natural waters	—	A	L	See Cu, ref. 948	F —	948
Mn	279.5	Water	Trace levels	A	L	—	Tantalum strip	1554

Mo	313.2	Natural waters	0.18–1.4 μg/ml	A	L	Acidify with HCl, filter, add KCNS + ascorbic acid and pass through Dowex 1-X8 column. Elute with $HCl/HClO_4$, evaporate and treat with H_2SO_4 (1:1)	F	Air/C_2H_2	992
Ni	—	Sea water	Trace levels	A	L	—	Graphite furnace		807
Ni	232.0	Atmospheric particulates	Trace levels	A	L,S	Collect on filter and take 4 mm diam. disc sample. Ash at 1200°C and atomize at 2600°C	Graphite furnace		614
Ni	—	Drinking water	Trace levels	A	L	See Co, ref. 1337	Graphite furnace		1337
P	526 (HPO band)	Air, waters	0.9–2 μg/ml (P in air) 0.003–120 μg/ml (H_3PO_4 in water)	E	L	Remove cations by ion-exchange. Sulphate (> 5 μg/ml) may interfere	F	Air/H_2	223
P	253.6	Waters	μg/ml levels	E	L	—	A	—	777
Pb	—	Atmospheric particulates	From 0.005 μg/m³	A	S	Filter air through porous graphite cup and introduce directly to furnace	Graphite furnace		31
Pb	—	Sea water	14–60 pg/g	A	L	Comparison with anodic stripping voltammetry and isotope dilution mass spectrometry	F	—	47
Pb	217.0	Atmospheric particulates	—	A	L	Extract membrane filter with $HNO_3/HClO_4$ and add $HClO_4$. Match standards for acid content (Study of critical pH range)	F	Air/C_2H_2	84
Pb	—	Atmospheric particulates	—	A	S	Collect particulate Pb (> 5 μ) on Micropore C disc and molecular Pb on carbon bed. Analyse separately	Graphite furnace		141
Pb	283.3	Air and gas particulates, automobile exhausts	1–250 mg/l	A	L	Filter and extract with HCl + small addition of HNO_3	F	Air/C_2H_2	212
Pb	283.3	Natural waters	2–14 μg/ml	A	L	Concentrate on Dowex-1X8 resin (Br^-, CNS^- or Cl^- form)	F	Air/C_2H_2	191
Pb	217.0	Natural and waste waters	Trace levels	A	L	(Use of Asiatic clam as Pb monitor). Extract shells by method of Yeager et al. (Env. Sci. Tech. 1971, **5**, 1020)	F	Air/C_2H_2	362
Pb	217.0	Sea and river waters	0–8 ng/ml	A	L	See Cd, ref. 365	Graphite furnace (FLA-1)		365
Pb	—	River water	Trace levels	A	L	Extract with dithizone/$CHCl_3$, evaporate and redissolve in aqueous medium	Graphite furnace		392

TABLE 2.7 AIR AND PARTICULATES, WATER, SEWAGE, EFFLUENTS — continued

Element	λ/nm	Matrix	Concentration	Tech.	Analyte form	Sample treatment	Atomization	Ref.
Pb	—	Street dust	0–3000 μg/g	A	L	Dry at 120°C, grind and treat with HNO_3	F —	402
Pb	—	Atmospheric particulates	—	A	S	—	Graphite furnace	433
Pb	—	Air	0.1–2.5 μg/m³	A	L	Bubble air through HCl/ICl solution, add EDTA and extract with dithizone into $CHCl_3$. Back extract with HNO_3/H_2O_2	Graphite furnace (HGA-70)	495
Pb	217.0	Air	0.6–13 μg/m³	A	L	Collect on membrane filter, wet-ash with HNO_3/H_2O_2, evaporate, dissolve in 0.5% tartaric acid solution	F Air/C_2H_2	522
Pb	283.3	Waters	Trace levels	A	L	See Cd, ref. 572	F Air/C_2H_2	572
Pb	—	Air	From 0.003 μg per m³, as organic Pb	A	L	—	F —	660
Pb	—	Air, foods	—	A	L	Review	F —	661
Pb	—	Effluents	Trace levels	A	L	See Cd, ref. 805	F —	805
Pb	—	Sea water	Trace levels	A	L	See Cd, ref. 808	Graphite furnace	808
Pb	—	Atmospheric particulates	Trace levels	A	S	See Cd, ref. 811	Graphite furnace	811
Pb	—	Environmental samples	Trace levels	A	S	See As, ref. 830	Tube furnace	830
Pb	217.0	Air	Trace levels (Pb-alkyls)	A	G	Trap volatile compounds at Pb, Se in U-tube at −70°C. Sweep by N_2 into GC column and pass effluent to heated (980°C) silica tube	Tube furnace	831
Pb	—	Waters, sludges, sediments	Trace levels	A	L	See Cd, ref 766	F N_2O/C_2H_2 + Mo cup	766
Pb	—	Natural waters	From 1 ng/ml	A,E	L	See Cd, ref. 237	F Air/C_2H_2 P —	237
Pb	—	Air	Trace levels	A	S,G	Absorb particulate Pb on graphite disc and collect any filterable Pb on graphite bed. Heat bed and/or disc to 1500°C to evolve Pb	Graphite furnace ("T"-furnace)	933
Pb	—	Air	—	A	L	—	F —	973
Pb	—	Environmental samples	ng/g levels	E	G	See Cd, ref. 1105	P —	1105

Pb	—	Drinking waters	0–200 μg/l	A	L	—	F Delves cup	1213
Pb	283.3	Water	0.3–4 ng/ml	A	L	See Cd, ref. 1293	Graphite furnace	1293
Pb	217.0 283.3	Airborne particulates	From 0.01 μg/m³	A	L	Collect on glass filter and extract with acid. (2000 m³ sample)	F —	1327
Pb	—	Airborne particulates	—	A	L	—	Graphite furnace	1343
Pb	217.0	Water	—	A	L	—	Graphite furnace	1549
Pb	217.0	Water	Trace levels	A	L	—	Tantalum strip	1554
S	—	Mill effluent vapours	—	E	G	Flame photometric S analyser adapted as continuous monitor.	F —	326
S (H_2SO_4)	—	Sulphuric acid aerosols	—	E	G	Filter to remove H_2SO_4 aerosol from other gaseous components. Volatilize from filter with inert gas purge at $<250°C$	F —	327
S (H_2SO_4)	—	Air	—	E	—	—	F —	937
Sb	217.6	Water	0.01–0.15 μg/ml	A	L	See As, ref. 380. Add small amount of Hg(II) to overcome interference by As, Co, Mg, Ni and Se	Graphite furnace (FLA-1)	380
Sb	—	Environmental samples	Trace levels	E	G	See As, ref. 1227	P —	1227
Se	204.0	Environmental samples	From 0.1 μg/l (waters) to 15 ng/g (solids)	E	L	Chelate Se (IV) with 5-nitro-o-phenylenediamine and extract into toluene. Separate by G.C.	P Microwave-excited plasma	221
Se	196.0	Waters, effluents	1–50 μg/l	A	L	Remove cations by treatment with ion-exchange resin (Rohm and Hass IR-124). Add Mo solution to enhance sensitivity	Graphite furnace (HGA-2000)	232
Se	196.1	Water	0.02–0.3 μg/ml	A	L	See As, Sb, ref. 380	Graphite furnace (FLA-1)	380
Se	196.1	Waters	μg/ml levels	A	L	—	F —	777
Se	—	Environmental samples	Trace levels	A	S	See As, ref. 830	Tube furnace	830
Se	196.1	Air	Trace levels (Se-alkyls)	A	G	See Pb, ref. 831	Tube furnace	831
Se	196.1	Waters, effluents, sludges, sediments	From 0.002 μg/ml (in final extract)	A	L	Waters, effluents — Treat with HNO_3/H_2O_2, evaporate to low volume and add Ni. Sludges, sediments — Treat with HNO_3, evaporate almost to dryness, digest with HNO_3/H_2O_2 and add Ni	Graphite furnace (HGA-2000)	1061

TABLE 2.7 AIR AND PARTICULATES, WATER, SEWAGE, EFFLUENTS — continued

Element	λ/nm	Matrix	Concentration	Tech.	Analyte form	Sample treatment	Atomization	Ref.
Se	—	Water	From 1 μg/l	A	L,G	See As, ref. 1207	F Air/Ar/H_2	1207
Si	288.1	Airborne particulates	—	E	S	See Al, ref. 366	P,S —	366
Si	250.1	Waters	μg/ml levels	E	L	—	A —	777
Sn	—	Drinking water	Trace levels	A	L	See Co, ref. 1337	Graphite furnace	1337
Tl	—	Sea and river water particulates	Trace levels	A,F	L	See Ag, ref. 334	Flameless method	334
V	318.4	Airborne particulates	From 25 ng/m^3 (flame) or 0.4 ng/m^3 (furnace)	A	L	For furnace method, dry at 100°C, ash at 1100°C and atomize at 2500°C	F N_2O/C_2H_2 Graphite furnace	1326
W	400.8	Natural waters	0.25–5 μg/ml	A	L	Adjust to pH 1.6–1.7 with HNO_3, add Na acetate to bring to pH 2.0 and extract with benzoin anti-oxine into MIBK. Solubilize W complex by addition of 10% solution of 1-ephedrine in MIBK	F N_2O/C_2H_2	167
Zn	213.9	Atmospheric particulates	—	A	L	See Pb, ref. 84	F Air/C_2H_2	84
Zn	213.9	Air and gas particulates	0.05–5 mg/l	A	L	See Pb, ref. 212	F Air/C_2H_2	212
Zn	—	Tobacco smoke	—	A	L	See Cu, ref. 272	F —	272
Zn	213.8	Natural waters	18–685 μg/ml	A	L	See Pb, ref. 191	F Air/C_2H_2	191
Zn	334.5	Airborne particulates	—	E	S	See Al, ref. 366	P,S —	366
Zn	—	Effluents	Trace levels	A	L	See Cd, ref. 805	F —	805
Zn	—	Sea water	Trace levels	A	L	See Cd, ref. 808	Graphite furnace	808
Zn	—	Atmospheric particulates	Trace levels	A	S	See Cd, ref. 811	Graphite furnace	811
Zn	—	Natural waters	—	A	L	See Cu, ref. 948	F —	948
Zn	213.9	Water	Trace levels	A	L	—	Tantalum strip	1554
Anionic detergents (indirect)	324.7 (Cu)	Waters	3 ng/ml — 2.5 μg/ml (as Na salt)	A	L	Form Cu phenanthroline complex and extract into MIBK	F Air/C_2H_2	258

Carbon disulphide (indirect)	324.7 (Cu)	Air	—	A	L	Absorb CS_2 on activated charcoal, extract with iso-amyl acetate. Complex Cu with pyrrolidine + $CuSO_4$, adjusted to pH 2.0–2.5 with HCl. Measure Cu in iso-amyl acetate layer	F	Air/C_2H_2	546
Various (9)	—	Power plant effluents	Trace levels	A	L	Filter (45 μ) and acidify to 0.2% HNO_3 strength. Soak polythene sample vessels in 2% HNO_3 for 24 hours, prior to use	Graphite furnace (HGA-2100)		3
Various (5)	—	Streams and lake waters	Trace levels, from 0.01 μg/ml	A	L	Study of both soluble and suspended particulate trace element variations. (Cd, Co, Cu, Pb, Zn)	F	—	26
Various (13)	—	Marine sediments	Trace levels	A	L	Treat successively with $HCl/HF/HNO_3/HClO_4$. Ash at 400°C and redissolve. Omit ashing stage for Cd determination	F	—	27
Various (23)	—	Atmospheric particulates	Minor and trace levels	A,E	L	Dry-ash filters at 400–425°C and dissolve residue in $HF/HCl/HNO_3$ in PTFE pressure vessel. Treat aliquots with (a) 1000 μg/ml K or (b) 1000 μg/ml each Cs, La. Dilute as required and match standards to diluted matrix. Determine Cs, Li, Rb by FES and other elements by AAS	F	Air/C_2H_2 N_2O/C_2H_2	54
Various (6)	—	Water	Trace levels	A	L	Study of recycling polluted water through soils and vegetation to improve quality. Analyse soils, plants and water for Ca, Mg, K, Pb, Zn, Cu	F	—	61
Various (10)	—	River (Danube) waters	μg/l levels	A	L	Treat with HCl/HNO_3 to dissolve any suspended matter	F	—	252
Various	—	Airborne pollutants	Trace levels	A	L	Review	F	—	389 390
Various (12)	—	Natural waters	Trace levels	E	L	Evaporate, as necessary, to salt concentration of 3–20 g/l and add $Ba(NO_3)_2$ or In_2SO_4 buffer	A	—	420
Various (7)	—	Environmental samples	Trace levels	A	G	Reduce with $NaBH_4$ (method for As, Sb, Se, Te, Bi, Sn, Ge)	Heated cell		428
Various	—	Industrial waste waters	Trace levels	E	S	Filter, concentrate by ion-exchange separation, evaporate eluate and ash residue to constant weight. Mix with dolomite and excite with a.c. arc	A	10 A a.c.	449

TABLE 2.7 AIR AND PARTICULATES. WATER, SEWAGE, EFFLUENTS — continued

Element	λ/nm	Matrix	Concentration	Tech.	Analyte form	Sample treatment	Atomization	Ref.
Various (7)	—	Waters	1 μg/ml	A	L	Adjust to pH 4 and extract Zn, Cd, Ni, Co, Fe, Mn, Cu with APDC/MIBK. (Multi-channel AAS, with vidicon detector).	F Air/C_2H_2	672
Various (11)	—	Atmospheric particulates	Trace levels	A	S,L	—	Graphite furnace	674
Various (7)	—	Dusts, fumes, effluents	Trace levels	A	L	Collect particulate matter on filter and acid-digest (Method for Fe, Mn, Cu, Zn, Ni, Cd, Pb)	F —	787
Various	—	Waters, environmental samples	ng/g levels	A	L	Review	Graphite furnace	792
Various (11)	—	Water	ng/g levels	A	L	Analysis of standard NBS sample for water analysis	F Air/C_2H_2 Graphite furnace Cold vapour (Hg)	806
Various (11)	—	Dusts and atmospheric particulates	Trace levels	A	L,S	Collect by filtration and dissolve in $HCl/HNO_3/HClO_4$. (Results given for Pb, Cu by solid sample method)	Graphite furnace (HGA-74)	809
Various	—	Atmospheric (oceanic) particulates	Trace levels	A	S	Analyse small area of filter disc directly	Graphite furnace	810
Various	—	Marine samples; waters, fish, sediments	μg/g levels	A	L	Methods cover Zn (flame); Hg (cold vapour); Cd, Co, Cr, Cu, Mn, Pb, Ni (furnace); As, Se (Hydride generation)	F Air/C_2H_2 Graphite furnace Cold vapour (Hg)	876
Various	—	Environmental samples	Trace levels	A	L,S	Review	F — Graphite furnace	893
Various (11)	—	Atmospheric particulates	—	A	S	—	Graphite furnace	703
Various	—	Water	—	A	L	(Study of accuracy and precision)	F —	704
Various	—	Industrial process waters	—	A,E	L	(Automated system)	F —	745
Various (11)	—	Atmospheric particulates	—	A	L	Collect on filter and extract with HNO_3 (Study of height and time variations, by AAS and XRF)	Graphite furnace (HGA-72) F Air/C_2H_2	622
Various	—	Waters	Trace levels	E	L	—	P ICP	906
Various (10)	—	Airborne particulates	Trace levels	E	S	Collect on glass-fibre filter, punch disc, transfer to electrode crater and treat with H_2SO_4/HF. Add CaF_2/graphite buffer mixture + In/Co as internal standards	A 10 A d.c.	928

Various (7)	—	Sea water	Trace levels	A	L	(High-speed continuous background correction method)	Graphite furnace (Model 63)		971
Various	—	Waters	Trace levels	E	S,L	Filter (0.45 μm) and acidify	Hollow-cathode		1134
Various	—	Environmental samples	Trace levels	E	L		P	ICP	1136
Various	—	Air and water	Trace levels	A	L,S	Review	Graphite furnace		1225
Various (19)	—	Airborne particulates	Trace levels	E	S	Collect on glass-fibre filter and digest with HF only, for best sensitivity	A	D.C. arc	1241
Various (39)	—	Water	Trace levels	E	S	Enrich metals by treatment of sample with soybean milk + δ-gluconic lactone. Boil to coagulate and centrifuge to separate. Ash coagulate in low-temperature plasma. (See also, refs. 1278, 1288, 1289)	A	—	1297
Various	—	Water	ng/ml levels	A	L	Stabilize with HCl addition	Graphite furnace		1322
Various (8)	—	Sea water	ng/ml levels	A	L	(a) Extract with hexamethylene-ammonium-hexamethylene-dithiocarbamate/butyl acetate (b) Evaporate with H_2SO_4/HNO_3	F	—	1356
Various (7)	—	Water	Trace levels	A	L	Acidify with HCl, buffer to pH 5.0 after filtration and extract with DDC into acetone/$CHCl_3$ (2:5). Evaporate, redissolve and absorb on cation-exchange resin. Elute, evaporate and redissolve in methanol/HBr	F	Air/C_2H_2	1362
Various (6)	—	Water	μg/l levels	A	L	Acidify, adjust to pH 3.5 and extract into MIBK	F	Air/C_2H_2	1385
Various	—	Water sediments	Trace levels	A	L	Treat with $HF/HNO_3/HClO_4$	F	—	1386
Various (20)	—	Lake water sediments	—	A	L	Digest with $HNO_3/HClO_4/HF$ at 140°C in PTFE bomb	F	—	1479
Various (8)	—	Waters	ng/g levels	E	L	Nebulize in argon	A	D.C. arc-plasma	1492
Various	—	Waters	ng/g levels	E	L	Add (1:1) to graphite powder, evaporate to dryness and mix (88:12) with LiF	A	10 A d.c.	1497
Various (13)	—	Environmental samples	Trace levels	E,A	L,S	Review	F,A	—	1513
Various (6)	—	Waters	Trace levels	A	L	Study of extraction procedure for Al, Cd, Cu, Fe, Pb, Zn	F	—	1556

2.8 FOODS

2.8.1 Introduction

It is interesting to note that the increasing use of electrothermal atomization techniques for trace element analysis of foods has not led to a general blind acceptance of results of analyses that were previously beyond the capability of alternative analytical techniques. Accuracy is often established by the analysis of standard reference materials and by inter-laboratory studies. Consequently, atomic spectroscopy is becoming a technique accepted by food analysts and official bodies.

2.8.2 Flame Methods

The direct flame AAS determination of Zn in decarbonated beers was recommended by the Institute of Brewing Analysis Committee following a satisfactory inter-laboratory study (262). Black (301) showed that a simple (1 + 4) dilution of soybean oil with MIBK was sufficient preparation to allow the determination by flame AAS of Cd, Cu, Fe, K, Mg, Mn and Zn. This preparation seems to be preferable to the 2 hour reflux extraction with acid and EDTA used by Jacob and Klevay (241) for the determination of Cu and Zn.

Air/C_2H_2 flames were used in the flame AAS determinations of high concentrations of As (1185) and of Se (454, 1188) in animal feed pre-mixes. Although the Ar/H_2/entrained air flame was more sensitive for Se, the matrix interferences were reported to be more severe than with the air/C_2H_2 flame. EDLs were used as light sources in both reports to overcome the effects of the considerable flame absorption at the low wavelengths used; one wonders whether the separated N_2O/C_2H_2 flame which is relatively transparent at these wavelengths would have been an improvement (see *ARAAS*, 1973, **3**, 10). Both analyses required minimal sample preparation—dissolution in 1% $(NH_4)_2CO_3$ for As (1185). and $HClO_4$/H_2O_2 oxidation for Se (1188). In this latter procedure the sample was heated with $HClO_4$ without prior oxidation which seems unnecessarily hazardous.

Hydride generation procedures were used to determine lower concentrations of As, Se, Sb and Te in foods after wet or dry ashing (400, 1186, 1316, 1449). AAS with N/H_2/entrained air flame was used for the determination of As (1316) and for Se (1186) in the 5–20 mg/kg range, whereas Thompson (400) determined much lower concentrations of Se of 0.07–0.60 mg/kg with a detection limit of 10^{-4} mg/kg by using AFS with Ar/H_2/entrained air flame. In this latter work, similar detection limits were reported for As, Sb and Te in pure solutions.

Franco and Holak (1179) used FES with an N_2O/H_2 flame to determine B in caviar, following ashing with HNO_3 and extraction with 2-ethyl-1,3 hexanediol in MIBK. The method was tested collaboratively on samples with H_3BO_3 added at 1–2 mg/g and on a sample containing 0.5% H_3BO_3. Recoveries of greater than 95% were obtained.

Solvent extraction procedures using quaternary ammonium salt ion association systems were reported for the determination of Cd after low-temperature ashing (667) or wet oxidation (1045). The accuracy of the methods was checked by the analysis of NBS reference materials (667) and by a collaborative study (1045). Another collaborative study (539) also reported excellent recoveries and agreement with theoretical or expected concentrations of Cd, Cu, Ni and Mn in soybean oil. In this report all four metals were extracted simultaneously from solutions of ashed samples at pH 7 using NaDDC and MIBK and then determined by flame AAS.

Baetz and Kenner (1410) used a chelating ion-exchange separation and concentration procedure to determine Cd, Co, Cu, Mn, Ni, Pb and Zn in eight different food commodities after oxidation with HNO_3/H_2SO_4/H_2O_2. The average recovery of added trace metals was 97.2% and the accuracy of the method was checked using NBS reference mtaerials of bovine liver and orchard leaves.

2.8.3 Electrothermal Methods

The determination of trace elements in foods and beverages by flame AAS either without sample pretreatment or with simple dilution procedures is limited to those matrices that have low solids contents and thus produce little matrix interference. Whereas for example, Fazakas (941) found no significant matrix interferences in the AAS determination of Cu and Cr in mineral waters using a graphite tube furnace. Calibration graphs were linear up to 10 $\mu g\ l^{-1}$ and the RSD ranged from 0.02–0.095. Matrix interferences in the graphite furnace AAS determination of Pb in wines diluted (1 + 9) with water were overcome by standard additions (1037) whereas the determination of Pb in milk samples required dry ashing and that of Pb in butter samples required HNO_3 extraction from samples dissolved in petroleum ether (270).

Wolf *et al* (144) found that the direct determination of Cr in sugars by furnace AAS gave low results due to pre-atomization losses of organically bound Cr, and that a prior low-temperature ashing (LTA) was necessary to determine all of the Cr present in unrefined sugars. These observations are supported by those of Black (301) who found direct analysis of Cr in soybean oil to be unsatisfactory by furnace AAS, and provide indirect confirmation of the observations of Mertz (*Clin. Chem.*, 1975, 21, 468) that the Cr-containing compound that affects glucose tolerance in humans contains organically bound Cr that is difficult to determine.

Additives to aid the sample decomposition within the furnace and to retain the analyte during this stage have been described (706, 959, 1448). Tsujino (959) mixed the sample with either W or TiO_2 to catalyse sample decomposition, whereas Slavin *et al* (706) added a Ni salt solution to the sample in the furnace which allowed the formation of NiSe thus enabling the ashing to proceed up to 1200°C without losses of Se. Similarly, the addition of Ni prevented losses of As up to 1400°C, and the addition of F allowed ashing at temperatures up to 1000°C without losses of Cd. Inhat (1448) claimed that the precision for the graphite furnace AAS determination of Se in foods was significantly improved when the analytical solutions contained 5000 $\mu g\ ml^{-1}$ of Ni. This approach of matrix modification should prove of great value to those using electrothermal atomization techniques.

2.8.4 Other Methods

Sobel *et al* (739) reported the use of an ICP emission method for the determination of 15 elements in solutions of ashed foods. They found no matrix interferences for a variety of foods even at low analyte concentrations, *i.e.* 0.08–0.40 $\mu g\ ml^{-1}$ of Mn, when the sample and standard solutions were prepared in $HClO_4$ (1 + 1). In view of the potential hazard with such high concentrations of $HClO_4$ disposal of nebulizer wastes should be carefully considered.

There were many reports on the cold vapour AAS determination of Hg. One new approach to sample oxidation which seems ideal for routine analysis was the use of high-temperature heating block (250°C) in which test tubes were placed such that the protruding length of tube acted as a reflux condenser (723). No losses of Hg were reported and the accuracy was quoted as 96–112% of the NBS standard certified value. The long term RSD (138 days) was 0.075–0.095.

TABLE 2.8 FOOD AND BEVERAGES

Element	λ/nm	Matrix	Concentration	Tech.	Analyte form	Sample treatment	Atomization	Ref.
As	—	Food products	Trace levels	A	L,S	Prepare solution in HNO_3 and add Ni, to allow ashing temperature of 1400°C. See also: Cd, Hg, Se, ref. 706	Graphite furnace	706
As	—	Fish tissue	Trace levels	A	L	Use low-temperature ashing technique and extract with HNO_3 (Study of P interference)	Graphite furnace	909
As	193.7	Feed premix	10–80% (as Roxarsone)	A	L	Extract with 1% $(NH_4)_2CO_3$	F Air/C_2H_2	1185
As	193.7	Foods	From 5 μg/g	A	L	Dry-ash with $Mg(NO_3)_2$ at 500°C, digest and convert As to AsH_3 by reaction with $NaBH_4$	F H_2/N_2	1316
As	—	Foods	From 0.1 μg/g	A	G	Digest with $HNO_3/H_2SO_4/HClO_4$ and form hydride by $NaBH_4$ reaction.	F Ar/H_2	1449
B	518 (band)	Caviar	0.1–0.5% (H_3BO_3)	E	S	Heat with HNO_3 in sealed PTFE vessel (150°C). Extract solution with 2-ethylhexane-1,3-diol into MIBK	F H_2/N_2	1179
Ca	554 (band)	Milk and dairy produce	—	E	L	Ash, redissolve and add 20% NH_4 molybdate solution + buffer solution of NH_4Cl/NH_4OH	F Air/propane	143
Ca	—	Animal feeds	—	A	L	Digest with $HNO_3/H_2SO_4/HClO_4$	F —	286
Cd	228.8	Rice	0–2 μg/g	A	L	Treat with 1% HNO_3 for 16 hours at room temperature. Nebulise supernatant liquid	F Air/C_2H_2 (+sample boat)	7
Cd	228.8	Fish, mussels, cereals	5–500 ng/ml in extract	A	L	Digest with HNO_3/H_2SO_4, dilute and (a) adjust to pH 10.0, for extraction with dithizone/$CHCl_3$ or (b) adjust to pH 3.8–4.0 for extraction with APDC/MIBK	F —	72
Cd	—	Wine	Up to 0.21 μg/g	A	L	Determine Cd, Cr, Pb by furnace method and Cu, Fe, Mn, Zn by flame method	Graphite furnace	119
Cd	—	Brown rice	—	A	L	Extract with 1 N HCl at 30°C for 24 hours (Collaborative study—11 participants	F —	324
Cd	228.8	Fats, oils	—	A	L	Digest with H_2SO/H_2O_2 and extract, at pH 7, with NaDDC/MIBK	F Air/C_2H_2	539

Cd	228.8	Foodstuffs	0.01–0.1 μg/g	A	L	Ash at low temperature and extract Cd from acid solution with Aliquot-336, into MIBK	F —	667
Cd	228.8	Fish	0.14–5.6 μg/g	A	L	Wet-digest with acid (a) $HNO_3/H_2SO_4/HClO_4$ or (b) HNO_3/H_2O	Ta ribbon	690
Cd	—	Canned oysters	1.84 μg/g	A	L	Wet-ash and extract Cd with dithizone/MIBK	F —	804
Cd	—	Food products	Trace levels	A	L,S	Convert Cd to fluoride, to permit ashing temperature of up to 1000°C. See also As, Hg, Se, ref. 706	Graphite furnace	706
Cd	228.8	Tomato soup	0.01–0.04 μg/g	A	L	Wet-ash with H_2SO_4/H_2O_2, cool, add KI, dilute, add 1% Amberlite LA 2 in MIBK and separate organic layer. Extract standards similarly.	F Air/C_2H_2	1045
Cd	—	Foods	0.1–0.9 μg/g (dry weight)	A	L	Dry-ash at 475°C and redissolve	F —	1414
Co	240.7	Feed grains, forages	5–200 ng/g	A	L	Ash at 450°C, extract with HNO_3, adjust to pH 3–5 and extract Co with 1-nitroso-2-naphthol into $CHCl_3$. Evaporate and redissolve in $CHCl_3$	Graphite furnace	1187
Co	—	Animal feeds	From 1 ng/g	A	L	Dry-ash at 450°C with NH_4NO_3, treat with 6 M HCl and extract Co with 1-nitroso-2-naphthol into $CHCl_3$, at pH 3.3. Wash extract with NaOH, evaporate $CHCl_3$ layer and redissolve in isoamyl acetate	F Air/C_2H_2 Graphite furnace	1204
Cr	—	Wine	Up to 0.17 μg/g	A	L	See Cd, ref. 119	Graphite furnace	119
Cr	—	Sugars	Up to 270 ng/g	A	S	Determine inorganic Cr directly. For organically-bound Cr, ash by O-plasma before introduction to furnace	Graphite furnace	144
Cr	357.9	Carbonated mineral waters	0.55–5.26 ng/g	A	L	Dry at 100°C, ash at 1100°C and atomize at 2400°C	Graphite furnace	941
Cu	—	Wine	Up to 1 μg/g	A	L	See Cd, ref. 119	F —	119
Cu	324.7	Edible fats and oils	0.03–0.35 μg/g	A	L	Extract under reflux with acid-EDTA	F Air/C_2H_2	241
Cu	324.7 327.4	Cheese	—	A	L	Digest with concentrated NH_3 solution + MIBK	Graphite furnace (HGA-72)	282
Cu	324.7	Fats, oils	—	A	L	See Cd, ref. 539	F Air/C_2H_2	539

TABLE 2·8 FOOD AND BEVERAGES — continued

Element	λ/nm	Matrix	Concentration	Tech.	Analyte form	Sample treatment	Atomization	Ref.
Cu	324.7	Carbonated mineral waters	0.30–7.4 ng/g	A	L	See Cr, ref. 941	Graphite furnace	941
Fe	—	Wine	Up to 21 μg/g	A	L	See Cd, ref. 119	F —	119
Fe	248.3	Wine	5–50 μg/g	A	L	—	F —	1260
Hg	253.7	Canned sea-foods	Trace levels	A	—	Comparison with GLC method, for MeHg+	Flameless method	116
Hg	253.7	Canned fish, mussels	Trace levels	A	G	Digest with $KMnO_4$, followed by addition of $NH_2OH.HCl$, extract fats with $CHCl_3$ and reduce Hg with $SnCl_2$	Cold vapour	147
Hg	253.7	Fish and fish products	—	A	G	Dry, homogenise and combust in O_2. Dissolve products in $HCl/KMnO_4$. For high sensitivity, extract Hg with dithizone, re-extract with HCl, reduce and collect Hg vapour on Ag wire before volatilisation	Cold vapour	261
Hg	—	Foods	From 0.5 ng/g	A	G	(a) Wet oxidise with HNO_3/H_2SO_4 or (b) combust with O_2. Reduce with $SNCl_2$ in acid medium	Cold vapour	457
Hg	—	Fish	—	A	G	Combust with Co oxide in air at 700°C and absorb vapour in $KMnO_4/H_2SO_4$. Reduce with $SnCl_2$	Cold vapour	459
Hg	—	Food products	Trace levels	A	L,S	Convert to sulphide, to allow ashing temperature up to 300°C. See also As, Cd, Se, ref. 706	Graphite furnace	706
Hg	253.7	Fish	Trace levels	A	G	Treat with HNO_3/H_2SO_4 in test-tube and digest in Al heating block at 250°C. Add $KMnO_4$, reduce with hydroxylamine sulphate	Cold vapour	723
Hg	253.7	Fish	Trace levels	A	G	Macerate tissue, extract with solvent, separate methyl-Hg by GC and pass through heated quartz tube containing CoO, to reduce to Hg	Cold vapour	724
Hg	—	Wine	—	A	G	Oxidise (a) $KMnO_4/H_2SO_4$; (b) 5% CrO_3 in H_2SO_4; (c) HNO_3 + CrO_3/H_2SO_4 (best)	Cold vapour	1339
Hg	253.7	Seeds, grain, food products	0.01–10 μg/g	A	G	Digest with HNO_3/H_2SO_4 at 60°C, filter, reduce with $SnCl_2/NaCl/NH_2OH.HCl$	Cold vapour	1413

Hg	—	Foods	0.02–9.4 μg/g (dry weight)	A	G	Combust in O_2-flask and analyse by Hatch and Ott method (Anal. Chem., 1968, **40**, 2085)	Cold vapour		1414
K	—	Animal feeds	—	A	L	See Ca, ref. 286	F	—	286
Mg	—	Animal feeds	—	A	L	See Ca, ref. 286	F	—	286
Mn	—	Wine	Up to 2.62 μg/g	A	L	See Cd, ref. 119	F	—	119
Mn	279.5	Fats, oils	—	A	L	See Cd, ref. 539	F	Air/C_2H_2	539
Na	—	Animal feeds	—	A	L	See Ca, ref. 286	F	—	286
Ni	232.0	Fats, oils	—	A	L	See Cd, ref. 539	F	Air/C_2H_2	539
P	213.6	Animal feeds	100–600 μg/ml (in extract)	E	L	—	P	ICP	1452
Pb	—	Evaporated milk	0.14–0.4 μg/g	A	L	Comparison of AAS and ASV methods (23 laboratories)	F	—	1451
Pb	283.3	Canned milk	0.1–.04 μg/g	A	L	Decompose with HNO_3 in sealed vessel at 140°C	Graphite furnace		1529
Pb	—	Wine	Up to 0.57 μg/g	A	L	See Cd, ref. 119	Graphite furnace		119
Pb	217.0	Milk, butter	Trace levels (from 4 ng/g)	A	L	Evaporate milk samples to dryness with HNO_3, ash at 480°C, digest with HNO_3, dilute with water and filter. Dissolve butter samples in petroleum (60–80° boiling-range), extract with 1:1 HNO_3, evaporate to dryness and dissolve in $H_2O/HNO_3/$ acetone (1:1:1)	Graphite furnace		270
Pb	217.0 283.3	Beverages, fruit drinks	0.5–20 μg/ml (in extract)	A	L	Study of Pb content of samples after storage (30 mins.) in glazed ceramic ware	F	Air/C_2H_2	396
Pb	283.3	Foods, diets	Trace levels	A	L	Add Mg acetate solution. evaporate to dryness at 140°C and ash at 500°C. Dissolve in HNO_3, add Na citrate solution, adjust to pH 3.5 with NH_3 and extract with APDC/MIBK. See also, Zn, ref. 515	F	Air/C_2H_2	515
Pb	—	Wine, fruit juice	Trace levels	A	L	Dilute (1:10) with H_2O. Calibrate by standard addition method	Graphite furnace		1037
Pb	—	Milk	—	A	L	Comparison of methods	F	—	1361

TABLE 2·8 FOOD AND BEVERAGES — continued

Element	λ/nm	Matrix	Concentration	Tech.	Analyte form	Sample treatment	Atomization	Ref.
Pb	—	Foods	1–11.5 μg/g (dry weight)	A	L	See Cd, ref. 1414	F —	1414
Se	196.0	Animal feeds	0.07–0.6 μg/g	F	G	Digest with $HNO_3/HClO_4$, then HCl, dilute, add $NaBH_4$ solution and pass evolved hydride to flame	F Ar/H_2	400
Se	196.0	Animal feed pre-mix	0.0057–0.42%	A	L	Digest with $HClO_4/H_2O_2$ and filter. (Comparison with XRF method)	F Air/C_2H_2 or Ar/H_2	454
Se	—	Food products	Trace levels	A	L,S	Add Ni, to allow ashing temperature of up to 1200°C. See also As, Cd, Hg, ref. 706	Graphite furnace	706
Se	196.0	Feed premix	0.002–0.4%	A	L	Wet-ash with $HClO_4/H_2O_2$, treat with HCl and reduce to H_2Se with $NaBH_4$	F Air/H_2/N_2	1186
Se	196.0	Feed premix	0.006–0.5%	A	L	Heat with H_2O_2/HCO_4 and dilute	F Air/C_2H_2 or Air/Ar/H_2	1188
Se	—	Foods	0.1–3 μg/g	A	L,G	(a) Digest with $HNO_3/H_2SO_4/HClO_4$, evaporate, redissolve in HCl/H_2SO_4, add Sn Cl_2 + Zn + KI and pass evolved H_2Se to flame; (b) digest with $HNO_3/H_2SO_4/HClO_4$. Precipitate Se with ascorbic acid, redissolve with $HNO_3/HClO_4$ and add Ni (5000 μg/ml) before introduction to furnace	F Air/Ar/H_2 Graphite furnace	1448
Se	—	Foods	From 0.1 μg/g	A	G	See As, ref. 1449	F Ar/H_2	1449
Sn	224.6 235.5 286.3		From 0.5 μg/g	A	L	Digest with HNO_3/H_2SO_4 and dilute to final solution containing 50% C_2H_5OH	F Air/C_2H_2 N_2O/C_2H_2	1450
Sr	—	Milk, hay	—	A	L	Dry-ash at 450°C and dissolve in 6 M HCl. Add La to mask Ca	F —	476
Zn	—	Wine	Up to 5 -g/g	A	L	See Cd, ref. 119	F —	119
Zn	213.9	Edible fats and oils	0.03–0.35 μg/g	A	L	See Cu, ref. 241	F Air/C_2H_2	241
Zn	—	Beer	0–1 mg/l	A	L	Decarbonate and aspirate directly	F —	262
Zn	—	Animal feeds	—	A	L	Add $LaCl_3$ to overcome general interferences. For low Zn levels, extract Zn and determine in HCl solution	F Air/C_2H_2	322

Zn	213.9	Foods, diets	Trace levels	A	L	Mix with solid Mg acetate, add $CaCO_3$ (to convert any P present into $Ca_3(PO_4)_2$, dry and ash. Extract with HCl, dilute and filter.	F	Air/C_2H_2	515
Various	—	Foods	Trace levels	A	L	Review	F	—	48
Various	—	Foods	Trace levels	A	L	Review of sample preparation methods	F	—	50
Various (8)	—	Oysters	Trace levels	A	L	Digest with HNO_3 or $HNO_3/HClO_4$. (Study of pond oyster culture, including analysis of oysters, plants and sediments for Cd, Co, Cu, Fe, Mn, Ni, Pb, Zn)	F	—	60
Various (7)	—	Animal feed	μg/g levels	A	L	Method covers Cd, Co, Cu, Pb, Mn, Ni, Zn	F	—	265
Various (12)	—	Soybean oil	μg/g levels	A	L	(a) Char, dry-ash at 600°C and dissolve in HCl; (b) dilute (1:4) with MIBK and aspirate; (c) volatilize directly (1 μl) in furnace. (Comparison of methods)	F Graphite furnace (3000°C, in N_2)	Air/C_2H_2	301
Various (8)	—	Granulated sugar	From <0.1% (Cu) to >20% (K) of ash. (Ash= 0.01–0.06%)	A	L	Study of mineral distribution in sugar crystals, for K, Ca, Na, Fe, Mg, Zn, Mn, Cu	F	—	329
Various (5)	—	River fish	μg/g levels	A	L	Comparison of AAS and XRF results for Co, Cu, Fe, Mn and Zn	Graphite furnace		475
Various (7)	—	Corn	μg/g levels	A	L	Wet-ash with $HNO_3/HClO_4$ and dissolve in HCl (Cd, Co, Cu, Mn, Pb, Zn) or aqua regia (Hg)	F Cold vapour (Hg)	—	477
Various (11)	—	Fruit juices	μg/ml levels	A	L	(a) Centrifuge, add HCl and dilute (Na, K, Mg, Ca); (b) extract solution with DDDC/MIBK (Mn, Cu, Ni, Pb); (c) digest aliquot of (a) with H_2SO_4 + HNO_3 and extract with DDDC/MIBK (Fe, Zn); (d) oxidise with $KMnO_4$ and extract with MIBK (Cr)	F	Air/C_2H_2	510
Various (9)	—	Meats	Trace levels	A	L	Treat with 1:1 HNO_3, digest at 80°C, filter and dilute	F Graphite furnace	—	548

TABLE 2·8 FOOD AND BEVERAGES — continued

Element	λ/nm	Matrix	Concentration	Tech.	Analyte form	Sample treatment	Atomization	Ref.
Various (15)	—	Foods	ng/ml levels (in extract)	E	L	Ash and acid-digest. Use unmatched standards, prepared in 50% $HClO_4$	P ICP	739
Various (12)	—	Cane juice, molasses, sugar	Trace levels	A	L	Study of 9 sample preparation treatments. Recommended: Dilute with 0.1 N HCl for Mg, K, Na, Fe, Zn; dilute with 0.1 N HCl + 1500 μg/ml La, for Ca; dilute with 10% citric acid for Cu, Mn; ash and dilute with 0.1 N HCl for Al, Co, Mo, Si	F —	957
Various	—	Rice	Trace levels	A	—	Mix with TiO_2 or W to accelerate sample decomposition	— —	959
Various	—	Foods	Trace levels	A	L	Study of sample treatment and interferences	Graphite furnace	1138
Various	—	Foods	Trace levels	A	L	Review of literature since 1972	F —	1177
Various	—	Foods	Trace levels	A	L	Review of sample treatments, covering wide range of foods	Graphite furnace	1206
Various (6)	—	Foods	Trace levels	A	L	(a) Extract with 1% HNO_3 (Cd in rice); (b) wet-digest with $HNO_3/H_2SO_4/HClO_4$ in PTFE pressure vessel; (c) ash with low temperature plasma and extract with HCl (Methods for Cd, Pb, Mn, Sr, Mo, Cu in various foods)	Graphite furnace (HGA-74)	1323
Various (7)	—	Foods	μg/g levels	A	L	Add Sr, digest with $HNO_3/H_2SO_4/H_2O_2$ + V_2O_5 catalyst, dilute, filter off $PbSO_4$/$SrSO_4$, convert to carbonates, dissolve in 2 N HNO_3 and analyse for Pb. Adjust filtrate to pH 6.5 ± 0.5, pass through Chelex-100 resin column, wash with $(NH_4)_2SO_4$ and elute with 2 N H_2SO_4 (Cd, Co, Cu, Mn, Ni, Zn)	F —	1410

2.9 BODY TISSUES AND TISSUE FLUIDS

2.9.1 Introduction

There has been some consolidation of existing principles and techniques, especially following the recent rapid increase in the application of electrothermal methods. The trend has been towards the improvement of established methods and their application to specific problems. To some extent, it has become evident that a single technique may not be suitable for the determination of all the metals required in a particular matrix.

Berman (86) has reviewed the application of flame AES and flame AAS techniques to biological samples. This paper provides valuable information on the determination of 21 elements and discusses known interferences and methods for their minimisation.

2.9.2 Sample Preparation

One of the greatest problems in trace-metal analysis is that of contamination. Franklin and Pickett (1137) have investigated the contamination of blood samples with Zn, Fe and Cd and other trace metals, during sampling with various commercially available blood-collection devices.

Szivos and Pungor (461) evaluated four sample preparation methods for Ca, Mg, Na and K in serum and found that a (1 + 49) dilution with aqueous 0.25% Sr was best for precision (RSD <0.02), accuracy and speed of analysis.

Wet ashing of large animal tissue samples can often be difficult and Rooney (500) has described three procedures using $HNO_3/HClO_4$ oxidation that are applicable to sample weights from 0.5 g to 10.0 g. Roos (393) investigated the use of a HNO_3/H_2SO_4 mixture with a pressure dissolution technique for the determination of Ca, Cu, Fe and Mg in whole blood. Hinners *et al* (655) found that quantitative recovery of Cd, Cu, Mn and Zn from bovine liver could be achieved using extraction with 1% HNO_3 at ambient temperatures and concluded that wet or dry oxidation of the sample was unnecessary for these determinations.

2.9.3 Flame Methods

Absorption—Baak, Heck and Van der Slik (985) considered that flame AAS resulted in serum Ca levels that were as close as possible to the "true" values and used this as a reference procedure for an evaluation of (1) continuous-flow automated colorimetric analysis with acid dialysis and complexing with cresolphthaleine; (2) as (1) without dialysis and using alizarin as the complexing reagent; (3) auto-titration with EGTA and fluorimetry; (4) manual EDTA titration using murexide as indicator. The interference effects from excessive amounts of bilirubin, protein and haemoglobin from haemolysed erythrocytes were studied. Method (1) was closest to flame AAS in its relative freedom from the interferences studied and with an RSD of 0.017 was comparable to that of flame AAS.

Herrman (871) reported a simple indirect method for F by flame AAS, in which the sample was heated with excess SiO_2/H_2SO_4 and the SiF_4 evolved is determined in a N_2O/C_2H_2 flame. The Si, which is proportional to the F concentration was measured by AA. Denton, Delves and Armstein (987) used flame AAS to study the release of Fe from stroma during erythroblast maturation, and found that digestion of the stroma with HNO_3 was adequate for sample preparation.

The analysis of hair for Ni (473) and for Pb (57) has been reported. The recovery of Ni added to hair was 99% with an RSD of 0.12 at the 0.22 $\mu g\ g^{-1}$ level (473). In the latter report (57) Pb contamination was removed by washing with EDTA and the authors then favoured a dry-ashing procedure that lasted for four days !

Flame methods for Pb in blood or urine were reported for screening or comparative purposes. The Delves cup method was used in 3 studies (62, 196, 1063). Cooke *et al* (62) used the technique for mass screening of children using capillary bood samples collected on to filter paper. Jackson *et al* (1063), who modified the technique to allow determinations of urinary Pb, studied various parameters that affected performance, and claimed that background correction was not required. Marcus *et al* (196) compared the micro-cup method with a macrotechnique and reported a mean RSD for the micro method of 0.0575. Mitchell, Kahl and Ward (761) devised a sampling cup for the determination of Mn in serum in a N_2O/C_2H_2 flame; this represents a useful advance on a technique hitherto restricted to volatile metals.

Reports of the determination of thallium in blood are not frequent but are of toxicological interest. Singh and Joselow (171) developed an APDC/MIBK extraction technique using the Delves cup method. Calibration by standard additions was essential. Although the reported detection limit of 5 μg l^{-1} is an improvement on existing techniques for blood Tl, further improvements in sensitivity will be required to discern early industrial exposure.

Stevens (874) determined Pb and Cd directly in serum with an RSD of 0.30 by aspirating microlitre volumes of samples diluted with detergent into an air/C_2H_2 flame. Thompson and Godden (585), reported a similar procedure in which whole blood was diluted with Triton XI 00 and haemolysed by ultrasonic radiation. Pulse nebulization of 200 μl volumes into a high solids burner gave detection limits for Pb and Cd of 0.1 and 0.01 μg ml^{-1} respectively.

Church and Robinson (474) determined as little as 0.04 μg of Se in bovine liver by a hydride generation flame AAS method. They obtained good recoveries for NBS reference material and an RSD of 0.15 at the 1 μg g^{-1} level.

Fluorescence—Kolihova and Sychra (41) used AFS to determine Zn in whole blood by using a nickel cup method with a circular shielded air/C_2H_2 flame. No interferences were reported. Werrell, Vickers and Williams (188) determined Cd in bovine liver by AFS using separated air/C_2H_2 or air/H_2 flames with detection limits of 0.005 μg ml^{-1}. Thompson and Godden (585) reported an improved detection limit for blood Cd of 0.001 μg ml^{-1} using AFS with their pulse nebulization method.

2.9.4 Electrothermal Methods

Improvement of precision and accuracy is notable especially with some metals such as Cu, Cd and Mn. Ross and Gonzalez (467) achieved RSDs of 0.023 and 0.048 for the determination of Mn in serum and urine respectively using a graphite tube furnace. The detection limit was 1.0 μg l^{-1}. Evenson and Warren (202) found that background correction was unnecessary for the determination of Cu in serum if the sample was diluted with 10 mmol l^{-1} HNO_3. The day-to-day RSD was 0.021 at 0.78 mg l^{-1} and 0.014 at 1.57 mg l^{-1}. Perry, Koirtyohann and Perry (203), determined Cd in blood and urine using a graphite tube furnace and AAS after wet ashing with HNO_3/H_2O_2. They obtained RSDs of 0.07 and 0.14 at the 2.8 and 0.18 μg l^{-1} levels; these are excellent results at realistic concentrations.

Cernik and Sayers (195) have extended the filter paper-graphite cup technique to the determination of Cd in blood and obtained a detection limit of 1.0 nmol l^{-1} with an RSD of 0.088 at 71 nmol l^{-1}. Posma *et al* determined Cd and Pb (245) and Cd, Zn, Pb and Cu (601) in serum and urine. In the former paper (245) the authors compared the graphite tube and cup with a routine flame method and in the latter (601) they evaluated analytical data obtained without the use of a background corrector. Serum samples were injected directly into the furnace and non-specific absorption signals were separated by ramp atomization.

An oxidative atmosphere during the ashing step reduced losses of Cd prior to atomization. This report discussed many other important aspects of electrothermal AAS.

Riner, Wright and McBeth (1) used an HNO_3 oxidation for the determination of Pb in faeces using a graphite tube furnace. Acid concentrations above 4% decreased the Pb signal.

Hurlbut (1035) described a useful technique for Be in urine that did not require a prior ashing stage, The sensitivity (for 1% absorption) was 0.2 $\mu g\ l^{-1}$ with a detection limit of 10 pg. A simple method for Pt and Pd in biological samples was devised by Miller and Doerger (558) in which NaCl was added to convert the metals to their chlorides. Zachariassen *et al* (200) found that it was necessary to separate Ni from blood samples by solvent extraction prior to its determination by electrothermal AAS. Kirkbright and Wilson (4) investigated the direct determination of I_2 and found that severe interferences necessitated continuum source background correction; the optimum atomization temperature was 2100°C.

Robinson, Woolcott and Rhodes (929) applied a "stop-flow" mode with a hollow T-shaped atomizer for the direct anaysis of Cu and Pb in blood, urine, sea-water, filter paper and polyethylene. A constant temperature of 2700°C was maintained thereby removing the need for a programmable heating supply. Continuous sample introduction into a heated graphite furnace for the AFS determination of Zn in biological systems has been described (481).

Sanz and Palacios (530), Brandenberger (264) and Delves (989) have reviewed the application of electrothermal atomization and AAS to the analysis of organic matrices.

2.9.5 Other Techniques

Grime and Bickers (233) determined pg amounts of Li in 5 μl volumes of serum combining a tantalum filament for sample vaporization and conventional flame emission spectroscopy in an air/C_2H_2 flame. Kawaguchi and Vallee (671) described the determination of 8 elements in pg quantities using a low pressure microwave induced helium plasma, with a tantalum filament vaporization system. Glick and Marich (1050) atomized biological samples with a laser microprobe. Detection limits were claimed to be 10^{-12} or 10^{-15}g according to element.

Inductively-coupled plasmas: Dahlquist and Knoll (854) have discussed the capabilities of ICP sources. Detection limits below 1$\mu g\ l^{-1}$ have been reported for Hg and As in urine and the freedom from matrix and inter-element effects has been emphasised. Nakashima Sasaki and Shibata (922) digested biological material with $HNO_3/HClO_4$ followed by concentration of Ag on a BiI_3 collector. Ag was detected down to 0.002 $\mu g\ ml^{-1}$ by plasma emission spectroscopy. Sb, Bi, Pb, Sn enhanced the Ag emission but Na, Fe and Mn depressed the emission and should be separated.

Cold vapour methods for Hg: Two reports have been concerned with the determination of Hg in hair (55, 346). Saishido and Suzuki (346) compared the AA method of Magos (*ARAAS*, 1972, **2**, 122) with neutron activation analysis for determining the organic Hg content of hair and found good correlation (correlation coefficient = 0.985). There is no doubt that Hg in hair is correlated to the body burden, thus indicating previous excessive ingestion and that this is important for exposure to organic Hg. However, blood Hg is considered to be a more appropriate index of current or recent organic Hg ingestion and the scarcity of direct and accurate techniques for this determination is an indication of the difficulties of this analysis. Gibbs, Jarosewich and Windom (59) have reported that preserved fish do not give reliable estimates of their heavy metal concentrations. Preservation in formalin, ethanol, or isopropanol produced a significant initial decrease of Hg. Carlsen (527) adapted the Hg cold vapour method for the indirect determination of thiol groups in proteins. *p*-Hydroxymercuribenzoate was complexed with protein *via* the thiol groups and the complex was separated, digested, and analysed for Hg. Toffaletti and Savoy (1248) have used aqueous sodium borohydride rather than $SnCl_2$ for reduction. All forms of Hg are rapidly reduced and they reported a detection limit of 1–2 ng, an RSD of 0.065, and a recovery of 103% for CH_3Hg.

2.9.6 Standardisation

Standardisation is very important in trace-metal analysis. Lauwerys *et al* (198) reported a programme in which 66 European laboratories participated in the analysis of Pb, Hg and Cd in blood, urine and aqueous solutions. The methods used were flame, discrete sampling and electrothermal AAS, spectrophotometry, polarography and anodic stripping voltammetry (ASV). Systematic errors appear responsible for high interlaboratory variation between laboratories where precise results were obtained. Browne, Ellis and Weightman (529) describing the results from two laboratories using AAS and one laboratory using ASV for the determination of blood Pb in 93 men. When any one of the laboratories reported $\geq$0.8 mg l^{-1} blood, they disagreed in 28 cases out of 35. In view of the last finding further work is necessary.

TABLE 2.9 BODY TISSUES AND TISSUE FLUIDS

Element	λ/nm	Matrix	Concentration	Tech.	Analyte form	Sample treatment	Atomization		Ref.
Ag	328.0	Animal tissue	0.1–100 μg/g	A	L	Wet oxidize with $HNO_3/HClO_4$, add tartaric acid + NH_4OH and dilute	F	Air/C_2H_2	500
Ag	328.1	Bovine liver	0.06 μg/g	A	S	Char at 280°C. Calibrate by addition of aqueous standards	Graphite furnace (HGA-74)		554
Ag	328.1	Biological materials	2–40 μg/g	E	L	Add $HClO_4/HNO_3$, heat to fuming, dissolve in H_2O, add 0.4 M NaI + standard Bi solution and centrifuge. Digest with HNO_3 and dilute	P	Argon ICP	922
Ag	—	Marine organisms	—	A	L	Comparison of AAS and NAA results	F	—	1236
Ag	328.1	Human skin	5–20 μg/g	E	S	Dry at 95–105°C and ash at 450°C. Mix with graphite powder	A	14 A a.c.	1567
As	—	Cattle blood, milk. hair	—	A	L	Char with HNO_3/H_2SO_4 and digest with H_2O_2 until clear. Reduce to AsH_3 and pass vapour to flame. (Environmental study).	F	Ar/H_2	361
As	—	Urine	ng/ml levels	E	L	Hydride generation method	P	Argon plasma (ICP)	854
As	—	Urine	Trace levels	A	L	—	P	ICP + graphite yarn atomizer	740
As	193.7	Urine	Trace levels	A	L	Treat with HCl + KI and extract with CCl_4. Re-extract with H_2O (inorganic As) or dichromate solution (total As)	Graphite furnace		1406
As	193.7	Biological materials	0–30 μg/g	A	L	Ash with $(Mg(NO_3)_2$, redissolve and form AsH_3 by $NaBH_4$ reaction	F	—	1552
Au	—	Serum	Trace levels	A	L	—	Graphite furnace		742
Be	—	Animal tissue	From 0.05 μg/ml (in extract)	A	L	Ash with low-temperature O-plasma	F	N_2O/C_2H_2	534
Be	—	Urine	From 0.2 ng/ml	A	L	—	Graphite furnace		1035
Ca	630 (band)	Blood serum	—	E	L	Centrifuge, treat with H_2O/5% butanol and precipitate with 10% TCA. For Ca, neutralise aliquot with NH_4OH and add 0.7% EDTA. Determine Na, K directly	F	Air/C_2H_2	52

TABLE 2.9 BODY TISSUES AND TISSUE FLUIDS — continued

Element	λ/nm	Matrix	Concentration	Tech.	Analyte form	Sample treatment	Atomization	Ref.
Ca	422.7	Serum	5 m.eq./l	E	L	—	F O_2/H_2	117
Ca	—	Dental enamel	—	A	L	Study of Ca loss from surface enamel	F —	278
Ca	—	Serum	—	A	L	Dilute (×100). Match standards for Na content. (Interference study)	F —	358
Ca	—	Serum	—	A	L	Comparison of 4 treatments : (1) Dilute with NaEDTA; (2) dilute with Sr solution (to 1% Sr); (3) addTCA, centrifuge and dilute with Sr solution; (4) dilute with Sr solution (to 0.25% Sr). (Method 4 preferred)	F Air/C_2H_2	461
Ca	—	Serum	2.3–2.9 mmol/l	A	L	Dilute (1:49) with $LaCl_3$ solution (Comparison with 3 other methods)	F Air/C_2H_2	985
Ca	422.7	Tumour cells	1.4–2.0 μmol/g	—	—	—	— —	1049
Ca	422.7	Biological samples	—	A	L	(Interference-filter system)	Graphite furnace	1257
Ca	—	Nerve cells	—	E	L	—	F —	1344
Cd	—	Blood	—	A	L	—	F —	28
Cd	228.8	Bovine liver	0.005 μg/ml	F	L	Digest with H_2SO_4/H_2O_2 and extract with HI + Aliquot 336-I/xylene. Discard H_2O layer, add ethylene-diamine solution to back-extract Cd	F Air/C_2H_2 (separated)	188
Cd	228.8	Blood	0–890 n.mol/l	A	L,S	Dry on filter paper and remove at 4 mm disc	Graphite cup	195
Cd	228.8	Liver tissue	300 μg/g	A	L	Digest with HNO_3 at 80°C	F Air/C_2H_2 (+ graphite tube)	197
Cd	—	Blood, urine, aqueous solutions	—	A	L	Comparison programme (66 laboratories) covering flame and non-flame AAS, absorptionmetry, polarography and anodic-stripping voltammetry	F — Graphite furnace	198
Cd	228.8	Blood, urine	0–5 μg/l	A	L	Digest with HNO_3/H_2O_2. Evaporate and redissolve in HNO_3	Graphite furnace (HGA-2000)	203

Cd	228.8	Blood	0.5–20 ng/g	A	L	Oxidise whole blood with HNO_3 + HF. Treat samples and standards with heparin. (Comparison of methods)	Graphite furnace F	Air/C_2H_2	245
Cd	—	Biological materials	—	A	L	Multi-channel AAS system for Cd, Cu, Zn, Add 8-hydroxyquinoline and extract into MIBK	F	—	263
Cd	—	Biological materials	Trace levels	A	L	Dry-ash at 420–450°C, moisten with HCl, evaporate and redissolve. Add NH_4 citrate solution and adjust to pH 9.2 with NH_4OH. Extract with dithizone/$CHCl_3$ and back-extract with 0.5 M HCl. (Review of errors due to ashing, sample treatment and matrix effects)	F	—	268 269
Cd	—	Animal tissues	From 0.05 μg/ml (in extract)	A	L	See Be, ref. 534	F	—	534
Cd	228.8	Blood, urine	—	A	L	Dilute (1:50) with H_2O. Calibrate by standard addition method. Atomize at 2150°C	Graphite furnace (HGA-2000)		552
Cd	—	Blood	0.02–0.8 μg/100 ml	A	L	Ash on paper disc at 270°C	Graphite furnace		875
Cd	—	Blood	Trace levels	A	L	—	Graphite furnace		742
Cd	228.8	Blood	From 1 ng/ml (F) or 10 ng/ml (A)	A,F	L	Dilute (1:1) with 0.2% Triton-X.100 and homogenize ultrasonically. Use pulse nebulization technique, on 200 μl sample	F	Air/C_2H_2 Air/H_2	585
Cd	—	Bovine liver	Trace levels	F	L	Study of automatic scatter correction	F	Air/C_2H_2 $Air/Ar/H_2$	773
Cd	—	Blood tissues	Trace levels	A	L	Ash, treat with HNO_3, dilute and add Na acetate buffer (pH 6.5–7.0). Extract with solvent mixture of diphenyl-carbazone/toluene/pyridine	F	Air/toluene	1163
Cd	228.8	Human tissue	0–0.5 μg/g	A	L	Buffer sample extract with acetate (pH 8.5) and extract with K-ethylxanthate into MIBK	F	Air/C_2H_2	1283
Cd	—	Animal tissue	—	A	L	Dry-ash at 475°C	F	—	1408
Cd	228.8	Biological tissue	1.5–150 μg/g	A	L	Mix (×4) with H_2O and homogenize	F	Air/C_2H_2 + Delves cup	1477

TABLE 2.9 BODY TISSUES AND TISSUE FLUIDS — continued

Element	λ/nm	Matrix	Concentration	Tech.	Analyte form	Sample treatment	Atomization	Ref.
Co	240.7	Blood, serum	5–10 ng/ml	A	L,S	Dilute with H_2O. For Co in serum, dry at 750°C for 5 minutes and introduce residue to furnace	Graphite furnace (HGA-70)	993
Co	—	Blood, tissue	Trace levels	A	L	See Cd, ref. 1163	F Air/toluene	1163
Co	240.7	Human tissue	Trace levels	A	L	Buffer sample solution with NH_4 acetate (pH 8) and extract with K-ethylxanthate into MIBK	F Air/C_2H_2	1286
Cr	—	Blood	Trace levels	A	L	—	Graphite furnace	742
Cr	—	Marine organisms	—	A	L	See Ag, ref. 1236	F —	1236
Cu	327.4 249.2	Nails	0–85 μg/g	A	S	Wash sample clippings with acetone and water and dry for 1½ hours at 115°C	Graphite furnace (HGA-72)	21
Cu	324.7	Liver tissue	26 μg/g level	A	L	See Cd, ref. 197	F Air/C_2H_2 (+ graphite tube)	197
Cu	324.7	Serum	0–2.5 mg/l	A	L	Dilute (1:10) with HNO_3 (to pH 3.0)	Graphite furnace (HGA-2000)	202
Cu	—	Biological materials	—	A	L	See Cd, ref. 263	F —	263
Cu	—	Serum	—	A	L	Study of effect of haemolysis	F —	357
Cu	—	Blood	—	A	L	Add Triton X-100 to whole blood sample. (Comparison with other mehtods)	Graphite furnace	432
Cu	—	Ocular fluids	0.1–0.2 μg/ml	A	L	—	F —	531
Cu	324.7	Bovine liver	210 μg/g	A	S	See Ag, ref. 554	Graphite furnace (HGA-74)	554
Cu	—	Biological tissues	Trace levels	A	L	Homogenize with H_2O (1:4), transfer to sample cup and oven dry	F Delves cup	705
Cu	—	Animal serum	—	A	L	Ash at 250°C (Description of new design of furnace tube)	Graphite furnace	746
Cu	—	Serum, urine, tissues	Trace levels	A	L	Dilute urine samples directly. Centrifuge serum samples, precipitate with TCA and dilute. Ash tissues and dissolve in HCl	F —	982

Cu	324.7	Urine	Trace levels	A	L	Add 0.2 ml conc H_2SO_4 to 4 ml urine. Determine Cu directly. Dilute (1:3) with H_2O for Zn determination	F	Air/C_2H_2	984
Cu	324.7	Blood, serum	1–2 μg/ml	A	L	See Co, ref. 993	Graphite furnace (HGA-70)		993
Cu	—	Tissues	—	A	L,S	Homogenize and dry-ash or acid-digest in microsampling cup	F	N_2O/C_2H_2 + cup	1146
Cu	—	Blood, tissue	—	A	L	See Cd, ref. 1163	F	Air/toluene	1163
Cu	—	Blood, serum	—	A	L	—	Graphite furnace		1234
Cu	324.7	Human tissue	0–4 μg/g	A	L	See Cd, ref. 1283	F	Air/C_2H_2	1283
Cu	—	Urinary concrements	—	E	S	—	A	A.C. arc	1499
Cu	—	Biological calcifications	—	A	L	Interference study	Graphite furnace		1505
Cu	327.4	Human skin	50 μg/g	E	S	See Ag, ref. 1567	A	14 A a.c.	1567
F	—	Biological materials	Trace levels	E,A	L	Heat with excess SiO_2 + H_2SO_4. Pass evolved SiF_4 to flame or furnace	F Graphite furnace	N_2O/C_2H_2	871
Fe	248.3	Serum	Up to 3 μg/ml	A	L	Mix serum with $FeCl_3$, add $MgCO_3$ and centrifuge. Determine Fe in supernatant liquid. (Method for measuring iron-binding capacity)	Graphite furnace		63
Fe	—	Serum	—	A	L	Add TCA; heat at 90°C for 15 min	F	—	462
Fe	248.3	Cell stroma, cytoplasm	0–1 μg/ml	A	L	Digest with HNO_3 (Study of Fe release)	F	Air/C_2H_2	987
Fe	—	Urinary concrements	—	E	S	—	A	A.C. arc	1499
Fe	253.7	Hair	—	A	—	—	—	—	55
Hg	253.7	Tissues (fish, insects)	—	E	G	Freeze grind, homogenize, shake with HCl and extract with C_6H_6	P	Microwave emission in Ar or He plasma	161
Hg	—	Blood, urine, aqueous solutions	—	A	L	See Cd, ref. 198	F Graphite furnace	—	198
Hg	253.7	Hair	—	A	G	Comparison with results by NAA	Cold vapour		346
Hg	—	Urine	—	A	G	Add 3 M H_2SO_4 + $KMnO_4$, boil and transfer to reaction vessel	Cold vapour		456
Hg	—	Blood, hair	—	A	—	Comparison of methods	—	—	509

TABLE 2.9 BODY TISSUES AND TISSUE FLUIDS — continued

Element	λ/nm	Matrix	Concentration	Tech.	Analyte form	Sample treatment	Atomization	Ref.
Hg	253.7	Fish tissue	1–3 μg/g	A	G	Combust in O_2, absorb product in $KMnO_4/H_2SO_4$, reduce with $SnCl_2$, absorb evolved Hg on Au particles and re-heat at 500°C for analysis	Cold vapour	637
Hg	—	Urine	Trace levels	A	L	—	P ICP + graphite yarn atomizer	740
Hg	—	Biological samples	—	A	—	(Application of Zeeman-effect correction)	Graphite furnace	770
Hg	—	Urine	—	A	G	Reduce buffered urine sample with aqueous $NaBH_4$ solution and pass vapour to heated quartz tube	Heated quartz absorption tube	1248
Hg	—	Animal tissue	—	A	G	Combust in oxygen flask (Gutenmann, W. H. and Lisk, D. J.—J. Agric. Fd. Chem. 1960, **8**, 306) and determine Hg (Hatch, W. R. and Ott, W. C.—Anal. Chem, 1968, **40**, 2085)	Cold vapour	1408
Hg	253.7	Fish tissue	—	A	G	Digest with $HNO_3/H_2SO_4/HCl$, dilute, and reduce with $H_2SO_4/NaCl/SnCl_2/NH_2OH.H_2SO_4$	Cold vapour	1412
I	183.0 206.1	Rat thyroid	—	A	L	Dry at 50°C over P_2O_5, digest with HCl/HNO_3 and evaporate. Repeat with portions of HNO_3 until dissolved. D_2-lamp background correction essential	Graphite furnace (HGA-2000)	4
K	766	Blood serum	—	E	L	See Ca, ref. 52	F Air/C_2H_2	52
K	—	Blood	—	E	L	—	F —	260
K	—	Blood serum	—	A	L	See Ca, ref. 461	F Air/C_2H_2	461
K	—	Serum	—	E	L	Simultaneous determination of K and Na with vidicon spectrometer	F Air/H_2 or Air/propane	471
K	—	Tissues	—	E	L	—	F —	541
K	766.0	Tumour cells	278–290 μmol/g	—	—	—	— —	1049
K	—	Nerve cells	—	E	L	—	F —	1344

K	—	Fish blood	2–3 μg/ml	E	L	Treat with HNO_3, evaporate, redissolve in HCl and dilute in polythene vessel. Calibrate by standard addition method	F	—	1380
Li	670.7	Water, serum	0.01–80 μg/ml	E	L	Calibrate by standard addition method or with artificial serum standards.	F	Air/C_2H_2 or Air/H_2 (+ Ta filament)	233
Li	670.7	Serum	30–40 ng/ml	E,A	L	Match standards for Na and K content. Comparison of AAS and FES methods; latter preferred	F	—	257
Li	670.7	Serum	—	A	L	Deproteinize with 5% TCA	F	—	1355
Li	—	Blood serum	—	A	L	See Ca, ref. 461	F	Air/C_2H_2	461
Mg	285.2	Serum	0.7-1 nmol/l	A	L	—	F	Air/C_2H_2	986
Mg	285.2	Tumour cells	14–16 μmol/g	—	—	—	—	—	1049
Mn	—	Blood, serum, urine	0–2.5 μg/100 ml (urine)	A	L	Take 20 μl sample, dry at 150°C, char at 1350°C and atomize at 2150°C (Results compared with APDC/MIBK extraction procedure)	Graphite furnace		467
Mn	279.8	Bovine liver	10.4 μg/g	A	S	See Ag, ref. 554	Graphite furnace (HGA-74)		554
Mn	—	Serum	—	A	L	—	F	N_2O/C_2H_2 + Mo cup	761
Mn	279.4	Blood, serum	3–21 ng/ml	A	L	See Co, ref. 993	Graphite furnace (HGA-70)		993
Mn	—	Biological calcifications	—	A	L	See Cu, ref. 1505	Graphite furnace		1505
Na	590	Blood serum	—	E	L	See Ca, ref. 52	F	Air/C_2H_2	52
Na	—	Blood	—	E	L	—	F	—	260
Na	—	Blood serum	—	A	L	See Ca, ref. 461	F	Air/C_2H_2	461
Na	—	Serum	—	E	L	See K, ref. 471	F	—	471
Na	—	Tissues	—	E	L	—	F	—	541
Na	589.0	Tumour cells	60–70 μmol/g	—	—	—	—	—	1049
Na	—	Nerve cells	—	E	L	—	F	—	1344
Na	—	Fish blood	2–3 μg/ml	E	L	See K, ref. 1380	F	—	1380

TABLE 2.9 BODY TISSUES AND TISSUE FLUIDS — continued

Element	λ/nm	Matrix	Concentration	Tech.	Analyte form	Sample treatment	Atomization	Ref.
Ni	232.0	Blood, urine	2–35 μg/l	A	L	Ash at 560°C, dissolve in HCl, adjust to pH 9 with NH_4OH and extract with DDDC or APDC into MIBK, or as dimethylglyoxime complex. Atomize at 2600°C	Graphite furnace	200
Ni	—	Hair	0.13–0.51 μg/g	A	L	—	F —	473
Ni	—	Human tissue	Up to 0.19–0.64 μg/g (bone)	A	L	—	F —	480
Ni	—	Blood, tissue	—	A	L	See Cd, ref. 1163	F Air/toluene	1163
Ni	232.0	Human tissue	Trace levels	A	L	See Co, ref. 1286	F Air/C_2H_2	1286
Pb	283.3	Cattle faeces	Up to 20 μg/g (dry weight)	A	L	Dry and digest with HNO_3	Graphite furnace (HGA 2000)	1
Pb	—	Blood	—	A	L	—	F —	28
Pb	217.0	Hair	—	A	L	Dry ash cleaned sample for 4 days at 450°C and dissolve in HNO_3. Evaporate to low bulk, add NH_4OH + $(NH_4)_2SO_4$ + Thymol blue indicator and adjust to pH 2.8. Extract Pb with APDC/MIBK	F —	57
Pb	—	Blood	—	A	L	Collect on filter paper, dry and transfer to Ni cup	F Delves cup	62
Pb	—	Blood	From 0.05 μg/ml	A	L	—	F —	118
Pb	—	Blood	—	A	L	—	F Air/C_2H_2 (Delves cup)	196
Pb	—	Blood, urine, aqueous solutions	—	A	L	See Cd, ref. 198	F — Graphite furnace	198
Pb	283.3	Blood	150–1000 μg/l	A	L	Comparison of methods	F Air/C_2H_2 Graphite furnace (HGA-2100)	199
Pb	283.3	Blood	0–16 ng/ml	A	L	—	Graphite furnace	204
Pb	—	Blood	10–90 μg/100 ml	A	L	Dilute (1:5) with Triton X.100	Graphite furnace	220
Pb	217.0	Blood	0–300 ng/g	A	L	See Cd, ref. 245	Graphite furnace F Air/C_2H_2	245

Pb	—	Blood	—	A	L	Study of effect of anti-coagulants on blood-Pb determinations	F Delves cup	267
Pb	—	Cattle blood, milk, hair	—	A	L	See As, ref. 361	F Delves cup	361
Pb	—	Blood	—	A	L	Study of accuracy and precision	Graphite furnace	431
Pb	—	Blood	—	A	L	See Cu, ref. 432	Graphite furnace (Model 63)	432
Pb	—	Blood	—	A	L	—	Graphite furnace	460
Pb	217.0 283.3	Urine	Up to 20 μg/100 ml	A	L	Mix urine (1:1) with 0.1 M EDTA (+ H_2O_2 (1:1). Dry at 100°C, ash at 330°C and atomize at 1100°C. Calibrate by standard addition method	Graphite furnace	502
Pb	—	Blood	—	A	L	Study of inter-laboratory variations	— —	529
Pb	217.0	Bovine liver	0.32 μg/g	A	S	See Ag, ref. 554	Graphite furnace (HGA-74)	554
Pb	—	Blood	—	A	L	Oxidise chemically, before ashing stage, to prevent Pb losses	Graphite furnace	904
Pb	—	Biological tissues	Trace levels	A	L	See Cu, ref. 705	F Delves cup	705
Pb	—	Blood	Trace levels	A	L	Dilute with H_2O. See also "Various", ref. 707	Graphite furnace	707
Pb	—	Urine	2–800 μg/100 ml	A	L	Automated system	F Delves cup	708
Pb	—	Urine	Trace levels	A	L	—	P ICP + graphite yarn atomizer	740
Pb	—	Blood	Trace levels	A	L	Dilute (1:2) with 0.1% solution of Triton X-100. (Comparison of methods) See also Au, Cd, Cr, ref. 742	Graphite furnace F Delves cup	742
Pb	217.0	Blood	From 0.1 μg/ml (A)	A,F	L	See Cd, ref. 585	F Air/C_2H_2	585
Pb	283.3	Blood	Trace levels	A	L	Mix (1:1) with $HClO_4$/trichloroacetic acid, stand for 1 hour and add (×8) 0.01 M HCl. Ash below 500°C. Use 287.5 nm line for background correction	Graphite furnace	918
Pb	217.0	Blood, urine	Trace levels	A	L	Atomize at 2700°C	"Hollow-T" atomizer	929

TABLE 2.9 BODY TISSUES AND TISSUE FLUIDS — continued

Element	λ/nm	Matrix	Concentration	Tech.	Analyte form	Sample treatment	Atomization	Ref.
Pb	217.0	Urine	0.006–1.50 μg/ml (in final solution)	A	L	Dilute if necessary, dry sample at 110°C in sample cup and introduce to flame below tube	F Air/C_2H_2 + Al_2O_3 tube	1063
Pb	—	Blood	0.2–1.0μ g/ml	A	L	Calibrate by standard addition method	Graphite furnace	1124
Pb	—	Animal tissue	—	A	L	Dry-ash at 475°C	F —	1408
Pb	—	Urinary concrements	—	E	S	—	A A.C. arc	1499
Pb	217.0	Blood	—	A	L	Digest with HNO_3 at 70°C	Graphite furnace	1530
Pb	283.3	Human skin	40 μg/g	E	S	See Ag, ref. 1567	A 14 A a.c.	1567
Pd	—	Tissues	From 0.2 μg/g	A	L	Lyophilize, treat with HNO_3, add NaCl + aqua regia	Graphite furnace (HGA-2000)	558
Pt	—	Tissues	From 0.2 μg/g	A	L	See Pd, ref. 558	Graphite furnace	558
Se	—	Animal tissue	From 0.04 μg (absolute)	A	L	Combust (Schoeniger method) and reduce to H_2Se vapour	F —	474
Sr	460.7	Biological samples	—	A	L	Interference study	F Air/C_2H_2	784
Te	—	Animal tissue	From 0.5 μg/ml (in extract)	A	L	See Be, ref. 534	F —	534
Tl	276.8	Whole blood	0–10 μg/100 ml	A	L	Dilute (1:1) with H_2O, add 5% Triton X-100 solution and extract with APDC/MIBK by centrifuge	F Delves cup	171
Zn	—	Blood	—	A	L	—	F —	28
Zn	213.9	Blood	From 0.1 μg/ml	F	L	Dilute (×5–×10) and oxidize in cup with H_2O_2	F Air/C_2H_2 + Delves cup	41
Zn	213.9	Liver tissue	6 μg/g	A	L	See Cd, ref. 197	F Air/C_2H_2 (+ graphite tube)	197
Zn	213.9	Serum	50–400 μg/l	A	L	Dilute (1:10) with H_2O	F Air/C_2H_2	201
Zn	213.9	Plasma	0–20 μg/100 ml	A	L	Dilute (1:10) with H_2O	Graphite furnace	205
Zn	—	Biological materials	—	A	L	See Cd, ref. 263	F —	263
Zn	—	Serum	25–300 μg/l	A	L	Dilute with non-ionic surfactant, a stearate ester of sucrose, and aspirate	F —	437
Zn	213.9	Biological fluids	—	A	L	Apply background correction, using Cd 228.8 nm or Co 240.7 nm	F —	325

Zn	—	Serum	—	A	L	See Cu, ref. 357	F	—	357
Zn	—	Rat tissues	—	A	L	Ash at 550°C for 16 hours. (Results compared with NAA)	F	—	466
Zn	—	Biological samples	—	F	L	—	Graphite furnace		481
Zn	—	Biological tissues	Trace levels	A	L	See Cu, ref. 705	F	Delves cup	705
Zn	213.9	Serum	1 μg/ml	A	L	Take 50 μl sample and use peak area integration method	F	Air/C_2H_2	632
Zn	—	Bovine liver	Trace levels	F	L	See Cd, ref. 773	F	Air/C_2H_2 $Air/Ar/H_2$	773
Zn	213.9	Urine	Trace levels	A	L	See Cu, ref. 984	F	Air/C_2H_2	984
Zn	213.9	Rat muscle, liver tissue	μg/g levels	A	L	Extract with 0.1 N HNO_3	F	Air/C_2H_2	988
Zn	213.9	Fingernails	Trace levels	A	L,S	Determine directly (20 μg) for small samples or wet-ash if larger samples available	Graphite furnace		1139
Zn	—	Tissues	—	A	L,S	See Cu, ref. 1146	F	N_2O/C_2H_2	1146
Zn	—	Blood, tissue	—	A	L	See Cd, ref. 1163	F	Air/toluene	1163
Zn	—	Marine organisms	—	A	L	See Ag, ref. 1236	F	—	1236
Zn	—	Nematodes	16–313 μg/g	A	L	Transfer individual nematode with 1:1 glycerol/H_2O solution	Graphite furnace		1360
Zn	328.0	Human skin	1000 μg/g	E	S	See Ag, ref. 1567	A	14 A a.c.	1567
Oxalic acid (indirect)	422.7 (Ca)	Urine	—	A	L	Precipitate oxalic acid from urine by addition of excess Ca at pH 5. Determine Ca in supernatant liquid and also total Ca by AAS (pH 2)	F	—	455
Thiol groups (indirect)	253.7 (Hg)	Proteins	—	A	G	Add excess of p-hydroxymercuri-benzoate to solution of protein in 0.1 M Na_2HPO_4 + 0.1 M NaH_2PO_4 (pH 7.5). Separate by gel filtration (Sephadex G-25) and digest portion of the eluted Hg-complex with HNO_3, followed by $H_2SO_4/KMnO_4$. Determine Hg by reduction and cold vapour AAS	Cold vapour		527
Various (6)	—	Museum fish specimens	—	A	L,G	Digest with HNO_3. Method for As, Cd, Cu, Hg, Pb, Zn	F Cold vapour (Hg)	Air/C_2H_2	59

TABLE 2.9 BODY TISSUES AND TISSUE FLUIDS — continued

Element	λ/nm	Matrix	Concentration	Tech.	Analyte form	Sample treatment	Atomization	Ref.
Various (9)	—	Whole blood, serum	μg/ml levels	E	L	—	Induction-coupled argon plasma	76
Various (21)	—	Blood, urine, serum	—	E,A	L	Review, with 105 refs	F —	86
Various	—	Biological materials	—	—	—	Review, 22 refs	— —	264
Various	—	Blood	—	A	L	Study of pressure dissolution methods	F —	393
Various	—	Blood serum	—	A	L	Comparison of AAS, XRF and X-ray excitation with protons	F —	465
Various	—	Serum, urine, hair	Trace levels	E	L	—	Induction-coupled argon plasma	484
Various	—	Biological materials	—	A	L	Review (12 refs)	— —	530
Various (6)	—	Fish tissues	μg/g levels	A	L	Treat with $HNO_3/HClO_4$, heat under pressure at 120°C, cool, dilute and extract with NaDDC into MIBK. Calibrate by standard addition method (Cd, Cu, Fe, Mn, Pb, Zn)	F Air/C_2H_2 + Ta boat	562
Various	—	Biological fluids	—	A	L	Review	F — Graphite furnace	564
Various	—	Urine	Trace levels	E	L	—	P Argon plasma (ICP)	647
Various	—	Biological tissues	Trace levels	A	L	Comparison of wet and dry ashing treatments. Extraction with 1% HNO_3 satisfactory for some elements e.g. Cd	F —	655
Various (8)	—	Micro-enzymes	ng/ml levels	E	L	Add KCl to sample extract	P Argon plasma (ICP)	671
Various	—	Biological samples	Trace levels	A	L	Study of Zeeman effect applied to background correction	Graphite furnace	799
Various	—	Biological materials	Trace levels	A	L	"Spike-height" method, on very small samples (50 μl), applied to Zn, Cu, Pb, Cd etc in blood and tissues	F —	874
Various (6)	—	Blood	Trace levels	A	L	Discussion of method for Cd, Cu, Au, Bi, Mn and Pb	Graphite furnace	707

Various	—	Plasma, serum	Trace levels	A	L	Take 5 μl sample and add, in furnace, 10 μl HNO_3 or 5–10 μl HF	Graphite furnace	601
Various	—	Biological tissues	Trace levels	A	L,S	Review	Graphite furnace	989
Various	—	Serum	Trace levels	A	L	—	Graphite furnace	1005
Various (8)	—	Biological material	Trace levels	E	S	—	Laser microprobe	1050
Various	—	Biological materials	Trace levels	E	L	Applications review	P ICP	1100
Various	—	Blood	Trace levels	E	L,S	Study of sample contamination	A —	1137
Various	—	Biological samples	Trace levels	A	L	Review	F — Graphite furnace	1223
Various (7)	—	Blood	Trace levels	A	L	Deep-freeze, ash by plasma oxidation and wet-ash with HNO_3. Determine Fe, Mg, Zn by flame and Cd, Cu, Mn, Pb by furnace	F Air/C_2H_2 Graphite furnace	1281
Various	—	Waterfowl feathers	—	A,E	L	—	F —	1352
Various (8)	—	Urinary concrements	Trace levels	E	S	—	Laser + spark	1565

NEW BOOKS

Analytical Atomic Spectroscopy, W. G. Schrenk, published by Plenum Press, New York, 1975.

Basic Atomic Absorption Spectroscopy—A Modern Introduction, M. D. Amos, P. A. Bennett, J. P. Matousek, C. R. Parker, E. Rothery, C. J. Rowe and J. B. Sanders, published by Varian Techtron Pty. Ltd., 1975.

Flame Emission and Atomic Absorption Spectrometry; Volume 3—Elements and Matrices, edited by J. A. Dean and T. C. Rains, published by Marcel Dekker, New York, 1975.

Analytical Emission Spectroscopy, J. Mika and T. Torok, published by Crane, Russak and Co. Inc., New York, 1974.

Emission Spectroscopy, T. Kantor, chapter in 'Comprehensive Analytical Chemistry, Volume V.' C. L. Wilson and D. W. Wilson, edited by G. Svehla, published by Elsevier, New York, 1975.

Atomic Fluorescence Spectroscopy, V. Sychra, V. Svoboda and I. Rubeska, published by Van Nostrand Reinhold, London, 1975.

Computers for Spectroscopists, edited by R. A. G. Carrington, published by Adam Hilger, London, 1975.

A Course on Analytical Techniques, M. S. Cresser, published by Ministry of Agriculture and Natural Resources, Iran, 1975.

Institute of Petroleum Standards for Petroleum and its Products; Part I—Methods for Analysis and Testing, Applied Science Publishers Ltd., Barking, 1975.

REVIEWS

Atomic absorption spectroscopy; flame atomization	446, 1056, 1057
Atomic absorption spectroscopy; general	1017
Atomic absorption spectroscopy; electrothermal atomization	446, 1057, 1171, 1173, 1174, 1521
Emission spectroscopy; general	1558
Emission spectroscopy; plasmas sources	104, 701, 1403
Chemiluminescence	498
Derivative spectroscopy	219
Tunable lasers	680, 681
TV-type multichannel detectors	669, 670
Multielement analysis; general	1379
Determination of beryllium	1197
Determination of mercury	153
Determination of precious metals	355
Pollution analysis; atomic absorption spectroscopy	1559
The progress of analytical chemistry, 1910–1970	697

MEETINGS 1975

The following major conferences are covered in this volume. The numbers in brackets after each conference signify the relevant reference numbers in the Reference Section

Australia — The 3rd Australian Symposium on Analytical Chemistry, Melbourne, 27–30 May 1975. (424–435).

The 5th International Conference on Atomic Spectroscopy, Melbourne, 25–29 August 1975. (792–889).

Canada — The 58th Chemical Conference and Exhibition, Toronto, 25–28 May 1975. (906–914).

Czechoslovakia — The 2nd Czechoslovak Seminar on Atomic Absorption Spectroscopy, Reka, 23–27 April 1975. (1543–1549).

France — The 18th Colloquium Spectroscopicum Internationale, Grenoble, 15–19 September 1975. (576–632, 1573).

Hungary — Euroanalysis II, Budapest, 25–30 August 1975. (1301–1323, 1482–1515).

The Netherlands — Dutch Atomic Spectroscopy Working Group, Amsterdam, 23 May 1975. (1075–1078).

United Kingdom — The 2nd Pye Unicam Analytical Conference, Stratford-on-Avon, 20–22 May 1975. (384–393).

U.S.A. — The 26th Pittsburgh Conference, Cleveland, 3–7 March 1975. (703–770).

The 169th American Chemical Society National Meeting, Philadelphia, 6–11 April 1975. (890–905).

The 28th Annual Summer Symposium on Analytical Chemistry, Knoxville, 18–20 June 1975. (484–493).

The 170th American Chemical Society National Meeting, Chicago, 24–29 August 1975. (644–658).

The Federation of Analytical Chemistry and Spectroscopy Societies 2nd National Meeting, Indianapolis, 6–10 October 1975. (1090–1150).

The 89th Annual Meeting of the Association of Official Analytical Chemists, Washington, 13–16 October 1975. (1448–1454).

The Eastern Analytical Symposium, New York, 21 November 1975. (1222–1225).

REFERENCES

1 RINER, J. C., WRIGHT, F. C., and McBETH, C. A., A technique for determining Pb in faeces of cattle by flameless AAS, *At. Absorpt. Newsl.*, 1974, **13**, 129. (U.S. Livestock Insects Laboratory, Agricultural Research Service, U.S. Department of Agriculture, Kerrville, Texas 78028, U.S.A.).

2 GANJE, T. J., and PAGE, A. L., Rapid dissolution of plant tissue for Cd determination by AAS, *At. Absorpt. Newsl.*, 1974, **13**, 131. (Dept. of Soil Science and Agricultural Engineering, University of California, Riverside, Calif. 92502, U.S.A.).

3 GUILLAUMIN, J. C., Determination of trace metals in power plant effluents, *At. Absorpt. Newsl.*, 1974, **13**, 135. (The Detroit Edison Co., 2000 Second Avenue, Detroit, Mich. 48226, U.S.A.).

4 KIRKBRIGHT, G. F., and WILSON, P. J., The direct determination of I by AAS with the graphite furnace, *At. Absorpt. Newsl.*, 1974, **13**, 140. (Dept. of Chemistry, Imperial College, London SW7 2AY, England).

5 MAY, L. A., and PRESLEY, B. J., The determination of trace metals in beach asphalts, *At. Absorpt. Newsl.*, 1974, **13**, 144. (Dept. of Chemistry, University of Alabama, Birmingham, Ala. 35294, U.S.A.).

6 ZIMMERMANN, H. G., Techniques in sample preparation for AA determinations, *At. Absorpt. Newsl.*, 1974, **13**, 145. (Research Centre QIT, Sorel, Que. Canada).

7 HINNERS, T. A., BUMGARNER, J. E., and SIMMONS, W. S., Extraction of Cd from rice, *At. Absorpt. Newsl.*, 1974, **13**, 146. (Bioenvironmental Laboratory Branch, Human Studies Laboratory, National Environmental Research Center, Environmental Protection Agency, Research Triangle Park, N.C. 27711, U.S.A.).

8 SLAVIN, S., and LAWRENCE, D. M., An AA bibliography for July–Dec. 1974, *At. Absorpt. Newsl.*, 1975, **14**, 1. (Perkin Elmer Corp., Norwalk, Conn. 06856, U.S.A.).

9 MACQUET, J. P., and THEOPHANIDES, T., AA and relation between stability cis–trans isomerism in Pt complexes, *At. Absorpt. Newsl.*, 1975, **14**, 23. (Dept. of Chemistry, University of Montreal, Montreal, Que. H3C 3V1, Canada).

10 SHERFINSKI, J. H., A graphite tube degradation study of Ba at gunshot residue levels, *At. Absorpt. Newsl.*, 1975, **14**, 26. (Perkin-Elmer Corp., Lombard, Ill. 60148, U.S.A.).

11 RUBESKA, I., MIKSOVSKY, M., and HUKA, M., A branched capillary for buffering in flame spectrometry, *At. Absorpt. Newsl.*, 1975, **14**, 28. (Geological Survey of Czechoslovakia, 17000 Prague 7, Kostelini 26, Czechoslovakia).

12 WARD, D. A., and BIECHLER, D. G., Rapid, direct determination of Ca in natural waters by AAS, *At. Absorpt, Newsl.*, 1975, **14**, 29. (Kerr McGee Technical Centre, Oklahoma City, Okla. 73125, U.S.A.).

13 SOMMERFELD, M. R., LOVE, T. D., and OLSEN, R. D., Trace metal contamination of disposable pipette tips, *At. Absorpt. Newsl.*, 1975, **14**, 31, (Dept. of Botany and Microbiology, Arizona State University, Tempe, Ariz. 85281, U.S.A.).

14 TIFFANY, W. B., Pyroelectric detectors, *Opt. Spectra*, 1975, **9**, 22. (Molectron Corp., Sunnyvale, Calif. U.S.A.).

15 DUNCAN, F. W., Silicon photodiodes, *Opt. Spectra*, 1975, **9**, 26. (United Detector Technology, Santa Monica, Calif., U.S.A.).

16 NEVILLE, G. J., The passive interference filter, *Opt. Spectra*, 1975, **9**, 31. (Corion Corp., U.S.A.).

17 BRECH, F., Spectroscopy, *Opt. Spectra*, 1975, **9**, 39. (Jarrell-Ash Div., Fisher Scientific Co., U.S.A.).

18 ANON., Spectrochemistry Group, Ames Laboratory – USAEC, Iowa State University, *Arcs and Sparks*, 1974, **19**, 10.

19 THOMPSON, K. C., and GODDEN, R. G., Determination of trace levels of Ba in $CaCO_3$ by AAS, *Analyst*, 1975, **100**, 198. (Shandon Southern Instruments Ltd., Frimley Road, Camberley, Surrey GU16 5ET, England).

20 GUTSCHE, B., KLEINOEDER, H., and HERRMANN, R., Device for trace analysis for F in reaction tubes by AAS, *Analyst*, 1975, **100**, 192. (Dept. of Medical Physics, University of Giessen, D–6300 Giessen, West Germany).

21 VAN STEKELENBURG, G. J., VAN DE LAAR, A. J. B., and VAN DER LAAG, J., Cu analysis of nail clippings: an attempt to differentiate between normal children and patients suffering from cystic fibrosis, *Clin. Chim. Acta*, 1975, **59**, 233. (Bio-

chemical Laboratory, Wilhelmina Kinderziekenhuis, State University of Utrecht, Utrecht, The Netherlands).

22 BURES, J., Degeneracy of light and the optimum accuracy of photoelectric measurements, *J. Opt. Soc. Am,* 1974, **64**, 1598. (Laboratoire d'Optique et de Spectroscopie, Dept. de Genie Physique, Ecole Polytechnique, Case Postale 6079–Succursale 'A', Montreal, Que. H3C 3A7, Canada).

23 POUEY, M., Second order focusing conditions for ruled concave gratings, *J. Opt. Soc. Am,* 1974, **64**, 1616. (Laboratoire des Interactions Moleculaires et des Haute Pressions, C.N.R.S., Bellevue, 92190 Meudon, France).

24 WESTWOOD, W. D., and LIT, J. W. Y., Sectioned thin-film grating, *J. Opt. Soc. Am,* 1974, **64**, 1631. (Bell-Northern Research, Ottawa, Ont., Canada).

25 POUEY, M., Design of simple rotating stigmatic concave grating monochromators, *Appl. Opt.,* 1974, **13**, 2739. (Address as in ref. 23).

26 KUBOTA, J., MILLS, E. L., and OGLESBY, R. T., Pb, Cd, Zn, Cu and Co in streams and lake waters of Cayuga Lake Basin, New York, *Environ. Sci. Technol.,* 1974, **8**, 243. (Agronomy Dept., U.S. Plant, Soil and Nutrition Laboratory, Cornell University, Ithaca, N.Y. 14850, U.S.A.).

27 BRULAND, K. W., BERTINE, K., KOIDE, M., and GOLDBERG, E. D., History of metal pollution in Southern California coastal zone, *Environ. Sci. Technol.,* 1974, **8**, 425. (Scripps Institution of Oceanography, La Jolla, Calif., U.S.A.).

28 BOGDEN, J. D., SINGH, N. P., and JOSELOW, M. M., Cd, Pb and Zn concentrations in whole blood samples of children, *Environ, Sci. Technol.,* 1974, **8**, 740. (Div. of Environmental Toxicology, Dept. of Preventive Medicine and Community Health, New Jersey Medical School, Newark, N.J. 07103, U.S.A.).

29 TAM., K. C., As in water by flameless AAS, *Environ. Sci. Technol.,* 1974, **8**, 734. (Freshwater Institute, Department of the Environment, 501 University Crescent, Winnipeg, Man. R3T 2N6, Canada).

30 BISOGNI, J. J., jun., and LAWRENCE, A. W., Determination of submicrogram quantities of monomethyl Hg in aquatic samples,*Environ. Sci. Technol.,* **1974**, **8**, 850. (Dept. of Environmental Engineering, Cornell University, Ithaca, N.Y. 14850, U.S.A.).

31 LECH, J. F.,* SIEMER, D.,† and WOODRIFF, R.,† Determination of **Pb** in atmospheric particulates by furnace AA. *Environ. Sci. Technol.,* 1974, **8**, 840. (*Varian Instrument Div., 611 Hansen Way, Palo Alto, Calif., U.S.A.; †Dept. of Chemistry, Montana State University, Bozeman, Mont. 59715, U.S.A.).

32 SHANNON, D. G., and FINE, L. O., Cation solubilities of lignite fly-ashes, *Environ. Sci. Technol.,* 1974, **8**, 1026. (Plant Science Dept., South Dakota State University, Brookings, S. Dak. 57006, U.S.A.).

33 DAVISON, R. L., NATUSCH, D. F. S., WALLACE, J. R., and EVANS, C. A., jun., Trace elements in fly-ash, *Environ. Sci. Technol.,* 1974, **8**, 1107. (School of Chemical Sciences, University of Illinois, Urbana, Ill. 61801, U.S.A.).

34 RAINS, T. C., and MENIS, O., Determination of Al by FES with repetitive optical scanning, *Anal. Lett.* 1974, **7**, 715. (Analytical Chemistry Div., Institute for Materials Research, National Bureau of Standards, Washington, D.C. 20234, U.S.A.).

35 RANJITKAR, K. P., and TOWNSHEND, A., Indirect pyrophosphate determination by AA using Cu(II) ions and a liquid ion exchanger, *Anal. Lett.,* 1974, **7**, 743. (Dept. of Chemistry, University of Birmingham, P.O. Box 363, Birmingham, England).

36 PRAUCHEVA, C., DELIJSKA, A., and TZOLOV, T., Spectrographic determination of Si and Fe in chromites, *Metalurgiya (Sofia),* 1974, **29**, 16. (Iron and Steel Research Institute, Sofia, Bulgaria).

37 KRASNOBAEVA, N., and NEDJALKOVA-DASKALOVA, N., The use of controllable atmosphere in the analysis of dry residues obtained from solutions. The effect of Ar and Ba $(NO_3)_2$ additive on the arc plasma parameters, *Izy. Otd. Khim. Nauki, Bulg. Akad. Nauk.,* 1974, **7**, 385. (Institute for General and Inorganic Chemistry, Bulgarian Academy of Sciences, Sofia 13, Bulgaria).

38 KRASNOBAEVA, N., and NEDJALKOVO-DASKALOVA, N., The use of controllable atmosphere in the analysis of dry residues obtained from solutions. The effect of Ar and Ba $(NO_3)_2$ additions on the relative intensities of atomic spectral lines, *Bulg. Akad. Nauk.,* 1974, **7**, 399. (Address as in ref. 37).

39 DELIJSKA, A., and PRAUCHEVA, C., A possibility of the application of organic compounds as a thermochemical agent in spectral analysis — Paper presented at the VIth National Conference on Spectroscopy, Slautchev, Bulgaria, 30 September–3 October, 1974. (Address as in ref. 36).

40 SVEHLA, A., Study of spectrochemical analysis conditions of slags from ferro-tungsten production, *Chem. Listy,* 1974, **68**, 1075. (Oranian Ferroalloying Plant, 02753 Istebne, Czechoslovakia).
41 KOLIHOVA, D., and SYCHRA, V., AA and AFS in biochemistry and medicine. I: Direct microdetermination of Zn in blood serum by AFS, *Chem. Listy,* 1974, **68**, 1091. (Flame Spectrometry Laboratory, Technical University, 16628, Prague 6, Czechoslovakia).
42 RUBESKA, I., and MIKSOVSKY, M., Analysis of sulphidic minerals by AAS, *Collect. Czech. Chem. Commun.,* 1974, **39**, 3485. (Address as in ref. 11).
43 SILAKOVA, V. G., MAKULOV, N. A., MANOVA, T. G., and BOZHEVOLNOV, E. A., Sensitivity of spectrographic analysis of dry residues of solutions, *Zh. Anal. Khim.,* 1974, **29**, 1683. (All Union Scientific Research Institute of Chemical Reagents and Special Purity Chemicals, Moscow, U.S.S.R.).
44 GRUSHKO, L. F., IVANOV, N. P., and CHUPAKIN, M. S., Analytical possibilities of a tantalum ribbon atomizer, *Zh. Anal. Khim.,* 1974, **29**, 1842. (Address as in ref. 43).
45 KARPENKO, L. I., FADEEVA, L. A., SHEVCHENKO, I. D., and VIDISHEVA, A. Y., Spectrographic analysis of solutions using solution nebulisation, *Zh. Anal. Khim.,* 1974, **29**, 1887. (Institute of General and Inorganic Chemistry, Ukranian S.S.R. Academy of Sciences, Odessa Lab., U.S.S.R).
46 MIKHAILOVA, T. P., and SYLADNEVA, N. A., AA determination of Zn and Cd in technological solutions, *Zh. Anal. Khim.,* 1974, **29**, 1912. (Institute of Physico-chemical Principles of Processing Mineral Raw Materials, U.S.S.R.).
47 PATTERSON, C., Pb in sea water, *Science,* 1974, **183**, 553. (Dept. of Geological and Planetary Sciences, California Institute of Technology, Pasadena, Calif., U.S.A.).
48 LICK, D. J., Recent developments in the analysis of toxic elements, *Science,* 1974, **184**, 1137. (Food Science Dept., Cornell University, Ithaca, N.Y. 14850, U.S.A.).
49 ANON., Safety practices for AA Spectrophotometers, *Amer. Lab.,* 1974, **6** (3), 49.
50 HOLAK, W., AA in food analysis special techniques, *Amer. Lab.,* 1974, **6** (8), 10. (Food and Drug Administration, New York District, N.Y., U.S.A.).
51 GRAHAM, T. F., Automated sample preparation for AA, *Amer. Lab.,* 1974, **6** (9), 77. (Technicon Industrial Systems, Tarrytown, N.Y., U.S.A.).
52 ROCSIN, M., and PLAUCHITHIU, M. G., The determination of electrolytes (Na, K and Ca) in erythrocytes by flame photometry, *Rev. Chim. (Bucharest),* 1974, **25**, 325. (Medical Institute, Timisoara, Rumania).
53 COBB, W. D., FOSTER, W. W., and HARRISON, T. S., Determination of Al at low levels in steel by AAS, *Lab. Pract.* 1975, **24**, 143. (British Steel Corp., Scunthorpe Group, P.O. Box No. 1, Scunthorpe, Lincs., England).
54 RANWEILER, L. E., and MOYERS, J. L., AA procedure for analysis of metals in atmospheric particulate matter, *Environ. Sci. Technol.,* 1974, **8**, 152. (Atmospheric Analysis Laboratory, Dept. of Chemistry, University of Arizona, Tucson, Ariz. 85721, U.S.A.).
55 GIOVANOLIJAKUBCZAK, T., and BERG, G. G., Measurement of Hg in human hair, *Arch. Environ. Health,* 1974, **28**, 139. (Dept. of Radiation, Biology and Bio-physics, University of Rochester, School of Medicine and Dentistry, Rochester, N.Y. 14642, U.S.A.).
56 HEMPHILL, D. D., MARIENFELD, C. J., REDDY, R. S., and PIERCE, J. O., Roadside Pb contamination in the Missouri Pb belt, *Arch. Environ. Health,* 1974, **28**, 190. (University of Missouri, Columbia, Mo. 65201, U.S.A.).
57 CLARKE, A. N., and WILSON, D. J., Preparation of hair for Pb analysis, *Arch. Environ. Health,* 1974, **28**, 292. (Dept. of Chemistry, Vanderbilt University, Nash-ville, Tenn. 37203, U.S.A.).
58 JACKWERTH, E., LOHMAR, J., and WITTLER, G., Trace enrichment with activated carbon. Determination of traces of elements in powdered W, *Z. Anal. Chem.,* 1974, **270**, 6. (Institut fur Spektrochemie und angewandte Spectroscopie, D–4600 Dortmund, West Germany).
59 GIBBS, R. H., JAROSEWICH, E., and WINDOM, H. L., Heavy metal concentra-tions in museum fish specimens: effects of preservatives, *Science,* 1974, **184**, 475. (Dept. of Vertebrate Zoology, Smithsonian Institute, Washington, D.C. 20560, U.S.A.).
60 BOYDEN, C. R., and ROMERIL, M. G., A trace metal problem in pond oyster

culture, *Mar. Pollut. Bull.*, 1974, **5**, 74. (Applied Geochemistry Research Group, Imperial College, London S.W.7, England).

61 CHADWICK, M. J., EDWORTHY, K. J., RUSH, D., and WILLIAMS, P. J., Ecosystem irrigation as a means of groundwater recharge and water quality improvement, *J. Appl. Ecol.*, 1974, **11**, 231. (Dept. of Biology, University of York, York, England).

62 COOKE, R. E., GLYNN, K. L., ULLMAN, W. W., LURIE, N., and LEPOW, M., Comparative study of a micro-scale test for Pb in blood, for use in mass screening programmes, *Clin. Chem.*, 1974, **20**, 582. (Connecticut State Dept. of Health, Laboratory Division, Hartford, Conn. 06106, U.S.A.).

63 YEH, Y. Y., and ZEE, P., Micromethod for determining total Fe binding capacity by flameless AAS, *Clin. Chem.*, 1974, **20**, 360. (Laboratory of Nutrition and Metabolism, St. Jude Children's Research Hospital, P.O. Box 318, Memphis, Tenn. 38101, U.S.A.).

64 BEKMUKHAMBETOV, E. S., and DONOV, V. A., AA method for investigating thermal diffusion in gas and vapour mixtures, *Zh. Prikl. Spektrosk.*, 1974, **20**, 385.

65 MENZINGER, M., Electronic chemiluminescence in $M+X_2$ reactions: dissociation energies of the alkaline earth monohalides MX, *Can. J. Chem.*, 1974, **52**, 1688. (Dept. of Chemistry, University of Toronto, Toronto, Ont., Canada).

66 YAMANE, T., MUKOYAMA, T., and SASAMOTO, T., Solvent extraction of Pb, Ag, Sb and Tl with Zn dibenzyl-dithiocarbamate and its application to the separation of Bi from large quantities of Pb, *Anal. Chim. Acta*, 1974, **69**, 347. (Dept. of Applied Chemistry, Faculty of Engineering, Yamanashi University, Kofu, Japan).

67 INGLE, J. D., jun., Sensitivity and limit of detection in quantitative spectrometric methods, *J. Chem. Educ.*, 1974, **51**, 100. (Oregon State University, Corvallis, Ore. 97331, U.S.A.).

68 SARBECK, J. R., and LANDGRAF, W. C., Automated peak discrimination and integration for non-flame AA analysis at nanogram levels, *J. Pharm. Sci.*, 1974, **63**, 929. (Applications Research Dept., Quality Control Div., Syntex Laboratories Inc., Palo Alto, Calif., U.S.A.).

69 ANON., Trace analysis applicable to determination of minor amounts of impurities in chemicals. I: General survey, *Pure Appl. Chem.*, 1974, **37**, 481. (I.U.P.A.C., Analytical Chemistry Division).

70 PATENT, Apparatus for AAS, British Patent 1,367,394. (Commissariat a l'Energie Atomique, France).

71 BACKHAUS, G., Low temperature plasma incineration of samples for trace element analysis, *G-I-T Fachz. Lab.*, 1974, **18**, 533. (Kontron Technik GmbH, Munich, Germany).

72 WOIDICH, H., and PFANNHAUSER, W., Determination of Cd in foods, *Z. Lebensm.-Unters.-Forsch.*, 1974, **155**, 72. (Forschungsinstitut Ernahrungswirtsch, Vienna, Austria).

73 BURGETT, C. A., and GREEN, L. E., Improved flame photometric dectection without solvent 'flame-out', *J. Chromatogr. Sci.*, 1974, **12**, 356. (Hewlett-Packard, Avondale, Pa., U.S.A.).

74 BONNE, R., Device for analysing a substance by AAS with background correction, British Patent, 1,361,736. (Address as in ref. 70).

75 PATENT, New and improved method and apparatus for sample analysis by AAS, British Patent 1,354,977. (Technicon Instruments Corp., U.S.A.).

76 KNISELEY, R. N., Analytical applications of inductively coupled plasma optical ES, *U.S. At. Energy Comm.*, 1974, IS-T-626. (Ames Laboratory, Iowa State University, Ames, Iowa, U.S.A.).

77 FOSTER, P., and GARDEN, J., Rapid determination of Si in ferromanganese by AAS, *Analusis*, 1974, **2**, 675. (Laboratoire de Chimie Analytique Mineralogie, U.E.R. Phys.–Chim. Mater., Saint-Martin d'Heres, France).

78 SACKS, R. D., and HOLCOMBE, J. A., Radiative and electrical properties of exploding silver wires, *Appl. Spectrosc.*, 1974, **28**, 518. (Dept. of Chemistry, University of Michigan, Ann Arbor, Mich. 48104, U.S.A.).

79 DALE, L. S., A direct carrier distillation procedure for the spectrographic determination of impurities in UF_4, *Appl. Spectrosc.*, 1974, **28**, 564. (Australian Atomic Energy Commission, Research Establishment, Lucas Heights, N.S.W. 2232, Australia).

80 JACKSON, K. W., ALDOUS, K. M., and MITCHELL, D. G., Simultaneous determination of trace wear metals in used lubricating oils by AAS using a silicon target

vidicon detector, *Appl. Spectrosc.*, 1974, **28**, 569. (Div. of Laboratories and Research, New York State Department of Health, Albany, N.Y. 12201, U.S.A.).

81 BARNES, R. M.,* and SLAVIN, M.,† The total energy technique; a historical note, *Appl. Spectrosc.*, 1974, **28**, 574. (*Dept. of Chemistry, University of Massachusetts, Amherst, Mass. 01002, U.S.A.; †Box 876, Setauket, N.Y. 11785, U.S.A.).

82 NAKAMURA, M., and SHALIMOFF, G. V., Four function calculator used to compute wavelength automatically, *Appl. Spectrose.*, 1974, **28**, 581. (Lawrence Berkeley Laboratory, University of California, Berkeley, Calif. 94720, U.S.A.).

83 SIEMER, D. D., WOODRIFF, R., and WATNE, B., A simple technique for coating carbon AA atomizer components with pyrolytic carbon, *Appl. Spectrosc.*, 1974, **28**, 582. (Chemistry Dept., Montana State University, Bozeman, Mont. 59715, U.S.A.).

84 BRACHACZEK, W. W., BUTLER, J. W., and PIERSON, W. R., Acid enhancement of Zn and Pb AA signals, *Appl. Spectrosc.*, 1974, **28**, 585. (Scientific Research Staff, Ford Motor Co., P.O. Box 2053, Dearborn, Mich. 48121, U.S.A.).

85 JOHNSON, G. W., and SKOGERBOE, R. K., A simple device for renewing contact surfaces on carbon rod atomizers, *Appl. Spectrosc.*, 1974, **28**, 590. (Dept. of Chemistry, Colorado State University, Fort Collins, Colo. 80521, U.S.A.).

86 BERMAN, E., Biochemical applications of FE and AAS, *Appl. Spectrosc.*, 1975, **29**, 1. (Division of Biochemistry, Cook County Hospital, Chicago, Ill. 60612, U.S.A.).

87 SCHRENK, W. G., and EVERSON, R. T., AA interferences using a tantalum boat atomizing system, *Appl. Spectrosc.*, 1975, **29**, 41. (Chemistry Dept., Kansas Agricultural Experiment Station, Manhattan, Kan. 66502, U.S.A.).

88 GOLEB, J. A., and MIDKIFF, C. R., The determination of Ba and Sb in gunshot residue by flameless AAS using a tantalum strip atomizer, *Appl. Spectrosc.*, 1975, **29**, 44. (Bureau of Alcohol, Tobacco and Firearms, Forensic Branch, U.S. Treasury, Washington, D.C. 20226, U.S.A.).

89 HORLICK, G., CODDING, E. G., and LEUNG, S. T., Automated d.c. arc time studies using a computer coupled photodiode array spectrometer, *Appl. Spectrosc.*, 1975, **29**, 48. (Dept. of Chemistry, University of Alberta, Edmonton, Alta T6G 2G2, Canada).

90 MUSCAT, V. I., VICKERS, T. J., RIPPETOE, W. E., and JOHNSON, E. R., Signal-to-noise ratio comparison for dispersive and non-dispersive flame AF measurements, *Appl. Spectrosc.*, 1975, **29**, 52. (Dept. of Chemistry, Florida State University, Tallahassee, Fla. 32306, U.S.A.).

91 UCHIDA, T., and IIDA, C., AAS with absorption tube technique, *Appl. Spectrosc.*, 1975, **29**, 58. (Laboratory of Analytical Chemistry, Nagoya Institute of Technology, Showa, Nagoya 466, Japan).

92 BRENNER, I. B.,* ELDAD, H.,* and ARGOV, L.,* HAREL, A.,† and ASSOUS, M.,† D.C. (central plasma region) spectrochemical analysis of standard silicate rocks and minerals, *Appl. Spectrosc.*, 1975, **29**, 82. (*Geological Survey of Israel, Jerusalem, Israel; †Nuclear Research Centre, Negev, Beersheva, Israel).

93 HOPP, H. U., AAS determination of Zn, Ba, Ca and Mg in mineral oil products: emulsion method, *Erdol Kohle, Erdgas, Petrochem. Brennst.-Chem.*, 1974, **27**, 435. (Fuchs Mineralolwerke GmbH, 68 Mannheim, West Germany).

94 MURTY, P. S., and KAMAT, M. J., Spectrographic determination of Ru in siliceous materials, *Indian J. Technol.*, 1974, **12**, 41. (Spectroscopy Division, Bhabha Atomic Research Centre, Trombay, Bombay, India).

95 NYAN'KOVSKAYA, N. M., and BOGAUTDINOVA, K. O., Determination of the ash content in streptomycin, neomycin and monomycin sulphates on a flame photometer, *Khim.-Farm. Zh.*, 1974, **8**, 61. (Mosk. Zavod. Med. Prep. No. 2, Moscow, U.S.S.R.).

96 SOSKIN, K., and DUMANSKI, J., Reproducibility of responses of the 'Atomspek' AA spectrophotometer in a long series of measurements, *Chem. Anal.* J*Warsaw*I, 1974, **19**, 697. (Instytut Ekspertyz, Sadowych, Ul. Westerplatte 9, 31–033 Krakow, Poland).

97 KAUFMANN, K. J., KINSEY, J. L., PALMER, H. B., and TEWARSON, A., Chemiluminescent emission spectra and possible upper state potentials of KCl and KBr, *J. Chem. Phys.*, 1974, **61**, 1865. (Dept. of Chemistry, Massachusetts Institute of Technology, Cambridge, Mass., U.S.A.).

98 GOLDSTEIN, S. A., D'SILVA, A. P., and FASSEL, V. A., X-ray excited optical fluorescence of gaseous atmopheric pollutants: analytical feasibility study, *Radiat. Res.*, 1974, **59**, 422. (Ames Laboratory, Iowa State University, Ames, Iowa, U.S.A.).

99 PALERMO, E. F., MONTASER, A., and CROUCH, S. R., Multi-element, non-dispersive AFS in the time division multiplexed mode, *Anal. Chem.*, 1974, **46**, 2154. (Dept. of Chemistry, Michigan State University, Mich., U.S.A.).

100 ROELANDTS, I., BOLOGNE, G., DUPAIN-KLERKX, L., and CZICHOSZ, R., Separation of microgram quantities of Al from Ag matrices prior to its determination by AAS, *Sep. Sci.*, 1974, **9**, 445. (University of Liege, Sart Tilman, Liege, Belgium).

101 ANON., Cryolite, natural and artificial: determination of Na content: FE and AAS methods — International Standard, ISO 2366: 1974.

102 STRONG, B., and MURRAY-SMITH, R., Determination of Au in Cu-bearing sulphide ores and metallurgical flotation products by AAS, *Talanta*, 1974, **21**, 1253. (Charter Consolidated, Metallurgical Laboratory, Ashford, Kent, England).

103 TOPPING, J. T., and MacCREHAN, W. A., Preconcentration and determination of Cd in water by reversed phase column chromatography and AA, *Talanta*, 1974, **21**, 1281. (Chemistry Dept., Towson State College, Maryland 21204, U.S.A.).

104 GREENFIELD, S., McGEACHIN, H. McD., and SMITH, P. B., Plasma emission sources in analytical spectroscopy. I, *Talanta*, 1975, **22**, 1. (Albright and Wilson Ltd., Oldbury, Warley, West Midlands B69 4LN, England).

105 ACKERMANN, G., and MUNX, M., Influence of additives on the spectrographic analysis of solutions with respect to the atomizer process. I: Influence of inorganic ions, *Talanta*, 1975, **22**, 107. (Lehrstuhl fur Analytische Chemie, Bergakademie Freiberg, 92 Freiberg, Saxony, East Germany).

106 CHILOV, S., Determination of small amounts of Hg, *Talanta*, 1975, **22**, 205. (Research Laboratory, Kodak (Australasia) Pty. Ltd., P.O. Box 90, Coburg, Vic. 3058, Australia).

107 DITTRICH, K., and ZEPPAN, W., Determination of Zn in hydrochloric acid, gallium arsenide and gallium aluminiumarsenide with flameless AA, *Talanta*, 1975, **22**, 299. (Sektion Chemie der Karl-Marx-Universitat, Liebigstrasse 18, Leipzig-701, East Germany).

108 DITTRICH, K., and MOTHES, W., Quantitative determination of Au in photographic film by flameless AA, *Talanta*, 1975, **22**, 318. (Address as in ref. 107).

109 GOODE, S. R., Experimental and theoretical investigation into non-flame atomization — Thesis, 1974. *Diss. Abstr. Int. B*, 1974, **35**, 1176. (Michigan State University, East Lansing, Mich., U.S.A.).

110 KANTOR, T., CLYBURN, S. A., and VEILLON, C., Continuous sample introduction with graphite atomization systems for AAS, *Anal. Chem.*, 1974, **46**, 2205. (Dept. of Chemistry, University of Houston, Houston, Texas, U.S.A.).

111 RAINS, T. C., OLSON, C. D., VELAPOLDI, R. A., WICKS, S. A., and MENIS, O., Preparation of reference materials for stationary source emission analysis: Be, *Report*, 1974, NBSIR–74–439. (National Bureau of Standards, Washington, D.C., U.S.A.).

112 DAGNALL, R. M., JOHNSON, D. J., and WEST, T. S., Determination of Mo by AAS using a carbon filament atom reservoir, pp. 93–96 *in* Mitchell, P. C. H., *Editor*, Chemistry and uses of Mo, proceedings of (1st) conference, Reading, England, 17–21 September, 1973 — Published by Climax Molybdenum Co. Ltd., London, 1974. (Dept. of Chemistry, Imperial College, London S.W.7, England).

113 FORD, A., YOUNG, B., and MELOAN, C., Determination of Pb in organic colouring dyes by AAS, *J. Agric. Food Chem.*, 1974, **22**, 1034. (Dept. of Chemistry, Kansas State University, Manhattan, Kan., U.S.A.).

114 SAPEK, A., AAS determination of Pb, Ni and Co in soil extracts, *Chem. Anal. (Warsaw)*, 1974, **19**, 687. (Inst. Reclam. Grassl. Farming, Falenty, Warsaw, Poland).

115 FISHKOVA, N. L., and KAZARINA, T. M., AA determination of Au in pyrite and arsenopyrite, *Tr. Tsentr. Nauchno-Issled. Geologorazved. Inst. Tsvetn. Blagorodn. Met.*, 1974, No. 111, 83.

116 HALL, E. T., Hg in commercial canned sea food, *J. Assoc. Off. Anal. Chem.*, 1974, **57**, 1068. (U.S. Army Medical Laboratory, Health Service Command, Fort George G. Meade, Md., U.S.A.).

117 SADEK, S. H., and ABDEL SALAM, A. A. A., Critical appraisal of Ca estimation by flame photometry, *Ain Shams Med. J.*, 1974, **25**, 363. (Dept. of Clinical Pathology, Ain Shams University, Cairo, Egypt).

118 MITCHELL, D., ALDOUS, K. M., and RYAN, F. J., Mass screening for Pb

poisoning: capillary blood sampling and automated Delves-cup AA analysis, *N.Y. State J. Med.*, 1974, **74**, 1599. (Div. Lab. Res., Albany, N.Y., U.S.A.).

119 CASTELLI, A., CAVALLARO, A., CERUTTI, G., and FITTIPALDI, M., Pb, Cr, Cd, Zn, Mn, Cu and Fe in national wine: analysis by AAS, *Riv. Vitic. Enol.*, 1974, **27**, 247. (Ist. Agrar., University of Milan, Milan, Italy).

120 PENZIAS, G. J., Temperature measurements and gas analysis in flames and plasmas using spectroscopic methods, pp. 321–347 *in* Palmer, H. B., *Editor,* Combustion technology: some modern developments — Book published by Academic Press, New York, 1974. (Norcon Instruments Inc., South Norwalk, Conn., U.S.A.).

121 CLYBURN, S. A., BARTSCHMID, B. R., and VEILLON, C., AFS with a continuum source, graphite atomisation and photon counting, *Anal. Chem.*, 1974, **46**, 2201. (Dept. of Chemistry, University of Houston, Houston, Texas, U.S.A.).

122 OLES, P. J., and SIGGIA, S., AA method for determining micromolar quantities of 1,2-diols, *Anal. Chem.*, 1974, **46**, 2197. (Dept. of Chemistry, University of Massachusetts, Amherst, Mass., U.S.A.).

123 VAN DER HURK, J., HOLLANDER, T., and ALKEMADE, C. J., Excitation energies of SrOH bands measured in flames, *J. Quant. Spectrosc. Radiat. Transfer,* 1974, **14**, 1167. (Fys. Lab., University of Utrecht, Utrecht, The Netherlands).

124 GLUSHKO, L. N., ZAITSEV, A. S., KOVALENKO, L. A., and TVERDOKHLEBOV, V. I., Spectroscopic and probe measurements of the plasma temperature of a flame, *Zh. Prikl. Spektrosk.*, 1974, **20**, 886.

125 KATO, A., OSUMI, Y., NAKANE, M., and MIYAKE, Y., D.C. arc ES determination of trace amounts of impurities in high purity TiO_2 using GaF_3, *Bunseki Kagaku,* 1974, **23**, 1036 (Government Industrial Research Institute, Osaka, Japan).

126 BRISTOW, Q., Solid state computer interface and update unit for existing Perkin Elmer double beam AA spectrophotometers, *Anal. Chem.*, 1974, **46**, 2246. (Geological Survey of Canada, Ottawa, Ont., Canada).

127 IWATA, R., and OGATA, I., Determination of Ru in a H_2SO_4 solution by AAS, *Bull. Chem. Soc. Jpn.*, 1974, **47**, 2611. (Nat. Chem. Lab. Ind., Tokyo, Japan).

128 AMETANI, K., AAS determination of rare earths in single crystals of magnetic garnets and sulphides, *Bull. Chem. Soc. Jpn.*, 1974, **47**, 2238. (R.C.A. Research Laboratories, Machida, Japan).

129 GOIKHMANN, V. Y., IZHAK, A. P., and FRIDMAN, B. S., Flame photometric determination of Na and K oxides in high Ca glasses, *Steklo Keram.*, 1974, **7**, 38.

130 BUSCH, K. W., HOWELL, N. G., and MORRISON, G. H., Elimination of interferences in flame spectroscopy using spectral stripping, *Anal. Chem.*, 1974, **46**, 2074. (Dept. of Chemistry, Cornell University, Ithaca, N.Y., U.S.A.).

131 KNAPP, D. O., Multichannel detection in flame spectrometry using a silicon intensifier camera — Thesis, 1973. *Diss. Abstr. Int. B,* 1974, **35**, 1177. (University of Florida, Gainesville, Fla., U.S.A.).

132 KAEGLER, S. H., AAS, *Erdol Kohle, Erdgas, Petrochem. Brennst.-Chem.*, 1974, **27**, 514. (Deutsche Shell AG, Hamburg, Germany).

133 NEMETS, A. M., NIKOLAEV, G. I., and FLISYUK, V. G., Calculation of some parameters of spectral lines during the measurement of collision broadening and optical density of metal vapours by the AA method, *Zh. Prikl. Spektrosk.*, 1974, **21**, 212.

134 CLYBURN, S. A., KANTOR, T., and VEILLON, C., Pyrolysis treatment for graphite atomization systems, *Anal. Chem.*, 1974, **46**, 2213. (Dept. of Chemistry, University of Houston, Houston, Tex., U.S.A.).

135 HERRMANN, R., Flame spectroscopic detectors for the analysis of F in gas chromatography, *Z. Klin. Chem. Klin. Biochem.*, 1974, **12**, 393. (Institute for Medical Physics, Giessen, Germany).

136 FISHKOVA, N. L., Effect of organic solvents on the sensitivity of AA determination of Au, Pt, Pd and Ag, *Tr. Tsentr. Nauchno-Issled. Geologorazved. Inst. Tsvetn. Blagorodn. Met.*, 1974, No. 111, 99.

137 LOVETT, R. J., and PARSONS, M. L., AA of Re using a Ne analysis line, *Anal. Chem.*, 1974, **46**, 2241. (Dept. of Chemistry, Arizona State University, Tempe, Ariz., U.S.A.).

138 INGLE, J. D., jun., Precision of AAS measurements, *Anal. Chem.*, 1974, **46**, 2161. (Dept. of Chemistry, Oregon State University, Corvallis, Ore., U.S.A.).

139 ROBBINS, W. K., Analysis of petroleum for trace metals. Determination of trace quantities of Mn in petroleum and petroleum products by heated vaporization AA,

Anal. Chem., 1974, **46**, 2177. (Analytical and Information Division, Exxon Research and Engineering Co., Linden, N.J., U.S.A.).

140 SETCHELL, R. E., Analysis of flame emissions by laser Raman spectroscopy, *West. States Sect. Combust. Inst., (Pap.),* 1974, WSS/CI–74–6. TTechnical Staff, Aerodynamics Division, Livermore, Calif., U.S.A.).

141 ROBINSON, J. W., and WOLCOTT, D. K., Simultaneous determination of particulate and molecular Pb in the atmosphere, *Environ. Lett.*, 1974, **6**, 321. (Dept. of Chemistry, Louisiana State University, Baton Rouge, La., U.S.A.).

142 MORRIS, R. M., Determination of Si in cured tobacco leaf by AAS, *Tob. Sci.*, 1974, **18**, 120. (Philip Morris Research Centre, Richmond, Va., U.S.A.).

143 FETISOV, E. A., and FATEEVA, V. V., Determination of Ca content in milk and other dairy products by flame photometry, *Molochn. Promst.*, 1974, **7**, 18. (Vses. Nauchno-Issled. Inst. Molochn. Prom., Moscow, U.S.S.R.).

144 WOLF, W., MERTZ, W., and MASIRONI, R., Determination of Cr in refined and unrefined sugars by oxygen plasma ashing flameless AA, *J. Agric. Food Chem.*, 1974, **22**, 1037. (Nutrition Institute, Agricultural Research Service, Beltsville, Md., U.S.A.).

145 DALL'AGLIIO, M., and VISIBELLI, D., FE scanning as an analytical tool in geochemical studies of alkali metals, *Rend. Soc. Ital. Mineral. Petrol.*, 1974, **30**, 239. (Cent. Studi Nucl. Casaccia, Cons. Naz. Energ. Nucl., Rome, Italy).

146 HURTUBISE, R. J., Determination of Si in streptomycin by AA, *J. Pharm. Sci.*, 1974, **63**, 1128. (Quality Control Dept., Pfizer Inc., Terre Haute, Ind., U.S.A.).

147 WOIDICH, H., and PFANNHAUSER, W., Quantitative analysis of Hg in biological material. II: Digestion and AA determination in the vapour phase, *Z. Lebensm.–Unters.–Forsch.*, 1974, **155**, 271. (Address as in ref. 72).

148 HURTUBISE, R. J., Determination of disodium edetate dihydrate in streptomycin by AAS, *J. Pharm. Sci.*, 1974, **63**, 1131. (Address as in ref. 146).

149 BROWN, A., and HUSAIN, D., Kinetics study of electronically excited Sn atoms by AAS, *J. Photochem.*, 1974, **3**, 37. (Dept. of Physical Chemistry, University of Cambridge, Cambridge, England).

150 BELOUSOVA, I. M., BOBROV, B. D., KISELEV, V. M., and KURZENKOV, V. N., Induced emission of atomic I in pulsed magnetic fields, *Kvantovaya Elektron. (Moscow),* 1974, **1**, 1389.

151 ANKERSMIT, R., *et al.*, Standardisation of methods for the determination of traces of Hg. Part I: Determination of total inorganic Hg in organic samples, *Anal. Chim. Acta,* 1974, **72**, 37. (BITC, Brussels), Belgium).

152 SENITZKY, B., Optical filter using a vapour mirror, *Appl. Phys. Lett.*, 1974, **24**, 68.)Polytechnique Institute of New York, Farmingdale, N.Y., U.S.A.).

153 URE, A. M., The determination of Hg by non-flame AA and AFS, *Anal. Chim. Acta,* 1975, **76**, 1. (The Macaulay Institute for Soil Research, Craigiebuckler, Aberdeen, Scotland).

154 MIKA, J., and TOROK, T., Analytical ES — Book published by Crane, Russak and Co. Ltd., New York, 1974.

155 BALTISBERGER, R. J., and KNUDSON, C. L., The differentiation of submicrogram amounts of inorganic and organo-Hg in water by flameless AAS, *Anal. Chim. Acta,* 1974, **73**, 265. (University of North Dakota, Grand Forks, N.Dak. 58201, U.S.A.).

156 HENN, E. L., Determination of trace metals in polymers by flameless AA with a solid sampling technique, *Anal. Chim. Acta,* 1974, **73**, 273. (Calgon Corporation, Box 1346, Pittsburg, Pa. 15230, U.S.A.).

157 ATSUYA, I., Determination of Cr in iron and steels by u.h.f. plasma-torch spectrometry, *Anal. Chim. Acta,* 1975, **74**, 1. (Kitami Institute of Technology, Kitami, Japan).

158 DHUMWAD, R. K.,* and RAMACHANDRAN, S.,† A special holder for the Jarrell-Ash spectrometer 750V for the analysis of steel rod samples, *Anal. Chim, Acta,* 1975, **74**, 11. (*Fuel Processing Division, Bhabha Atomic Research Centre, Trombay, Bombay, India; †Steel Melt Shop, Mukand Iron and Steel Works Ltd., Kalwe, Thana, India).

159 BROOKS, R. R., and SMYTHE, L. E., Trends in AAS, *Anal. Chim. Acta,* 1975, **74**, 35. (School of Chemistry, University of New South Wales, Sydney, N.S.W., Australia).

160 ROBINSON, J. W., and WOLCOTT, D. K., A hollow-T carbon atomizer for AAS,

Anal. Chim. Acta, 1975, **74**, 43. (Dept. of Chemistry, Louisiana State University, Baton Rouge, La. 70803, U.S.A.).

161 TALMI, Y., The rapid sub-picogram determination of volatile organo-Hg compounds by gas chromatography with a microwave emissive spectrometric detector system, *Anal. Chim. Acta,* 1975, **74**, 107. (Analytical Chemistry Div., Oak Ridge National Laboratory, Oak Ridge, Tenn. 37830, U.S.A.).

162 HOLAK, W., Analysis of paints for Pb by AAS, *Anal. Chim. Acta,* 1975, **74**, 216. (Food and Drug Administration, 850 Third Avenue, Brooklyn, N.Y. 11232, U.S.A.).

163 ANON., Furnace atomizer for AAS, *Lab. Equip. Dig.*, 1975, **13**, 97.

164 DRESNER, S., It's superspray, *Pop. Sci.,* 1975, **1**, 45.

165 JULEFF, C. M., Veryfying purity of chemicals, *Electron. Prod.,* 1975, (Feb.), 34. (Micro-Image Technology Ltd.).

166 JOHANSEN, G., Elimination of the effect of acetone vapour in AA analysis, *At. Absorpt. Newsl.,* 1975, **14**, 44. (Dept. of Clinical Chemistry, Glostrup Hospital, DK–2600, Glostrup, Denmark).

167 KORREY, J. S., and GOLDEN, P. D., The determination of microgram quantities of W in natural waters by solvent extraction and AAS, *At. Absorpt. Newsl.,* 1975, **14**, 33. (Canada Centre for Inland Waters, Department of Environment, Burlington, Ont. L7R 4A6, Canada).

168 HARRINGTON, D. E., and BRAMSTEDT, W. R., The determination of Sn, Sb and Ta in the presence of precious metals by AAS, *At. Absorpt. Newsl.,* 1975, **14**, 36. (T. R. Evans Research Center, Diamond Shamrock Corp., Painesville, Ohio 44077, U.S.A.).

169 BRACA, G.,* SBRANA, G.,* SCANDIFFIO, G.,* and CIONI, R.,† A simple and rapid determination of Ru in organic compounds by AAS, *At. Absorpt. Newsl.,* 1975, **14**, 39. (*Instituto di Chimica Organica Industriale, and †Instituto di Mineralogia e Petrografia, Universita di Pisa, Pisa, Italy).

170 MURPHY, J.,* and STOCKTON, H.,† Mg interference in the AA determination of Pb, *At. Absorpt. Newsl.,* 1975, **14**, 40. (*Geology Dept., and †Chemistry Dept., University of Wyoming, Laramie, Wyo. 82071, U.S.A.).

171 SINGH, N. P., and JOSELOW, M. M., Determination of Tl in whole blood by Delves cup AAS, *At. Absorpt. Newsl.,* 1975, **14**, 42. (Address as in ref. 28).

172 PRICE, W. J., AA: an essential tool in modern metallurgy, *Met. Mater.,* 1974. **8**, 485. (Pye Unicam Ltd., York Street, Cambridge, England).

173 GREENFIELD, S., JONES, I. L., McGEACHIN, H. M., and SMITH, P. B., Automatic multi-sample, simultaneous, multi-element analysis with a h.f. plasma torch and direct reading spectrometer, *Anal. Chim. Acta,* 1975, **74**, 225. (Address as in ref. 104).

174 EL-KHOLY, H. K., BURRIDGE, J. C., and SCOTT, R. O., A triple flow gas-sheathed d.c. arc for spectrochemical analysis, *Anal. Chim. Acta,* 1975, **74**, 247. (Macaulay Institute for Soil Research, Craigiebuckler, Aberdeen AB9 2QJ, Scotland).

175 EBDON, L., HUBBARD, D. P., and MICHEL, R. G., Studies in AFS, III: The determination of Sn in steels, *Anal. Chim. Acta,* 1975, **74**, 281. (Dept. of Chemistry and Biology, Sheffield Polytechnic, Sheffield S1 1WB, England).

176 CHOW, A., and LIPINSKY, W., The determination of Ga by AAS, *Anal. Chim. Acta,* 1975, **75**, 87. (Dept. of Chemistry, University of Manitoba, Winnipeg, Man. R3T 2N2, Canada).

177 THOMPSON, K. C., GODDEN, R. G., and THOMERSON, D. R., A method for the formation of pyrolytic graphite coatings and enhancement by Ca addition techniques for graphite rod flameless AAS, *Anal. Chim. Acta,* 1975, **74**, 289. (Address as in ref. 19).

178 JACKWERTH, E., and BERNDT, H., Determination of trace heavy metals in alkali and alkaline salts by AAS after concentration with active carbon, *Anal. Chim. Acta,* 1975, **74**, 299. (Address as in ref. 58).

179 MURUGAIYAN, P., and NATARAJAN, S., Varian-Techtron burners in AFS, *Anal. Chim. Acta,* 1975, **75**, 217. (Analytical Chemistry Div., Bhabha Atomic Research Centre, Bombay–400 085, India).

180 MURUGAIYAN, P., NATARAJAN, S., and VENKATESWARLU, C., Direct determination of Zn in high-purity materials by AFS, *Anal. Chim. Acta,* 1975, **75**, 221. (Address as in ref. 179).

181 SCOTT, R. H.,* and KOKOT, M. L.,† Application of inductively coupled plasmas to the analysis of geochemical samples, *Anal. Chim. Acta,* 1975, **75**, 257. (*National

Physical Research Laboratory, C.S.I.R., Pretoria, South Africa; †U.S. Steel International (New York) Inc., Raw Materials Investigation Div., Sandton, South Africa).

182 MERMET, J.–M., and ROBIN, J., Interference studies in a h.f. plasma, *Anal. Chim. Acta,* 1975, **75**, 271. (Laboratoire de Chimie Industrielle et Analytique, Bat. 401, I.N.S.A., 69621–Villeurbanne, France).

183 WATLING, R. J., The determination of Hg at picogram per litre levels in H_2O with a microwave-induced argon plasma emission system, *Anal. Chim. Acta,* 1975, **75**, 281. (Applied Spectroscopy Div., National Physical Research Laboratory, C.S.I.R., P.O. Box 395, Pretoria, South Africa).

184 GRIES, W. H., and NORVAL, E., New solid standards for the determination of trace impurities in metals by flameless AAS, *Anal. Chim. Acta,* 1975, **75**, 289. (National Physical Research Laboratory, C.S.I.R., P.O. Box 395, Pretoria, South Africa).

185 AGEMIAN, H., and CHAU, A.S.Y., A method for the determination of Hg in sediments by the automated cold vapour AA technique after digestion, *Anal. Chim. Acta,* 1975, **75**, 297. (Water Quality Laboratory, Canada Centre for Inland Waters, Burlington, Ont. L7R 4A6, Canada).

186 NAKAHARA, T., and MUSHA, S., The determination of Ga by AAS in premixed inert gas (entrained-air)/hydrogen flames, *Anal. Chim. Acta,* 1975, **75**, 305. (Dept. of Applied Chemistry, College of Engineering, University of Osaka Prefecture, Mozu-umemachi, Sakai 591, Japan).

187 BEYER, M. E., and BOND, A. M., Simultaneous determination of Cd, Cu, Pb and Zn in Pb and Zn concentrates by a.c. polarographic methods. Comparison with AAS, *Anal. Chim. Acta,* 1975, **75**, 409. (Dept. of Inorganic Chemistry, University of Melbourne, Parkville, Vic. 3052, Australia).

188 WORRELL, G. J., VICKERS, T. J., and WILLIAMS, F. D., A solvent extraction AF system for the determination of Cd in complex samples, *Anal. Chim. Acta,* 1975, **75**, 453. (Address as in ref. 90).

189 RADCLIFFE, D. B., BYFORD, C. S., and OSMAN, P. B., The determination of As, Sb and Sn in steels by flameless AAS, *Anal. Chim. Acta,* 1975, **75**, 457. (Central Electricity Generating Board, Marchwood Engineering Laboratories, Marchwood, Southampton, Hampshire, England).

190 KALRA, Y. P., and RADFORD, F. G., Suitability of ammonium EDTA extraction procedure for determining Ca in tree foliage, *Commun. Soil Sci. Plant Anal.,* 1975, **6**, 13. (Northern Forest Research Centre, Environment Canada, 5320–122 Street, Edmonton, Alta., T6H 3S5, Canada).

191 KORKISCH, J., and SORIO, A., Application of ion exchange methods to the determination of trace elements in natural waters. Part V: Pb, *Talanta,* 1975, **22**, 273. (Analytisches Institut der Universitat, Abteilung Rohmaterialanalyse Nuklearer Brennstoffe, Wahringerstrasse 38, A–1090 Vienna, Austria).

192 KORKISCH, J., GODL, L., and GROSS, H., Applications of ion-exchange methods to the determination of trace elements in natural waters. Part VI: Zn, *Talanta,* 1975, **22**, 281, (Address as in ref. 191).

193 KORKISCH, J., GODL, L., and GROSS, H., Applications of ion-exchange methods to the determination of trace elements in natural waters. Part VII: Cu, *Talanta,* 1975, **22**, 289. (Address as in ref. 191).

194 ODDO, N., and VITALI, A., New instrumental systems in AAS applied to toxic elements analysis. Determination of As with more advanced sampling systems, *Chim. Ind. (Milan),* 1975, **57**, 93. (Perkin Elmer, Milan, Italy).

195 CERNIK, A. A., and SAYERS, M. H. P., Application of blood Cd determination to industry using a punched disc technique, *Br. J. Ind. Med.,* 1975, **32**, 155. (Central Reference Laboratory, Health and Safety Executive, Chepstow Place, London, W.2, England).

196 MARCUS, M., HOLLANDER, M., LUCAS, R. E., and PFEIFFER, N. C., Micro-scale blood Pb determinations in screening: evaluation of factors affecting results, *Clin. Chem.,* 1975, **21**, 533. (Dept. of Pathology and Pediatrics, Fordham Hospital, Bronx, N.Y., U.S.A.).

197 EVENSON, M. A., and ANDERSON, C. T., Ultramicro analysis for Cu, Cd and Zn in human liver tissue by use of AAS and the heated graphite tube atomizer, *Clin. Chem.,* 1975, **21**, 537. (Dept. of Medicine, Pathology and Toxicology, University of Wisconsin, Madison, Wis. 53706, U.S.A.).

198 LAUWERYS, R., BUCHET, J. P., ROELS, H., BERLIN, A., and SMEETS, J.,

Intercomparison programme of Pb, Hg and Cd. Analysis in blood, urine and aqueous solutions, *Clin. Chem.*, 1975, **21**, 551. (Unite de Toxicologie Industrielle et Medicale, University of Louvain, Brussels, Belgium).

199 FERNANDEZ, F. J., Micromethod for Pb determination in whole blood by AAS with use of the graphite furnace, *Clin. Chem.*, 1975, **21**, 558. (Perkin-Elmer Corp., Norwalk, Conn., U.S.A.).

200 ZACHARIASEN, H., ANDERSON, I., KOSTOL, C., and BARTON, R., Technique for determining Ni in blood by flameless AA, *Clin. Chem.*, 1975, **21**, 562. (Falconbridge Health Centre, Falconbridge Nikkelverk Aktieselskap, Kristiansand, Norway).

201 MOMCILOVIC, B., BELONJE, B., and SHAH, B. A., Effect of the matrix of the standard on results of AAS of Zn in serum, *Clin. Chem.*, 1975, **21**, 588. (Bureau of Nutritional Sciences, Health Protection Branch, Tunney's Pasture, Ottawa, Ont. K1A 0L2, Canada).

202 EVENSON, M. A., and WARREN, B. L., Determination of serum Cu by AA with the use of the graphite cuvette, *Clin. Chem.*, 1975, **21**, 619. (Address as in ref. 197).

203 PERRY, E. F., KOIRTYOHANN, S. R., and PERRY, H. M., Determination of Cd in blood and urine by graphite furnace AAS, *Clin. Chem.*, 1975, **21**, 626. (Environmental Trace Substances Center, University of Missouri, Columbia, Mo. 65201, U.S.A.).

204 KILROE-SMITH, T. A., Linear working graphs in blood Pb determinations with the Beckman flameless AA cuvette, *Clin. Chem.*, 1975, **21**, 630. (National Research Institute, Occupational Diseases, South African Medical Research Council, P.O. Box 4788, Johannesburg 2000, South Africa).

205 CHOOI, M. K., TODD, J. K., and BOYD, N. D., Effect of carbon cup ageing on plasma Zn determination by flameless AAS, *Clin. Chem.*, 1975, **21**, 632. (Chemical Pathology Division, Foothills Hospital, Calgary, Alta. T2N 2T9, Canada).

206 BRASSEN, P., and MAESSEN, F. J. M. J., Electron temperatures and electron concentrations at low pressure in microwave induced plasmas, *Spectrochim. Acta, Part B*, 1974, **29*B***, 203. (Laboratory for Analytical Chemistry, University of Amsterdam, Nieuwe Achtergracht 166, Amsterdam, The Netherlands).

207 WAGENAAR, H. C., PICKFORD, C. J., and De GALAN, L., The interferometric measurement of AA line profiles in flames, *Spectrochim. Acta, Part B*, 1974, **29*B***, 211. (Laboratorium voor Instrumentele Analyse, Technische Hogeschool, Delft, The Netherlands).

208 HOHN, R., JACKWERTH, E., and KOOS, K., Determination of trace elements in high purity Al by AAS: trace enrichment by partial solution of the matrix in the presence of Hg, *Spectrochim. Acta, Part B*, 1974, **29*B***, 225. (Address as in ref. 58).

209 SZABO, Z. L., Oxidation and reduction tendencies on the electrode surfaces during arc excitation, *Spectrochim. Acta, Part B*, 1974, **29*B***, 231. (Institute of Inorganic and Analytical Chemistry, L. Eatvos University, Budapest, Hungary).

210 MACQUET, J. P., and THEOPHANIDES, T., Cis–trans effect and the influence of the structure of Pt complexes on AA, *Spectrochim. Acta, Part B*, 1974, **29*B***, 241. (Address as in ref. 9).

211 DECKER, R. J., and McFADDEN, P. A., Radial particle distribution in a d.c. arc, *Spectrochim. Acta, Part B*, 1975, **30*B***, 1. (University of Rhodesia, Salisbury, Rhodesia).

212 HERMANN, P., Determination of Pb and Zn in dust by AA, *Spectrochim. Acta, Part B*, 1975, **30*B***, 15. (Rheinisch–Westfalisher Technischer, Uberwachungs–Verein, Essen, West Germany).

213 PERIC, M. N., TODOROVIC, P. S., and VUKANOVIC, V. M., Determination of the diffusion coefficients of substances in the plasma of a d.c. arc in air using a photometric method, *Spectrochim, Acta, Part B*, 1975, **30*B***, 21. (Prirodno–matematicki Fakultet, Tehnolosko–metalurski Fakultet, Institut za Fiziku, Belgrade, Yugoslavia).

214 VUKANOVIC, V. M., A new equipment for mass transport study in a free-burning d.c. arc in air, *Spectrochim. Acta, Part B*, 1975, **30*B***, 31. (Address as in ref. 213).

215 NEUFELD, L., and SCHRENK, W. G., An extended analysis of the arc spectrum of Sc, *Spectrochim. Acta, Part B*, 1975, **30*B***, 45. (Address as in ref. 87).

216 STEPHENS, R., and STEVENSON, R. G., jun., Some observations on the interdependence of CN concentration and the atomization efficiency in the nitrous

oxide–acetylene flame, *Spectrochim. Acta, Part B*, 1975, **30***B*, 61. (Trace Analysis Research Centre, Dept. of Chemistry, Dalhousie University, Halifax, N.S., Canada).

217 URBAIN, H., and DESQUESNES, W., Some observations on the background continuum in spectrophotometric analysis with some Na rich flames, *Spectrochim. Acta, Part B*, 1975, **30***B*, 71. (C.N.R.S., Institut de Recherches sur la Catalyse, 69626–Villeurbanne, France).

218 THOMAS, C. P., An integrated-intensity method for ES computer analysis, *J. Res. U.S. Geol. Surv.*, 1975, **3**, 181. (957 National Center, U.S. Geological Survey, Reston, Va. 22092, U.S.A.).

219 O'HAVER, T. C., and GREEN, G. L., Derivative spectroscopy, *Am. Lab.*, 1975, **7**(3), 15. (Dept. of Chemistry, University of Maryland, College Park, Md. 20742, U.S.A.).

220 KERBER, J. D., AA grows up: for rapid analysis of trace materials in small samples *Ind. Res.*, 1975, **17**, 73. (Perkin–Elmer Corp., Norwalk, Conn., U.S.A.).

221 TALMI, Y., and ANDREW, A. W., Determination of Se in environmental samples using gas chromatography with a microwave ES detection system, *Anal. Chem.*, 1974, **46**, 2122. (Address as in ref. 161).

222 MONTASER, A., and CROUCH, S. R., New methods for programmed heating of electrically heated non-flame atom vapour cells, *Anal. Chem.*, 1975, **47**, 38. (Address as in ref. 99).

223 PRAGER, M. J.,* and SEITZ, W. R.,† FE photometer for determining P in air and natural waters, *Anal. Chem.*, 1975, **47**, 148. (*Nucor Corporation, Denville, N.J. 07834, U.S.A.; †N.E.R.C. Corvallis, Environmental Protection Agency, Athens, Ga. 30602, U.S.A.).

224 FREED, D. J., Flame photometric detector for liquid chromatography, *Anal. Chem.*, 1975, **47**, 186. (Bell Laboratories, Murray Hill, N.J. 07974, U.S.A.).

225 BOSCH, F. M., and BROEKAERT, J. A. C., Alternate method for the calculation of the detection limit in ES, *Anal. Chem.*, 1975, **47**, 188. (Laboratory for Inorganic Technical Chemistry, State University of Ghent, Grote Steenweg Noord 12, B–9710 Zwijnaarde, Belgium).

226 LAYMAN, L. R., and HIEFTJE, G. M., New, computer-controlled microwave discharge emission spectrometer employing microarc sample atomization for trace and micro elemental analysis, *Anal. Chem.*, 1975, **47**, 194. (Dept. of Chemistry, Indiana University, Bloomington, Ind. 47401, U.S.A.).

227 GRIFFIN, H. R., HOCKING, M. B., and LOWERY, D. G., As determination in tobacco by AAS, *Anal. Chem.*, 1975, **47**, 229. (Dept. of Chemistry, University of Victoria, Victoria, B.C. V8W 2Y2, Canada).

228 LARSON, G. F., FASSEL, V. A., SCOTT, R. H., and KNISELEY, R. N., Inductively coupled plasma–optical emission analytical spectrometry. A study of some inter-element effects, *Anal. Chem.*, 1975, **47**, 238. Address as in ref. 76).

229 NOGA, R. J., Determination of total Cr in coatings by AAS, *Anal. Chem.*, 1975, **47**, 332. (Research Center, De Soto Incorporated, Des Plaines, Ill. 60018, U.S.A.).

230 COKER, D. T., Determination of individual and total Pb alkyls in gasoline by a simple rapid gas chromatography/AAS technique, *Anal. Chem.*, 1975, **47**, 386. (Esso Research Centre, Abingdon, Oxfordshire, England).

231 HILDERBRAND, D. C., and PICKETT, E. E., Determination of the volatility of metal chelates by AA, *Anal. Chem.*, 1975, **47**, 424. (Dept. of Chemistry, University of Missouri, Columbia, Miss. 65201, U.S.A.).

232 HENN, E. L., Determination of Se in water and industrial effluents by flameless AA, *Anal. Chem.*, 1975, **47**, 428. (Address as in ref. 156).

233 GRIME, J. K., and VICKERS, T. J., Determination of Li in microlitre samples of blood serum using FES with a tantalum filament vaporizer, *Anal. Chem.*, 1975, **47**, 432. (Address as in ref. 90).

234 RIPPETOE, W. E., JOHNSON, E. R., and VICKERS, T. J., Characterization of the plume of a d.c. plasma arc for ES analysis, *Anal. Chem.*, 1975, **47**, 436. (Address as in ref. 90).

235 HOLCOMBE, J. A., BRINKMAN, D. W., and SACKS, R. D., Fortran-based photographic emulsion calibration procedure for use in quantitative spectrometry, *Anal. Chem.*, 1975, **47**, 441. (Address as in ref. 78).

236 SIMMONS, W. J., and LONERAGAN, J. F., Determination of Cu in small amounts of plant material by AAS using a heated graphite atomizer, *Anal. Chem.*,

1975, **47**, 566. (Dept. of Soil Science and Plant Nutrition, Institute of Agriculture, University of Western Australia, Nedlands, Western Australia).

237 SKOGERBOE, R. K., DICK, D. L., PAVLICA, D. A., and LICHTE, F. E., Injection of samples into flames and plasmas by production of volatile chlorides, *Anal. Chem.*, 1975, **47**, 568. (Address as in ref. 85).

238 LYTLE, F. E., ENG, J. F., HARRIS, T. D., and SANTINI, R. E., Spectroscopic excitation source with variable frequencies and shapes of modulation, *Anal. Chem.*, 1975, **47**, 571. (Dept. of Chemistry, Purdue University, Lafayette, Ind. 47907, U.S.A.).

239 HAWLEY, J. E., and INGLE, J. D., jun., Improvements in cold-vapour AA determination of Hg, *Anal. Chem.*, 1975, **47**, 719. (Address as in ref. 67).

240 SCHLEICHER, R. G., and BARNES, R. M., Remote coupling unit for r.f. inductively coupled plasma discharges in spectrochemical analysis, *Anal. Chem.*, 1975, **47**, 724. (Address as in ref. 81).

241 JACOB, R. A.,* and KLEVAY, L. M.,† Determination of trace amounts of Cu and Zn in edible fats and oils by acid extraction and AAS, *Anal. Chem.*, 1975, **47**, 741. (*Dept. of Biochemistry, University of North Dakota; † U.S. Department of Agriculture, Agricultural Research Service, Human Nutrition Laboratory, Grand Forks, N.Dak. 58201, U.S.A.).

242 FUJIWARA, K., HARAGUCHI, H., and FUWA, K., Response surface and atomization mechanism in air-acetylene flames, *Anal. Chem.*, 1975, **47**, 743. (Dept. of Agricultural Chemistry, University of Tokyo, Tokyo 113, Japan).

243 BUTLER, C. C., KNISELEY, R. N., and FASSEL, V. A., Inductively coupled plasma optical ES: application to the determination of alloying and impurity elements in low and high alloy steels, *Anal. Chem.*, 1975, **47**, 825. (Address as in ref. 76).

244 GUTZLER, D. E., and DENTON, M. B., Improvements in FES through the use of ultrasonic nebulization into a premixed oxygen/hydrogen flame, *Anal. Chem.*, 1975, **47**, 830. (Dept. of Chemistry, University of Arizona, Tucson, Ariz. 85721, U.S.A.).

245 POSMA, F. D., BALKE, J., HERBER, R. F. M., and STUIK, E. J., Microdetermination of Cd and Pb in whole blood by flameless AAS using carbon-tube and carbon-cup as sample cell and comparison with flame studies, *Anal. Chem.*, 1975, **47**, 834. (Laboratory for Analytical Chemistry, University of Amsterdam, Amsterdam, The Netherlands).

246 TORSI, G., and TESSARI, G., Time resolved distribution of atoms in flameless spectrometry: recovery of the source parameters from the response function, *Anal. Chem.*, 1975, **47**, 839. (Instituto di Chimica, Universita'degli Studi di Bari, Via G. Amendola, 173–70126, Bari, Italy).

247 TESSARI, G., and TORSI, G., Time resolved distribution of atoms in flameless spectrometry: experimental, *Anal. Chem.*, 1975, **47**, 842. (Address as in ref. 246).

248 CULVER, B. R., and SURLES, T., Interference of molecular spectra due to alkali halides in non-flame AAS, *Anal. Chem.*, 1975, **47**, 920. (Varian Instrument Div., 611 Hansen Way, Palo Alto, Calif. 94303, U.S.A.).

249 BOTHNER, M. H.,* and ROBERTSON, D. E.,† Hg contamination of sea water samples stored in polythene containers, *Anal. Chem.*, 1975, **47**, 592. (*Dept. of Oceanography, University of Washington, Seattle, Washington, D.C., U.S.A.; †Battelle–North West Laboratory, Richland, Washington, D.C., U.S.A.).

250 RONCO, A. E., and MERODIO, J. C., Interference in the determination of Ba by AAS: application in the analysis of calcareous rocks, *An. Asoc. Quim. Argent.*, 1974, **62**, 223. (Fac. Cienc. Nat. Mus., University Nac. La Plata, La Plata, Argentina).

251 THISTLEWAITE, P. J., and TREASE, M., Determination of Hg by a simple AA method, *J. Chem. Educ.*, 1974, **51**, 687. (Melbourne University, Parkville, Vic. 3052, Australia).

252 HEYN, A.,* OTTENDORFER, L. J.,† EBNER, F.,† and GAMS, H.,† Determination of heavy metals in the Danube near Vienna using flame AA, *Oesterr. Abwasser–Rundsch.*, 1974, **19**, 41. (*Dept. of Chemistry, Boston University, Boston, Mass., U.S.A.; †Bundesanstalt fur Wasserbiologie und Abwasserforschung, A–1223 Vienna, Postfach 7, Austria).

253 KNEIP, T. J., *et al.*, Tentative method of analysis for elemental Hg in ambient air by collection on Ag wool and AAS, *Health Lab. Sci.*, 1974, **11**, 342.

254 KERSTAN, W., LIMPENS, M. J., and HEUSER, H. J., Si determination in

ceramics by AA analysis, *Ber. Dtsch. Keram. Ges.*, 1974, **51**, 256. (Zentrallab., Wessel–Werk GmbH, Bonn, West Germany).

255 FAZAKAS, J., Suggestions for the nomenclature of AAS in the Roumanian Language, *Rev. Chim. (Bucharest)*, 1974, **25**, 399.

256 HURLBUT, J. A., GILBERT, L. K., BUDDINGTON, B. N., and REES, T. F., Determination of trace quantities of Cr and Mn in steel, ore and liquid samples by AAS, *J. Chem. Educ.*, 1974, **51**, 734. (Metropolitan State College, Denver, Colo., U.S.A.).

257 MAEHATA, E., KANOHDA, Y., HIRAI, T., and NAKA, H., Determination of serum Li by flame AA, *Eisei Kensa*, 1974, **23**, 779. (Mitsui Memorial Hospital, Tokyo, Japan).

258 LE BIHAN, A., and COURTOT-COUPEZ, J., Determination of traces of anionic detergents in fresh water using AAS, *Analusis*, 1974, **2**, 695. (Laboratoire de Chimie Analytique, Universite de Bretagne Occidentale, 6 Avenue Le Gorgeu, 29283 Brest–Cedex, France).

259 FREI, J., Flame photometry, pp. 254-259 *in* Curtius, H. C., and Roth, M., *Editors*, Clinical biochemistry: principles and methods, Vol. 1 — Book published by Walter de Gruyter, Berlin and New York, 1974. (Lab. Cent., Hop. Cantonal Univ., Lausanne, Switzerland).

260 ROMANOV, I. P., Determination of K and Na in blood using a flame photometer, *Lab. Delo*, 1974, (7), 438. (Sverdlovsk. Nauchno-Issled. Inst. Kurortol., Sverdlovsk, U.S.S.R.).

261 BOEK, K., Determination of Hg compounds in fish and fish products by application of the flask combustion Schoeniger method, *Z. Lebensm.–Unters. Forsch.*, 1974, **155**, 209. (Chem. Lebensmitteluntersuchungsanst., Hamburg, West Germany).

262 WEINER, J., Institute of Brewing analysis committee: determination of Zn in beer by AAS, *J. Inst. Brew.*, 1974, **80**, 486.

263 FALCHUK, K. H., EVENSON, M., VALLEE, B. L., and MATTHEWS, J. M., Multichannel AA instrument: Simultaneous analysis of Zn, Cu and Cd in biological materials, *Anal. Biochem.*, 1974, **62**, 255. (Dept. of Biological Chemistry, Harvard Medical School, Boston, Mass., U.S.A.).

264 BRANDENBERGER, H., AA, pp. 260-273 *in* Curtius, H. C., and Roth, M., *Editors*, Clinical biochemistry: principles and methods, Vol. 1 — Book published by Walter de Gruyter, Berlin and New York, 1974. (Chem. Dept., Inst. Forensic Med., Zurich, Switzerland).

265 HASAN, M. Z., SETH, T. D., and SHARMA, R., Determination of Cd, Co, Cu, Pb, Mn, Ni and Zn in animal feed by AAS, *Chemosphere*, 1974, **3**, 241. (Indian Toxicological Research Centre, Lucknow, India).

266 ONIANI, O. G., and EGORASHVILI, N. V., Determination of available Mn in krasnozem soils of the western Georgian S.S.R., *Agrokhimiya*, 1974, (9), 137. (Tbilis. Fil., Tsentr. Inst. Agrokhim. Obsluzhivaniya Sel'sk. Khoz., Tbilis, U.S.S.R.).

267 ELFBAUM, S. G., JULIANO, R., MacFARLAND, R. E., and PFEIL, D. L., Blood Pb determination by AAS with use of the Delves sampling-cup technique: effect of various coagulants, *Clin. Chem.*, 1974, **18**, 316. (Boston Medical Laboratory, Boston, Mass., U.S.A.).

268 OELSCHLAEGER, W., and BUEHLER, E., Determination of Cd in biological and other materials using AAS. II: Sources of error and their elimination in the AAS determination, *Landwirtsch. Forsch.*, 1974, **27**, 70. (Abt. Tierernachr., University of Hohenheim, Stuttgart–Hohenheim, West Germany).

269 OELSCHLAEGER, W., and BESTENLEHEHNER, L., Determination of Cd in biological and other materials using AAS. I: Sources of error and their elimination in ashing and preparing of the analytical solutions, *Landwirtsch. Forsch.*, 1974, **27**, 62. (Address as in ref. 268).

270 VELGHE, G., VERLOO, M., and COTTENIE, A., Determination of Pb in milk and butter by flameless AA, *Z. Lebensm.–Unters. Forsch.*, 1974, **156**, 77. (Faculty of Agricultural Sciences, State University, Ghent, Belgium).

271 MUECK, G., Continuous emission of an $AlCl_3$ arc plasma, *Z. Naturforsch., Teil A*, 1974, **29**, 1643. (Osram–Forsch., Munich, West Germany).

272 MORIE, G. P., and MORRISETT, P. E., Determination of transition metals in cigarette smoke condensate by solvent extraction and AAS, *Beitr. Tabakforsch.*, 1974, **7**, 302. (Research Laboratory, Tennessee Eastman Co., Div. Eastman Kodak Co., Kingsport, Tenn., U.S.A.).

273 MANOLIU, C., and TOMI, B., Ti behaviour in a nitrous oxide acetylene flame in its determination by AAS, *Rev. Chim. (Bucharest)*, 1974, **25**, 582.
274 SADYKOV, R. S., and AIDAROV, T. K., Apparatus for AAS, U.S.S.R. Patent 432,351.
275 BALLA, K. Z., HARSANYI, E. G., POLOS, L., and PUNGOR, E., Determination of metals of low concentration in high purity Cu by AAS, *Mikrochim. Acta*, 1975, 107. (Cent. Res. Inst. Phys., Technical University of Budapest, Budapest, Hungary).
276 OTTO, J., and CZYGAN, W., X-ray fluorescence and AA analyses of geochemical reference samples, *Neues Jahrb. Mineral., Monatsh.*, 1974, (11), 481. (Mineral. Inst., University of Freiburg, Freiburg, West Germany).
277 COROMINAS, L. F., Addendum to flame photometric determination of Na in fertilizers, *J. Assoc. Off. Anal. Chem.*, 1974, **57**, 1402. (Guanos y Fertilizantes de Mexico, S.A., Mexico 12, Mexico).
278 WHITE, G. E., COONEY, C. L., SINSKEY, A. J., and MILLER, S. A., In vitro assay to measure early Ca loss from surface enamel, *J. Dent. Res.*, 1974, **53**, 481. (Dept. of Nutrition and Food Science, Massachusetts Institute of Technology, Cambridge, Mass., U.S.A.).
279 WOODWARD, P. W., and PEMBERTON, J. R., Analysis of Hg preservatives in bacterins, vaccines and antiserums by AAS, *Appl. Microbiol.*, 1974, **27**, 1094. (Veterinary Service Laboratory, A.P.H.I.S., Ames, Iowa, U.S.A.).
280 OPALOVSKII, A. A., TYULENEVA, N. I., and ZEMSKOV, S. V., Useof ClF_3 during the analysis of lean Au-containing ores, *Zh. Prikl. Khim.*, 1974, **47**, 2157. (Inst. Neorg. Khim., Novosibirsk, U.S.S.R.).
281 JOHANSON, R., Determination of Cu, Cr, As and Zn in H_2SO_4/HNO_3 digested extracts of preserved wood by AAS with reference to As in Karri rail sleepers, *Holzforschung*, 1974, **28**, 117. (For. Prod. Lab., C.S.I.R.O., South Melbourne, Australia).
282 MAURER, L., Rapid, simple procedure for the determination of Cu in cheese with a graphite furnace, *Z. Lebensm.–Unters. Forsch.*, 1974, **156**, 284. (Institut fur Milchwirtschaft und Mikrobiologie, Hochschule fur Bodenkultur, Vienna, Austria).
283 WALTHALL, F. G., Spectrochemical computer analysis: programme description, *J. Res. U.S. Geol. Surv.*, 1974, **2**, 61.
284 ARPADJAN, S., DOERFFEL, K., HOLLAND-LETZ, K., MUCH, H., and PANNACH, M., Statistical optimisation of analytical chemistry conditions, *Fresenius' Z. Anal. Chem.*, 1974, **270**, 257. (Technische Hochschule fur Chemie 'Carl Schorlemmer', Leuna–Merseburg, East Germany).
285 ILLNER, E., Use of flame spectrophotometric analysis for trace-element determinations, *Z. Chem.*, 1974, **14**, 351. (V.E.B. Jenapharm, Jena, East Germany).
286 LORENZ, K., REUTER, F. W., and SIZER, C., Mineral composition of triticales and triticale milling fractions by x-ray fluorescence and AA, *Cereal Chem.*, 1974, **51**, 534. (Dept. of Food Science and Nutrition, Colorado State University, Fort Collins, Colo., U.S.A.).
287 YUDELEVICH, I. G., and SHABUROVA, V. P., Extraction–AA determination of Mo, W and Re, *Chem. Anal. (Warsaw)*, 1974, **19**, 941. (Institute for Inorganic Chemistry, Acad. Sci. U.S.S.R., Siberian Dept., Novosibirsk, U.S.S.R).
288 KING, H. G., In situ characterisation of common impurities in U metal by spark spectrochemistry, *U.S. At. Energy Comm.*, 1974, Y–1937. (Oak Ridge Y–12 Plant, Oak Ridge, Tenn., U.S.A.).
289 ALVAREZ-ALDUAN, F., and CAPDEVILA PEREZ, C., Spectrographic determination of Zr, Nb, Rh, Ru, Ta and W in U and its compounds, *An. Quim.*, 1974, **70**, 601. (Junta de Energia Nucl., Div. Quim. Analit., Madrid, Spain).
290 BEGAK, O. Y., NICKOLAEV, G. I., and POKROVSKAYA, K. A., Determination of Cu in steel by AA, *Zh. Prikl. Khim.*, 1974, **47**, 1711.
291 CHANDOLA, L. C., KARANJIKAR, N. P., and DIXIT, V. S., Intermittent a.c. arc method for spectrographic analysis of K fluorotantalate, *Chem. Anal. (Warsaw)*, 1974, **19**, 749. (Spectroscopy Div., Bhabha Atomic Research Centre, Trombay, Bombay, India).
292 JOSHI, B. D., BANGIA, T. R., and DALVI, A. G. I., Spectrographic determination of trace impurities in S, *Mikrochim. Acta*, **1974**, 829. Radiochemistry Div., Bhabha Atomic Research Centre, Trombay, Bombay, India).
293 GORYCZKA, J., Spectrographic determination of low Nb contents in low and high

alloy steels, *Chem. Anal. (Warsaw),* 1974, **19**, 949. (Abt. Analyt. Chem., Eisenforschungsinst., Gliwice, Poland).

294 DITTRICH, K., and NEIBERGALL, K., ES determination of Si, Al and Sn in scheelite ($CaWO_4$), *Chem. Anal. (Warsaw),* 1974, **19**, 921. (Address as in ref. 107).

295 FAITHFULL, N. T., Conversion of the Technicon Model II flame photometer to pre-mix burner operation, *Lab. Pract.,* 1974, **23**, 429. (Agricultural Science Analytical Laboratory, Institute of Rural Science, Penglais, Aberystwyth, Wales).

296 BAKER, K. F., and VARLEY, J. A., Improvements in or relating to chemical analysis apparatus, British Patent 1,371,787.

297 PATENT, Chemical analyser, British Patent 1,369,146. (Rohe Scientific Corp.).

298 WINTER, E., Improvements in monochromators, British Patent 1,370,414.

299 GRAY, A. L., and HOSKIN, W. J., Improvements relating to the ES analysis of materials, British Patent 1,368,810.

300 PATENT, Apparatus for and method of analysis of liquid samples, British Patent 1,372,004. (Philps' Electronic and Associated Industries Ltd.).

301 BLACK, L. T., Comparison of three AA techniques for determining metals in soybean oil, *J. Amer. Oil Chem. Soc.,* 1975, **52**, 88. (Northern Regional Research Laboratory, Peoria, Ill. 61604, U.S.A.).

302 CROSSWHITE, H. M., The Fe-Ne hollow-cathode spectrum, *J. Res. Nat. Bur. Stand., Sect. A.,* 1975, **79A**, 17. (The John Hopkins University, Baltimore, Md. 21218, U.S.A.).

303 WELSCH, E. P., and CHAO, T. T., Determination of trace amounts of Sb in geological materials by AAS, *Anal. Chim. Acta,* 1975, **76**, 65. (U.S. Geological Survey, Denver, Colo. 80225, U.S.A.).

304 SCOTT, R. H., and STRASHEIM, A., Determination of trace elements in plant materials by inductively coupled plasma optical ES, *Anal. Chim. Acta,* 1975, **76**, 71. (Address as in ref. 181).

305 NAGULIN, Y. S., SMOLYOK, E. L., AFANAS'EV, V. A., and GIMUSHIN, I. F., Three-channel spectrometer for studying high-temperature flames, *Zh. Prikl. Spektrosk.,* 1974, **21**, 1070.

306 KUZNETSOV, A. P., KUKUSHKIN, Y. N., and MAKAROV, D. F., Use of Ni matte as a collector for the noble metals in analysis of poor materials, *Zh. Anal. Khim.,* 1974, **29**, 2155. (Norilsk Min. Metall. Group Enterp., Norilsk, U.S.S.R.).

307 NAKAGAWA, R., and OHYAGI, Y., Determination of Cr in soil and sea-sediment by AAS, *Nippon Kagaku Kaishi,* 1974, (12) 2331. (Faculty of Science, Chiba University, Chiba, Japan).

308 BODROV, N. V., and NIKOLAEV, G. I., Use of AAS to study the reaction of Al_2O_3 with C and to study high-temperature evaporation and diffusion processes of Al in high-melting metals, *Zh. Prikl. Spektrosk.,* 1974, **21**, 400.

309 BELCHER, R., BOGDANSKI, S. L., and TOWNSHEND, A., Cavity molecular emission spectroscopy: flame analysis technique, *Inf. Chim.,* 1974, No. 137, 197. (Address as in ref. 35).

310 SUBBER, S. W. FIHN, S. D., and WEST, C. D., Simplified apparatus for the flameless AF determination of Hg, *Am. Lab.,* 1974, **6**(11), 38. (Dept. of Chemistry, Occident College, Los Angeles, Calif., U.S.A.).

311 MALLETT, R. C., and KELLERMAN, S., Assessment of the carbon rod atomizer for the determination of Ag, *Natl. Inst. Metall., Repub. S. Afr., Rep.* No. 1669, 1974 (Natl. Inst. Metall., Milner Park, Johannesburg, S. Africa).

312 PARKER, C. R., ERICKSON, J. O., and CULVER, B. R., Prevention of flashbacks in flame AA, *Am. Lab.,* 1974, **6** (11), 85. (Address as in ref. 248).

313 HWANG, J. Y., and THOMAS, G. P., New generation of flameless AA atomizer, *Am. Lab.,* 1974, **6**(11), 42. (Applications Laboratory, Instrumentation Laboratory Inc., Lexington, Mass. 02173, U.S.A.).

314 SUZUKI, T., KONDO, I., TSUTSUMI, K., and TSUGI, N., Determination of traces Ca, Na and Mg in Zr alloys by AAS, *Tokai Jigyo-sho, Doryoku-do, Kakunenryo Kaihatsu Jigyo-dan, (Rep.),* 1974 PNCT 831–74–01, 62.

315 TAMM, R. G. A., and WITTE, W. W. F., Tubular, electrically conducting sampler and heating apparatus for flameless AAS, German Offen. 2,323,774. (Bodenseewerk Perkin-Elmer Co., GmbH).

316 SAUPE, K., Annual review of methods of chemical analysis, *Giesserei,* 1974, **61**, 784.

317 NIKOLAEV, G. I., and PODGORNAYA, V. I., Effectiveness of using a substance in a graphite cuvette during AA analysis, *Zh. Prikl. Spektrosk.,* 1974, **21**, 593.

318 ENDO, Y., and NAKAHARA, Y., AAS analysis of Fe and Steel, *Tetsu To Hagane,* 1974, **60**, 1787. (Mizushima Works, Kawasaki Steel Corp., Mizushima, Japan).
319 FOURNIER, D., and RAMIREZ-MUNOZ, J., Analytical performance of the Beckman Autolam Burner II, *Flame Notes,* 1974, **6**, 18. (Beckman Instruments Inc., Fullerton, Calif. 92634, U.S.A.).
320 TOYOTA, T., IZAWA, K., and TOMIOKA, N., Determination of Total Hg in fertilizers by flameless AA photometry, *Hiken Kaiho,* 1974, **27**, 19. (Tokyo Fertilizer and Feed Inspection Office, Ministry of Agriculture, Tokyo, Japan).
321 DANIELSON, B. G., and OBERG, P. A., Integrating AAS for ultramicroanalysis of metals, *Anal. Biochem.,* 1974, **62**, 327. (Institute of Physiology, Medicine and Biophysics, University of Uppsala, Uppsala, Sweden).
322 OELSCHLAEGER, W., and SCHMIDT, S., Determination of Zn in vegetable, animal materials and in mineral fodders using AAS, *Landwirtsch. Forsch.,* 1974, **27**, 85. (Address as in ref. 268).
323 MOELLER, G., Flame spectrometric measurement of K, Mg and Ca in HCl solutions of plant ash using the Unicam SP90, *Z. Pfanzenernaehr. Bodenkd.,* 1974 **137**, 31. (Institut Pflanzenbau Pflanzenzuecht., University of Kiel, Kiel, West Germany).
324 ASANO, J., and TOKUNAGA, Y., Simple method of estimation of the Cd concentration in brown rice, *Tokai Kinki Nagyo Shikenjo Kenkyu Hokoku,* 1974, **27**, 55. (Tokai-Kinki National Agricultural Experimental Station, Tsu, Japan).
325 ARROYO, M., and PALENQUE, E., Analytical determination of Zn in biological fluids by AAS. I: Nonspecificity in measurements and a spectrophotometric method for its correction, *Rev. Clin. Esp.,* 1974, **133**, 211. (Lab. Cent., Hosp. Clin. San Carlos, Madrid, Spain).
326 ECKSTEIN, N. A., Continuous total S measurement at Harmac, *Pulp Pap. Can.,* 1974, **75**, 95. (MacMillan Bloedel Ltd., Nanaimo, B.C., Canada).
327 RICHARDS, L. W., and MUDGETT, P. S., Apparatus and method for H_2SO_4 aerosol analysis, U.S. Patent 3,838,972.
328 GRABNER, E., Comparison of two decomposition methods for the determination of Na and K, *Mikrochim. Acta,* 1975, 65. (Eidgenossische Anstaltfur Wasserversorgung Abwasserreinigung und Gewasserschutz, Dubendorf/Zurich, Switzerland).
329 TSEREVITINOV, O. B., ANDREEVA, E. V., and GOLUBEVA, I. L., Mineral composition of granulated sugar, *Sakh. Promst.,* 1974, (11), 21.
330 PRUDNIKOV, E. D., Relative sensitivity of the direct AA determination of traces of elements in a flame, *Izv. Vyssh. Uchebn. Zaved., Khim. Khim. Tekhnol.,* 1974, **17**, 1489. (Zhdanov Leningrad State University, Leningrad, U.S.S.R.).
331 OLES, P. J., Applications of AAS and x-ray fluorescence spectrometry in organic functional group analysis — Thesis, 1974. *Diss. Abstr. Int. B,* 1974, **35**, 2062. (Address as in ref. 122).
332 GRENGG, W. M., Hg determination system and method, U.S. Patent 3,852,604.
333 KARAMIAN, N. A., Separatory funnel, U.S. Patent 3,836,334.
334 BELYAEV, Yu. I., and ORESHKIN, V. N., Determination of Cd, Ag and Tl in suspended matter of sea and river waters by AA and AF methods with a non-flame electrical contact atomizer of solid samples, *Okeanologiya,* 1974, **14**, 917. (Geochim. Acad. of Sciences, Moscow, U.S.S.R.).
335 RAMIREZ-MUNOZ, J., and FOURNIER, D., Simple mirror device for FES with Beckman AA spectrophotometers, *Flame Notes,* 1974, **6**, 21. (Address as in ref. 319).
336 KOVATSITS, K., and VAJDA, F., Analysis of mineral materials by AAS, *Publ. Hung. Min. Res. Inst.,* 1974, **17**, 247.
337 MERODIO, J. C., Interference of Sr in the determination of Li by AAS, *An. Soc. Cient. Argent.,* 1974, **197**, 43. (Address as in ref. 250).
338 GADAEV, A. Y., Use of an electrocontact atomizer for AA determination of impurity elements in rocks, *Uzb. Geol. Zh.,* 1974, **18**, 68. (Inst. Geol. Geofiz. im. Abdullaeva, Tashkent, U.S.S.R.).
339 VARGIN, A. N., GOLUBEV, O. A., and MALKIN, O. A., Radiation of a molecular gas plasma studied in the vacuum u.v. region, *Teplofiz. Vys. Temp.,* 1974, **12**, 940. (Mosk. Inzh. Fiz. Inst., Moscow, U.S.S.R.).
340 SHAPKINA, Y. S., and PRUDNIKOV, E. D., Determination of small quantities of elements in microsamples by AE and AA flame spectrophotometric methods, *Vestn. Leningr. Univ., Geol., Geogr.,* 1974, (2), 147. (Address as in ref. 330).
341 VORONOV, B. G., Spectrum excitation arc source, U.S.S.R. Patent 428,228.

342 D'SILVA, A. P., and FASSEL, V. A., Radiographic intensifying screens, German Offen. 2,404,422.
343 FEDOSEEV, V. A., and KALINCHAK, V. V., Temperatures of binary-solution flames, *Fiz. Aerodispersnykh Sist.*, 1973, **8**, 73. (Odessa University, Odessa, U.S.S.R.).
344 NEMETS, A. M., ad NIKOLAEV, G. I., Determination of the concentration and saturated vapour pressure of the metals Cr, Mn and Fe in the solid phase by AA, *Zh. Prikl. Spektrosk.*, 1974, **21**, 405.
345 ZOLOTOVITSKAYA, E. S., FIDEL'MAN, B. M., and BONDAREVA, N. V., Nonselective light absorption by alkali metal halides during AA analysis, *Zh. Prikl. Spektrosk.*, 1974, **21**, 410.
346 SHISHIDO, S., and SUZUKI, T., Hg in human hair. Comparative study by using two different methods of AA and neutron activation analysis, *Tohoku J. Exp. Med.*, 1974, **113**, 351. (School of Medicine, Tohoku University, Sendai, Japan).
347 RATHJE, A. O., MARCERO, D. H., and DATTILO, D., Personal monitoring technique for Hg vapour in air and determination by flameless AA, *J. Am. Ind. Hyg. Assoc.*, 1974, **35**, 571. (Environment Control Operation, General Electric Co., Cleveland, Ohio, U.S.A.).
348 KATSKOV, D. A., KRUGLIKOVA, L. P., L'VOV, B. V., ORLOV, N. A., and POLZIK, L. K., Device for the separation of weak spectral lines against a background of extraneous radiation, *Zh. Prikl., Spektrosk.*, 1974, **21**, 366. (State Institute of Applied Chemistry, Leningrad, U.S.S.R.).
349 PASSAMANTE, A. P., and FERGUSON, G. D., Pulsed inert gas plasma model, *U.S. Nat. Tech. Inf. Serv., AD Rep.*, 1974, No. 780717/5GA. (Aero-Electron. Technol. Dept., Nav. Air Dev. Cent, Warminster, Pa., U.S.A.).
350 HUSAIN, D., MITRA, S. K., and YOUNG, A. N., Kinetic study of electronically excited N atoms by attenuation of atomic resonance radiation in the vacuum ultraviolet, *J. Chem. Soc., Faraday Trans.* 2, 1974, **70**, 1721. (Address as in ref. 149).
351 MASON, W. B., Flame photometry, pp. 49-63 *in* Henry, R. J., Cannon, D. C., and Winkelman, J. W., *Editors,* Clinical chemistry: principles and technics, 2nd edition — Book published by Harper and Row, New York, 1974. (Affiliated Laboratory, Bioscience Enterprises, Westwood, Calif., U.S.A.).
352 KHRAPAI, V. P., STARTSEVA, E. A., ZHURAVLEVA, L. E., and POPOVA, N. M., Determination of impurities in pure Ag by AAS, *Zh. Anal. Khim.*, 1974, **29**, 2137.
353 MAKAROV, D. F., KUKUSHKIN, Y. N., and EROSHEVICH, T. A., Buffer properties of Na, Cd, Cu and La sulphate solutions in the AA determinations of Ru, Ir and Rh in solutions of complex composition, *Zh. Anal. Khim.*, 1974, **29**, 2128. (Address as in ref. 306).
354 HORTON, R., and LYNCH, J. J., Geochemical field laboratory for the determination of some trace elements in soil and water samples, *Geol. Surv. Can., Pap.*, 1975, No. 75-1, Pt. A, 213. (Resources, Geophysics and Geochemistry Division, Geological Survey of Canada, Ottawa, Ont., Canada).
355 FISHKOVA, N. L., Determination of the Pt metals, Au and Ag by AAS, *Zh. Anal. Khim.*, 1974, **29**, 2121.
356 HWANG, J. Y., and THOMAS, G. P., An AA spectrophotometer for semi-flame and non-flame techniques, *Am. Lab.*, 1974, **6**(8), 55. Address as in ref. 313).
357 LOFBERG, R. T., and LEVRI, E. A., Analysis of Cu and Zn in hemolysed serum samples. *Anal. Lett.*, 1974, **7**, 775. (Div. of Biochemistry, Walter Reed Army Institute of Research, Washington, D.C. 20012, U.S.A.).
358 BONDO, P., and ACKART, T., Ca in serum by AAS without La, *Clin. Chem.*, 1974, **20**, 908. (South Bend Medical Foundation Inc., North Main Street, South Bend, Ind. 46601, U.S.A.).
359 HOYT, P. B.,* and WEBBER, M. D.,† Rapid measurement of plant-available Al and Mn in acid Canadian soils, *Can. J. Soil Sci.*, 1974, **54**, 53. (*Research Station, Agr. Can., Beaverlodge, Alta. Canada; †Soil Research Institute, Agr. Can., Ottawa, Ont., Canada).
360 ARDEN, W. M., HIRSCHFIELD, T. B., KLAINER, S. M., and MUELLER, W. A., Studies of gaseous flame combustion products by Raman spectroscopy, *Appl. Spectrosc.*, 1974, **28**, 554. (Block Engineering Inc., Blackstone Street, Cambridge, Mass. 02139, U.S.A.).
361 ORHEIM, R. M., LIPPMAN, L., JOHNSON, C. J., and BOVEE, H. H., Pb and As levels of dairy cattle in proximity to a Cu smelter, *Environ. Lett.*, 1974, **7**, 229.

(Dept. of Environmental Health, University of Washington, Seattle, Wash., 98175, U.S.A.).

362 CLARKE, A. N., and CLARKE, J. H., A static monitor for Pb in natural and waste waters, *Environ. Lett.*, 1974, **7**, 251. (Address as in ref. 57).

363 BELCHER, R., BOGDANSKI, S. L., KASSIR, Z. M., STILES, D. A., and TOWNSHEND, A., A preliminary study of halide determination by molecular emission cavity analysis, *Anal. Lett.*, 1974, **7**, 751. (Address as in ref. 35).

364 FULLER, C. W., Applications of AAS in the paint industry — Paper presented at an Oil and Colour Chemists Association meeting, Newcastle upon Tyne, England, 6 March 1975. (Tioxide International Ltd., Billingham, Cleveland, England).

365 YAMAMOTO, Y.,* KUMAMARU, T.,* KAMADA, T.,† TANAKA, T.,* and KAWABE, M.,* Determination of p.p.b. level of Cd, Pb and Cu in water by a carbon tube flameless AAS combined with dithizone–CCl_4 extraction, *Nippon Kagaku Kaishi,* 1975, 841. (*Dept. of Chemistry, Faculty of Science, Hiroshima University, 1–1–89, Higashisenda-machi, Hiroshima-shi 730, Japan; †Dept. of Hygiene, Faculty of Medicine, Hiroshima University, 1–2–3, Kasumi-cho, Hiroshima-shi 734, Japan).

366 UCHIDA, H., ADACHI, F., MORI, O., and NEGISHI, R., Multi-element analysis of airborne particulates by laser microprobe spectroscopy, *Bunseki Kagku,* 1975, **24**, 325. (Industrial Research Institute of Kanagawa Prefecture, 3173, Showa-machi, Kanazawa-ku, Yokohama-shi, Kanagawa, Japan).

367 FUJIWARA, K., HARAGUCHI, H., and FUWA, K., Profiles of the distribution of atoms in the nitrous oxide/acetylene flame, *Bull. Chem. Soc. Jpn.,* 1975, **48**, 857. (Address as in ref. 242).

368 AMETANI, K., AAS of Ti in the Fe sulphides of the NiAs type, *Bull. Chem. Soc. Jpn.,* 1975, **48**, 1047. (Address as in ref. 128).

369 KOIZUMI, H., and YASUDA, K., A new method of AAS using the Zeeman effect, *Bunko Kenkyu,* 1974, **23**, 290. (Naka Works, Hitachi Ltd., 882 Ichige, Katsuta-shi, Ibaraki 312, Japan).

370 OKAZAKI, K., A vacuum spark discharge source for producing highly ionized Al lines in the wavelength region of 50–100 Å, *Bunko Kenkyu,* 1974, **23**, 285. (Institute of Physical and Chemical Research, Honkomagome, Bunkyo-ku, Tokyo 113, Japan).

371 KAWAGUCHI, H., SAGA, T., and MIZUIKE, A., ES analysis of solutions by d.c. arc excitation with an ultrasonic nebulizer, *Bunko Kenkyu,* 1975, **24**, 99. (Faculty of Engineering, Nagoya University, Furo-cho, Chikusa-ku, Nagoya 464, Japan).

372 KUBOTA, M., and ISHIDA, R., Studies on the spectrochemical analysis of metal samples by means of the laser microprobe. IV: Effects of atmospheres on spectral line intensity, *Bunko Kenkyu,* 1975, **24**, 89. (National Chemical Laboratory, 1–1–5, Honmachi, Shibuya-ku, Tokyo 151, Japan).

373 ZENITANI, F., EBISU, T., and MINAMI, S., An automated spectrofluorometer system with an on-line minicomputer, *Bunko Kenkyu,* 1975, **24**, 13. (Dept. of Applied Physics, Faculty of Engineering, Osaka University, Yamada-kami, Suita, Osaka 565, Japan).

374 NAGANUMA, K., and KATO, M., Effect of pores in sample electrode prepared by powder metallurgy in ES analysis. I, *Bunko Kenkyu,* 1975, **24**, 79. (Government Industrial Research Institute, Nagoya 1 Hirate-machi, Kita-ku, Nagoya 462, Japan).

375 KASHIMA, J., and UMEMURA, F., Effect of the electrode temperature on the spark discharge. V: The decrease of the matrix effect on spectrographic analysis of of cast Fe, *Bunko Kenkyu*, 1975, **24**, 21. (Casting Research Laboratory, Waseda University, 1–500, Totsuka, Shinjuku-ku, Tokyo, Japan).

376 TOMINGA, M., KIMURA, A., MIYAZAKI, A., and UMEZAKI, Y., Determination of Cd by flameless AAS using a heated graphite atomizer, *Bunseki Kagaku,* 1975, **24**, 61. (National Research Institute for Pollution and Resources, 26–10, Ukima 4-chome, Kita-ku, Tokyo, Japan).

377 SAKURAI, H., Extraction and AAS determination of micro amounts of Re with NH_4 pyrrolidine dithiocarbamate, *Bunseki Kagaku,* 1975, **24**, 52. (Central Research Laboratory, Mitsubishi Petrochemical Co. Ltd., 1315, Wakaguri, Ami-machi, Inashiki-gun, Ibaraki, Japan).

378 YASUDA, S., and KAKIYAMA, H., Study of absorption spectra for alkali and alkaline earth metal salts in flameless AAS using a carbon tube atomizer, *Bunseki Kagaku,* 1975, **24**, 377. (National Research Institute, Kyushu, Shuku-machi, Tosushi, Saga, Japan).

379 KIDANI, Y.,* OSUGI, N.,* INAGAKI, K.,* and KOIKE, H.,† Indirect determination of anthanilic acid by AAS, *Bunseki Kagaku,* 1975, **24**, 218. (*Faculty of Pharmaceutical Sciences, Nagoya City University, 3–1, Tanabe-dori, Mizuho-ku, Nagoya-shi, Aichi, Japan; †School of Medical Technology and Nursing, Fugita Gakuen University, Katsutake-cho, Toyoake-shi, Aichi, Japan).

380 KAMADA, T.,† KUMAMARU, T.,* and YAMAMOTO, Y.,* Rapid determination of a trace amount of As(III, V), Sb(III, V) and Se(IV) in water by AAS with a carbon-tube atomizer, *Bunseki Kagaku,* 1975, **24**, 89. (*† Both addresses as in ref. 365).

381 KANDA, M., HIRI, Y., and MATSUMOTO, I., Rapid determination of Pb in TiO_2 by AAS using a graphite atomizer, *Bunseki Kagaku,* 1975, **24**, 299. (Shiseido Laboratories, 1050, Nippa-cho, Kohoku-ku, Yokohama-shi, Kanagawa, Japan).

382 HIIRO, K.,* KAWAHARA, A.,* TANAKA, T.,* and HIRAI, A.,† Interference of alkali salts in the AAS determination of Cu and Fe, *Bunseki Kagaku,* 1975, **24**, 275. (*Government Industrial Research Institute, Osaka, 1–8–31, Midorigaoka Ikeda-shi, Osaka, Japan; †Industrial Research Institute of Hyogo Prefecture, 3–1, Yukihira-cho, Suma-ku, Kobe-shi, Hyogo, Japan).

383 OOGURO, H., Correction of the sensitivity difference in AAS of Cr, *Bunseki Kagaku,* 1975, **24**, 361. (Central Research Laboratory, Matsushita Electric Industrial Co. Ltd., 1006, Kadoma, Kadoma-shi, Osaka, Japan).

The following papers (*) were presented at the Pye Unicam Analytical Conference, 20–22 May 1975, Stratford-on-Avon, England.

384* NALL, W. R., Analysis of complex steels and other alloys by AAS. (Ministry of Defence, Bragg Laboratory, Sheffield, England).

385* WHITESIDE, P. J., Methods of dissolution of siliceous samples for AAS. (Pye Unicam Ltd., Cambridge, England).

386* POWELL, R. J. W., Pb/Sn alloys: some analytical problems in AA and x-ray fluorescence. (G.E.C. Hirst Research Centre, Wembley, London, England).

387* CLAY, A. F., EVANS, F. E., and HUNT, D. W., Analysis by AAS of materials bearing precious metals. (Daniel C. Griffith & Co. Ltd., Witham, Essex, England).

388* HUMPHREYS, G. T. P., Use of AAS in the analysis of Ni based materials. (International Nickel Ltd., European Research and Development Centre, Birmingham, England).

389* DODGSON, J., Spectrophotometric methods for analysing particulate pollutants. (Institute of Occupational Medicine, Edinburgh, Scotland).

390* ELLIS, D. J., Some factors affecting the role of AA in air pollution studies. (Department of the Environment, Warren Spring Laboratory, Stevenage, Bucks., England).

391* BALLINGER, P. J., The use of AA in the determination of toxic elements in agricultural products. (Agricultural Development and Advisory Service, Ministry of Agriculture, Fisheries and Food, S.W. Region, Bristol, England).

392* VAN KOLLENBURG, L. W. J., Comparison of some flameless AA systems for the determination of Pb in river water. (Philips Research Laboratories, Eindhoven, The Netherlands).

393* ROOS, J. T. H., Experiences with a simple pressure dissolution technique for blood samples prior to AAS. (Dept. of Chemistry, The University, St. Andrews, Scotland).

394 SHAW, F., and OTTAWAY, J. M., The determination of trace amounts of Al and other elements in Fe and steel by AAS with carbon furnace atomization, *Analyst,* 1975, **100**, 217. (Dept. of Pure and Applied Chemistry, Strathclyde University, Glasgow G1 1XL, Scotland).

395 FULLER, C. W., A kinetic theory of atomization for non-flame AAS with a graphite furnace. II: Analytical applications of kinetic information for Cu, *Analyst,* 1975, **100**, 229. (Address as in ref. 364).

396 GEGIOU, D., and BOTSIVALI, M., AAS determination of Pb in beverages and fruit juices and of Pb extracted by their action on glazed ceramic surfaces, *Analyst,* 1975, **100**, 234. (Research Dept., State Chemical Laboratories, 16A Tsoha Street, Athens, Greece).

397 AGEMIAN, H., ASPILA, K. I., and CHAU, A. S. Y., A comparison of the extraction of Hg from sediments by using HCl/HNO_3, H_2SO_4/HNO_3 and HF/aqua regia mixtures, *Analyst,* 1975, **100**, 253. (Address as in ref. 185).

398 GRAY, A. L., Mass spectrometric analysis of solutions using an atmospheric pressure ion source, *Analyst,* 1975, **100**, 289. (Applied Research Laboratories Ltd., Wingate Road, Luton, Beds., England).

399 SMITH, A. E., Interference in the determination of elements that form volatile hydrides with $NaBH_4$ using AAS and the argon/hydrogen flame, *Analyst,* 1975, **100**, 300. (Imperial Chemical Industries Ltd., Mond Div., Research Dept., Northwich, Cheshire, England).
400 THOMPSON, K. C., The AF determination of Sb, As, Se and Te by using the hydride generation technique, *Analyst,* 1975, **100**, 307. (Address as in ref. 19).
401 BOYLE, E., and EDMOND, J. M., Cu in surface waters south of New Zealand, *Nature,* 1975, **253**, 106. (Dept. of Earth and Planetary Sciences, Massachusetts Institute of Technology, Cambridge, Mass. 01239, U.S.A.).
402 DAY, J. P., HART, M., and ROBINSON, M. S., Pb in urban street dust, *Nature,* 1975, **253**, 343. (Chemistry Dept., Manchester University, Manchester M13 9PL, England).
403 STUPP, H. J., and OVERHOFF, T., AES with laser excitation of graphite under elevated temperature. I: Plasma properties, *Spectrochim. Acta, Part B,* 1975, **30***B*, 77. (Kernforschungsanlage Julich, Institute fur Reaktorentwicklung, Postfach 365, 517 Julich, West Germany).
404 STUPP, H. J., and OVERHOFF, T., AES with laser excitation of graphite under elevated temperature. II: Characteristics and peculiarities of the emission spectra, *Spectrochim. Acta, Part B,* 1975, **30***B*, 89. (Address as in ref. 403).
405 SCHWARZ, W., and MILKOWITS, A., Influence of thermal conductivity and steel alloy structure with quantitative ES analysis, *Spectrochim. Acta, Part B,* **1975**, **30***B*, 101. (Versuchanstalt der Gebr. Bohler & Co. AG, Edelstahlaverke, Kapfenberg, A–8605, Austria).
406 BARNES, R. M., and SCHLEICHER, R. G., Computer simulation of r.f. induction-heated argon plasma discharges at atmospheric pressure for spectrochemical analysis. I: Preliminary investigation, *Spectrochim. Acta, Part B,* 1975, **30***B*, 109. (Address as in ref. 81).
407 EPSTEIN, M. S., and O'HAVER, T. C., Improvements in repetitive scanning techniques for reducing spectral interferences in FES, *Spectrochim. Acta, Part B,* 1975, **30***B*, 135. (Address as in ref. 219).
408 STRASHEIM, A., and BLUM, F., A study of sparked Al samples with a scanning electron microscope and energy-dispersive x-ray analyser, *Spectrochim. Acta, Part B,* 1975, **30***B*, 147. (Address as in ref. 304).
409 NIKITIN, E. E., Theory of elementary atomic and molecular processes in gases — Book published by Clarendon Press, Oxford, 1974.
410 MENSHIKOVA, V. I., MALYKH, V. D., and SHESTAKOVA, T. D., Direct determination of Au in solids by AA, *Zh. Anal. Khim.,* 1974, **29**, 2132. (Irkutsk State Institute of Rare and Nonferrous Metals, Irkutsk, U.S.S.R.).
411 APOLITSKII, V. N., YUDINA, L. P., LATYSHEVA, Y. P., and PRONIN, V. A., Studies on AA determination of Pt and Rh, *Zh. Anal. Khim.,* 1975, **30**, 141. (Irkutsk Scientific Research Institute of Rare and Nonferrous Metals, Irkutsk, U.S.S.R.).
412 KRIVCHIOVA, E. P., Third-element effect in spectrographic analysis of Fe-based alloys using a laser, *Zh. Anal. Khim.,* 1975, **30**, 187. (Physiotechnical Institute, Ukranian S.S.R. Academy of Sciences, Kharkov, U.S.S.R.).
413 GRINZAID, E. L., NADEZHINA, L. S., KOLOSOVA, L. P., and LISNYANSKAYA, M. G., Error statistics of the fire assay: spectrometric determination of Pt, Pd, Au and Rh, *Zh. Anal. Khim.,* 1975, **30**, 190. (M. I. Kalinin Leningrad Polytechnic Institute, Leningrad, U.S.S.R.).
414 PRUDNIKOV, E. D., Use of a microsound–flame adaptor atomizer in AA determination of small amounts of elements, *Zh. Anal. Khim.,* 1975, **30**, 232. (Address as in ref. 330).
415 KATSOV, D. A., KRUGLIKOVA, L. P., and L'VOV, B. V., New technique of AA analysis of solids by using a graphite capsule-flame atomizer, *Zh. Anal. Khim.,* 1975, **30**, 238. (Address as in ref. 348).
416 L'VOV, B. V., KRUGLIKOVA, L. P., POLZIK, L. K., and KATSKOV, D. A., Theory of flame AA analysis. I: Distribution of aerosol in flame from slot burners, *Zh. Anal. Khim.,* 1975, **30**, 645. (Address as in ref. 348).
417 L'VOV, B. V., KRUGLIKOVA, L. P., POLZIK, L. K., and KATSKOV, D. A., Theory of flame AA analysis. II: Distribution of atoms of the analyte in flames from slot burners, *Zh. Anal. Khim.,* 1975, **30**, 652. (Address as in ref. 348).
418 LAKTIONOVA, N. V., KARYAKIN, A. V., and AGEEVA, L. V., The mechanism of action of NaCl as carrier on intensity of rare-earth element lines in their

chemical–spectrochemical determination, *Zh. Anal. Khim.*, 1975, **30**, 703. (Vernadskii Institute of Geochemistry and Analytical Chemistry, U.S.S.R. Academy of Sciences, Moscow, U.S.S.R.).

419 PLESKACH, L. I., and CHIRKOVA, G. D., Flame photometric determination of sulphate sulphur in superphosphates, *Zh. Anal. Khim.*, 1975, **30**, 729. (S. M. Kirov State University, Alma-Ata, U.S.S.R.).

420 KARYAKIN, A. V., PAVLENKO, L. I., and SAFRONOVA, N. S., Effect of natural water macrocomponents on the determination of microelements, *Zh. Anal. Khim.*, 1975, **30**, 775. (Address as in ref. 418).

421 FISHKOVA, N. L., ZDOROVA, E. P., and POPOVA, N. N., Determination of low Au and Ag contents in mineral raw materials by fire assay–AA, *Zh. Anal. Khim.*, 1975, **30**, 806.

422 LENC, J., The determination of Cr in corundum materials by AAS, *Sklar Keram.*, 1975, **25**, 48. (Research Institute on Electrotechnical Ceramics, Hradec Kralove, Czechoslovakia).

423 AMOS, M. D., BENNETT, P. A., MATOUSEK, J. P., PARKER, C. R., ROTHERY, E., ROWE, C. J., and SANDERS, J. B., Basic AAS: a modern introduction — Book published by Varian Techtron Pty. Ltd., Australia, 1975.

The following papers (*) were presented at the Third Australian Symposium on Analytical Chemistry, 27–30 May 1975, Melbourne, Australia.

424* DOOLAN, K. J., and SMYTHE, L. E., Automatic scatter correction in the analysis of environmental samples by AF. (Address as in ref. 159).

425* LIDDELL, P. R., Factors influencing precision and detection limits in flame AAS. (Varian Techtron Pty. Ltd., P.O. Box 222, Springvale, Vic. 3171, Australia).

426* STEINER, J. W., KENNARD, C. H. L., and WOOD, B. J., Efficiency of the nebulization process using AAS for ultra-trace analysis. (Dept. of Chemistry and A. B. Baker Dept. of Surgery, University of Queensland, St. Lucia, Qld. 4067, Australia).

427* PAKALNS, P., and FARRAR, Y. J., The effects of surfactants on the extraction–AAS determination of Cu, Fe, Mn and Pb. (Atomic Energy Commission, Research Establishment, Private Mail Bag, Sutherland, N.S.W. 2232, Australia.

428* CHAPMAN, J. F., and DALE, L. S., AAS analysis of some metals by hydride evolution using a heated absorption tube. (Address as in ref. 79).

429* STUX, R. L., PRADHAN, N. K., FINCH, A., and SANDERS, J. B., Matrix management in the carbon rod atomizer. (Varian Techtron Pty. Ltd., P.O. Box 222, Springvale, Vic. 3171, Australia).

430* PRADHAN, N. K., FINCH, A., and ROWE, C. J., Application of the carbon rod atomizer to trace metal analysis in coal. (Address as in ref. 429).

431* GARNYS, V. P., and SMYTHE, L. E., The graphite tube atomizer as a survey instrument for trace metals in blood: some fundamental studies. (Address as in ref. 159).

432* NOLLER, B. N., and BLOOM, H., Determination of trace metals in blood by non-flame AAS. (Chemistry Dept., University of Tasmania, Box 252C, G.P.O., Hobart, Tas. 7001, Australia).

433* NOLLER, B. N., and BLOOM, H., Direct determination of trace metals in atmospheric particulate material using low-volume sampling and non-flame AAS. (Address as in ref. 432).

434* BELCHER, R., BOGDANSKI, S. L., KNOWLES, D. J., and TOWNSHEND, A., The determination of S by molecular emission cavity analysis. (Address as in ref. 35).

435* STEINER, J. W., KENNARD, C. H. L., and WOOD B. J., Comments on flame atomization in AAS. (Address as in ref. 426).

436 I.P. standards for petroleum and its products. Part I: Methods for analysis and testing, 34th edition — Book published by Applied Science Publishers Ltd., England, 1975.

437 MAEHATA, E., KANOHDA, Y., and NAKA, H., Determination of Zn in serums by AAS, *Eisei Kensa*, 1974, **23**, 687. (Address as in ref. 257).

438 MIZUIKE, A., and FUKUDA, K., Solid–liquid extraction of trace impurities from K and Pb nitrates and from $PbSO_4$, *Mikrochim. Acta*, 1975, 281. (Address as in ref. 371).

439 WATSON, G. H. R., Quantitative determination of elements in alloys, British Patent 1,366,074.

440 GORDANIER, D. E., and WINTERS, B. L., Spectrographic impurity determination after partition chromatography of U, *U.S. At. Energy Comm., Rep.*, 1974, KY–653, (Paducah Gaseous Diffusion Plant, Paducah, Ky., U.S.A.).
441 RIGIN, V. I., Determination of traces of B in graphite, *Zavod. Lab.*, 1974, **40**, 1195. (Krasnoyarsk State University, U.S.S.R.).
442 DITTRICH, K., and WENNRICH, R., Spectrographic determination of Si in FeY garnet, Fe_2O_3 and Y_2O_3, *Chem. Anal. (Warsaw).* 1974, **19**, 475. (Address as in ref. 107).
443 LOCKWOOD, T. H., Analysis of asbestos for trace metals, *Am. Ind. Hyg. Assoc. J.*, 1974, **35**, 245. (National Institute for Occupational Safety and Health, Cincinnati, Ohio, U.S.A.).
444 CHANDOLA, L. C., and DIXIT, V. S., Simultaneous spectrographic determination of volatile and refractory impurities in Ta_2O_5 using d.c. arc excitations. *Curr. Sci.*, 1974, **43**, 372. (Address as in ref. 291).
445 GREEN, R. B., and LATZ, H. W., Detection and quantitation of several atomic species by intra-cavity quenching of laser emission, *Spectrosc. Lett.*, 1974, **7**, 419. (Clippinger Graduate Research Laboratory, Dept. of Chemistry, The University, Athens, Ohio 45701, U.S.A.).
446 SYTY, A., Developments in methods of sample injection and atomization in atomic spectrometry, *C.R.C. Crit. Rev. Anal. Chem.*, 1974, **4**(2), 155. (Dept. of Chemistry, Indiana University of Pennsylvania, U.S.A.).
447 HEINRICHS, H., Determination of Hg in water, rocks, coal and petroleum with flameless AAS, *Fresenius' Z. Anal. Chem.*, 1975, **273**, 197. (Geochemistry Institute, University of Gottingen, West Germany).
448 SIXTA, V., MIKSOVSKY, M., and SULCEK, Z., Determination of Ba in strongly mineralised waters by AAS, *Fresenius' Z. Anal. Chem.*, 1975, **273**, 193. (Address as in ref. 11).
449 BATCA, A., and IONESCU, M., Simultaneous determination of heavy-metal cations in industrial waste water, *Rev. Chim. (Bucharest)*, 1974, **25**, 585.
450 ABBEY, S., LEE, N. J., and BOUVIER, J. L., Analysis of rocks and minerals using an AA spectrophotometer. V: Improved lithium fluoroborate scheme for 14 elements, *Geol. Surv. Pap. Can., Pap.*, No. 74–19, 1974.
451 BOSCH, H., BUECHEL, E., GRYGIEL, H., and LOHAU, K., Automation of AAS; determination of Mn in the structural constituents of steel, *Arch. Eisenhuettenwes.*, 1974, **45**, 699. (Chemistry Laboratory, August Thyssen-Huette AG, Duisburg-Hamborn, West Germany).
452 PEARTON, D. C. G., and MALLETT, R. C., Use of carbon-rod devices in the determination of noble metals by AAS, *Natl. Inst. Metall., Repub. S. Afr., Rep.* No. 1598, 1974. (National Institute for Metallurgy, Milner Park, Johannesburg, South Africa).
453 CARRINGTON, R. A. G., Computers for spectroscopists — Book published by Adam Hilger, London, 1975.
454 NG, S., MUNROE, M., and McSHARRY, W., Determination of Se in animal-feed pre-mix by AAS, *J. Assoc. Off. Anal. Chem.*, 1974, **57**, 1260. (Analytical Research Laboratory, Quality Control Dept., Hoffman-La Roche Inc., Nutley, N.J., U.S.A.).
455 MENACHE, R., Routine micromethod for determination of oxalic acid in urine by AAS, *Clin. Chem.*, 1974, **20**, 1444. (Laboratory of Clinical Chemistry, Hasharon Hospital, Petah, Tikva, Israel).
456 LIEU, V. T., CANNON, A., and HUDDLESTONE, W. E., Non-flame AA attachment for trace Hg determination, *J. Chem. Educ.*, 1974, **51**, 752. (California State University, Long Beach, Calif., U.S.A.).
457 *Deleted.*
458 HUBER, B. W., TAMM, R. G. A., and BRAUN, K. J., Apparatus for atomizing a sample for flameless AA measurement, German Offen. 2,314,207.
459 TANAKA, Y., IKEBE, K., TANAKA, R., and KUNITA, N., Microanalysis of Hg in fish by quartz tube combustion and AAS, *Shokuhin Eiseigaku Zasshi*, 1974, **15**, 386. (Osaka Prefecture Institute for Public Health, Osaka, Japan).
460 EALY, J. A., BOLTON, N. E., McELHENY, R. J., and MORROW, R. W., Determination of Pb in whole blood by graphite furnace AAS, *J. Am. Ind. Hyg. Assoc.*, 1974, **35**, 566. (Industrial Hygiene Dept., Oak Ridge National Laboratory, Oak Ridge, Tenn., U.S.A.).
461 SZIVOS, K., and PUNGOR, E., Determination of Ca, Na, K and Mg in human

blood serums by AA, *Acta Pharm. Hung.*, 1974, **44**, 253. (Address as in ref. 275).
462 NAKANE, K., IGUCHI, K., KOSAKA, A., and SAITO, H., Determination of Fe in serum by AAS, *Rinsho Byori*, 1974, **22**, 741. (School of Medicine, Nagoya University Nagoya, Japan).
463 VANPEE, M., VIDAUD, P., and CASHIN, K. D., Emission spectra and burning velocity of the premixed cyanogen-fluorine flame, *Combust. Flame*, 1974, **23**, 227. (Dept. of Chemical Engineering, University of Massachusetts, Amherst, Mass., U.S.A.).
464 HERRMANN, H., New methods for preparation of beads for x-ray fluorescence and AA, *Bull. Soc. Fr. Ćeram.*, 1974, **103**, 13. (Soc. Herrmann-Moritz, Chassant, France).
465 WHEELER, R. M., LIEBERT, R. B., ZABEL, T., CHATURVEDI, R. P., VALKOVIC, V., PHILLIPS, G. C., ONG, P. S., CHENG, E. L., and HRGOVCIC, M., Techniques for trace element analysis. X-ray fluorescence, x-ray excitation with protons and flame AA, *Med. Phys.*, 1974, **1**, 68. (T. W. Bonner Nuclear Laboratory, Rice University, Houston, Tex., U.S.A.).
466 BERGMAR, B., SJOSTROM, R., and WING, K. R., Variation with age of tissue Zn concentrations in albino rats determined by AAS, *Acta Physiol. Scand.*, 1974, **92**, 440. (Dept. Prosthet. Dent., University of Umea, Umea, Sweden).
467 ROSS, R. T., and GONZALEZ, J. G., Direct determination of trace quantities of Mn in blood and serum samples using selective volatilization and graphite tube reservoir AAS, *Bull. Environ. Contam. Toxicol.*, 1974, **12**, 470. (National Enviraonmental Research Centre, Environmental Protection Agency, Research Triangle Park, N.C., U.S.A.).
468 CALLANDER, R. H., GERSTEN, J. L., LEIGH, R. W., and YANG, J. L., Role of metastable dimers in the theory of induced AF, *Phys. Rev. Lett.*, 1974, **33**, 1311. (City College, City University, New York, N.Y., U.S.A.).
469 BALL, G. A., USOL'TSEVA, M. V., PRONIN, V. A., SHIPITSIN, S. A., SKUDAEV, Y. D., and YUDELEVICH, I. G., Extraction–AA determination of Au in lean products using a furnace-flame atomizer, *Izv. Sib. Otd. Akad. Nauk. SSSR, Ser. Khim. Nauk*, 1974, (5), 71. (Address as in ref. 287).
470 PATENT, AAS of Sb, Se, Te and As, Japan Kokai 74 77,684.
471 COOK, T. E., MILANO, M. J., and PARDUE, H. L., Application of a vidicon spectrometer for simultaneous FE analysis for Na and K in serum, *Clin. Chem.*, 1974, **20**, 1422. (Dept. of Chemistry, Purdue University, West Lafayette, Ind., U.S.A.).
472 SORENSON, J. R. J., Evalution of interferences: reply to comments, *Arch. Environ. Health*, 1974, **29**, 297.
473 NECHAY, M. W., and SUNDERMAN, F. W., jun., Measurements of Ni in hair by AAS, *Ann. Clin. Lab. Sci.*, 1973, **3**, 30. (School of Medicine, University of Connecticut, Farmington, Conn., U.S.A.).
474 CHURCH, M. R., and ROBISON, W. H., Rapid routine AAS method for the determination of Se at submicrogram levels in animal tissue, *Int. J. Environ. Anal. Chem.*, 1974, **3**, 323. (Wildilfe Research Centre, U.S. Bureau of Sport, Fisheries and Wildlife, Denver, Colo., U.S.A.).
475 HARMS, U., Determination of the transition metals Mn, Fe, Co, Cu and Zn in river fish using x-ray fluorescence analysis and flameless AAS, *Arch. Fischereiwiss.*, 1974, **25**, 63. (Isotopenlab., Bundesforschungsanst. Fisch., Hamburg, West Germany).
476 BARTH, J., and BRUCKNER, B. H., Simplified AA determination of stable Sr in milk and hay: comparison of methods and stepwise procedure, *Report*, 1974, EPA–680–4–73–2. (National Environmental Research Center, Las Vegas, Nev., U.S.A.).
477 GARCIA, W. J., BLESSIN, C. W., and INGLETT, G. E., Heavy metals in whole kernel dent corn determined by AA, *Cereal Chem.*, 1974, **51**, 788. (N.R.R.L., A.R.S., Peoria, Ill, U.S.A.).
478 TESTA, J. F., and WYATT, J. L., General chemistry laboratory project utilizing AA, *J. Chem. Educ.*, 1975, **52**, 50. (Essex Community College, Baltimore, Md., U.S.A.).
479 HINNERS, T. A., and SIMMONS, W. S., Evaluation of interferences: comments, *Arch. Environ. Health*, 1974, **29**, 296. (Address as in ref. 7).
480 NOMOTO, S., Determination and pathophysiological study of Ni in humans and animals II: Measurement of Ni in human tissues by AAS, *Shinshu Igaku Zasshi*, 1974, **22**, 39. (Central Clinical Laboratory, Shinshu University Hospital, Nagano, Japan).
481 CLYBURN, S. A., SERIO, G. F., BARTSCHMID, B. R., EVANS, J. E., and

VEILLON, C., Sensitive AF system for metals in biological systems, *Anal. Biochem.*, 1975, **63**, 231. (Address as in ref. 110).

482 LANBINA, T. V., YUDELEVICH, I. G., and POLUBOYAROV, V. A., AA determination of Pt metals in nonaqueous media. III: Influence of organic solvents on AA determination of Pt metals, *Izv. Sib. Otd. Akad. Nauk SSSR, Ser. Khim. Nauk*, 1974, (5), 78. (Address as in ref. 287).

483 FIRKINS, J. L., Tyre cord dip pick-up by AAS, *Rubber Chem. Technol.*, 1974, **47**, 448. (Celanese Fibers Mark. Co, Charlotte, N.C., U.S.A.).

The following papers (*) were presented at the 28th Annual Summer Symposium on Analytical Chemistry, 18–20 June, 1975, Knoxville, Tenn., U.S.A.

484* FASSEL, V. A., Simultaneous multi-element determinations at the major, minor, trace and ultratrace levels by inductively coupled plasma–optical emission spectroscopy. (Address as in ref. 98).

485* HORLICK, G., Photodiode arrays for multi-element spectroscopy. (Address as in ref. 89).

486* PARDUE, H. L., Applications of a custom-designed vidicon spectrometer for analytical spectroscopy. (Address as in ref. 471).

487* WINEFORDNER, J. D., Multi-element AFS. (Dept. of Chemistry, University of Florida, Gainesville, Fla. 32611, U.S.A.).

488* ALDOUS, K. M., Multi-element AAS with image tubes. (Address as in ref. 80).

489* HOWELL, N. G., and MORRISON, G. H., Multi-element AES with image tubes. (Address as in ref. 130).

490* HIRSCHFELD, T, Fourier-transform spectroscopy. (Address as in ref. 360).

491* DECKER, Jr., J. A., Hadamard-transform spectroscopy. (Spectral Imaging, Inc., Concord, Mass., U.S.A.).

492* HAGER, R. N., jun., Second-derivative spectroscopy. (L.S.I. Spectrometrics, Pinellas Park, Fla 33565, U.S.A.).

493* O'HAVER, T. C., Wavelength modulation in analytical spectroscopy. (Address as in ref. 219).

494 BELCHER, R., RANJITKAR, K. P., and TOWNSHEND, A., Determination of nanogram amounts of Bi by means of candoluminescence emission following application to the surface of a CaO based matrix, *Analyst,* 1975, **100**, 415. (Address as in ref. 35).

495 HANCOCK, S., and SLATER, A., A specific method for determination of trace concentrations of tetramethyl- and tetraethyl-Pb vapours in air, *Analyst,* 1975, **100**, 422. (The Associated Octel Co. Ltd., Ellesmere Port, Cheshire, England).

496 LAU, O. W., and LI, K. L., Determination of Pb and Cd in paint by AAS utilising the Delves microsampling technique, *Analyst,* 1975, **100**, 430. (Chemistry Dept., Chinese University of Hong Kong, Shatin, N.T., Hong Kong).

497 OTTAWAY, J. M., and SHAW, F., Carbon furnace AES; a preliminary appraisal, *Analyst,* 1975, **100**, 438. (Address as in ref. 394).

498 GLOVER, J. H., Chemiluminescence in gas analysis and FES, *Analyst,* 1975, **100**, 449. (75 Craven Gardens, Wimbledon, London SW19 8LU, England).

499 HOLDING, S. T., and ROWSON, J. J., The determination of Ba in unused lubricating oil by means of AAS, *Analyst,* 1975, **100**, 465. (Shell Research Ltd., Thornton Research Centre, P.O. Box 1, Chester CH1 3SH, England).

500 ROONEY, R. C., The determination of Ag in animal tissues by a wet-oxidation process followed by AAS, *Analyst,* 1975, **100**, 471. (Rooney and Ward Ltd., Blackwater Station Estate, Camberley, Surrey, England).

501 BELCHER, R., BOGDANSKI, S. L., HENDEN, E., and TOWNSHEND, A., Elimination of interferences in the determination of As and Sb by hydride generation using MECA, *Analyst,* 1975, **100**, 522. (Address as in ref. 35).

502 EBERT, J., and JUNGMANN, H., Rapid and sensitive determination of Pb in urine by flameless AAS, *Fresenius' Z. Anal. Chem.*, 1974, **272**, 287. (Analyt. Abt., Hoechst AG, Knapsack bei Koeln, West Germany).

503 LAPINSKAYA, M. E., KUZNETSOVA, G. V., and SHITT, E. S., Spectrographic analysis of slags of Cr/Mo/V steel, *Zavod. Lab.*, 1974, **40**, 1478. (Ural Branch, Dzerzhinsk Thermo-Tech. Inst., U.S.S.R.).

504 MURTY, P. S., and MARATHE, S. M., Spectrographic determination of Nd, Sm, Gd, Tb and Dy in high purity Eu_2O_3, *Fresenius' Z. Anal. Chem.*, 1974, **272**, 341. (Address as in ref. 94).

505 SHERSTYUK, A. I., VOVK, V. N., and VLASOVA, L. I., Spectrographic analysis

of high-speed steel with preliminary transfer of sample,*Zavod, Lab.*, 1974, **40**, 1471. (Zaporozhe Automobile Works, Ukraine, U.S.S.R.).

506 EFIMENKO, N. I., and KUSHNAREVA, V. N., Determination of Hf in solutions by the aerosol-spark method, *Zavod. Lab.*, 1974, **40**, 1473. (Upper Dnieper Mining-Metall. Combine, Vol'nogorsk, U.S.S.R.).

507 ENGLEMAN, R., New source of atomic U for absorption spectroscopy and other applications, *Spectrosc. Lett.*, 1974, **7**, 547. (University of California, Los Alamas Science Laboratory, N.Mex. 87544, U.S.A.).

508 BRAMAN, R. S., and JOHNSON, D. L., Selective absorption tubes and emission technique for determination of ambient forms of Hg in air, *Environ. Sci. Technol.*, 1974, **8**, 996. (Dept. of Chemistry, University of South Florida, Tampa, Fla. 33620, U.S.A.).

509 WIADROWSKA, B., and SYROWATKA, T., Determination of total Hg in blood or hair by AAS, *Roczn. Panst. Zakl. Hig.*, 1974, **25**, 701. (Dept. of Sanitation and Toxicology, State Institute of Hygiene, Warsaw, Poland).

510 CHIRICOSTA, S., BRUNO, E., and CLASADONTE, M. T., Determination of the mineral constituents of Sicilian citrus-fruit juices by AAS, *Essenze Deriv. Agrum.*, 1974, **44**, 259. (Istituto Merceol., University of Messina, Italy).

511 OL'KHOVICH, P. F., and VOROPAI, V. P., Rapid flame-photometric determination of K and Na in Ti bronze, *Ukr. Khim. Zh.*, 1974, **40**, 1336. (Institute of General and Inorganic Chemistry, Academy of Science Ukraine SSR, Kiev, U.S.S.R.).

512 KONDRACHOFF, W., and PUJADE-RENAUD, J. M., Determination of soluble Pb in Pb CrO_4 pigments, *Double-Liaison*, 1974, **21**, 593. (Society Cappelle Freres, Rue de la Lys, Halluin, France).

513 NASER, M. I., BASILY, A. B., and RAAFAT, A. M., Spectrographic estimation of Ga and Ge in coal and coke, *Indian J. Technol.*, 1974, **12**, 359. (National Research Centre, High Polytechnic Institute, Cairo, Egypt).

514 VITKUN, R. A., ZELYUKOVA, Y. V., and POLUEKTOV, N. S., Flameless AA determination of Hg in selenides and tellurides using formaldehyde as reducing agent, *Ukr. Khim. Zh.*, 1974, **40**, 1304. (Institute of General and Inorganic Chemistry, Academy of Sciences Ukraine, SSR, Odessa Laboratory, U.S.S.R.).

515 REITH, J. F., ENGELSMA, J., and VAN DITMARSCH, M., Pb and Zn contents of food and diets in the Netherlands, *Z. Lebensm.-Unters. Forsch.*, 1974, **156**, 271. (Dept. of Food Chemistry, State University of Utrecht, Utrecht, The Netherlands).

516 Fertiliser analysis (1974 to 1966), Annotated Bibliography No. 1683, published by Commonwealth Agricultural Bureaux, Slough, England, 1974.

517 Chemical determination of Fe and Mn in soil and plant material (1974 to 1968), Annotated Bibliography No. 1681, published by Commonwealth Agricultural Bureaux, Slough, England, 1974.

518 Chemical determination of Ca and Mg in soil and plant material (1973 to 1968), Annotated Bibliography No. 1675, published by Commonwealth Agricultural Bureaux, Slough, England, 1974.

519 Chemical determination of Cu and Zn in soil and plant material (1974 to 1969), Annotated Bibliography No. 1679, published by Commonwealth Agricultural Bureaux, Slough, England, 1974.

520 Chemical determination of B in soil and plant material (1974 to 1969), Annotated Bibliography No. 1673, published by Commonwealth Agricultural Bureaux, Slough, England, 1974.

521 Chemical determination of Al in soil and plant material (1974 to 1967), Annotated Bibliography No. 1674, published by Commonwealth Agricultural Bureaux, Slough, England, 1974.

522 SZIVOS, K., POLOS, L., FEHER, I., and PUNGOR, E., AA method for determination of Pb in air, *Period. Polytech., Chem. Eng.*, 1974, **18**, 281. (Address as in ref. 275).

523 ALDUAN, F. A., and ROCA ADELL, M., Spectrographic determination of traces of B on steel, *An. Quim.*, 1974, **70**, 821.

524 BARTOE, J. D. F., and BRUECKNER, G. E., New stigmatic, coma-free, concave-grating spectrograph, *J. Opt. Soc. Am.*, 1975, **65**, 13. (E. O. Hulburt Center for Space Research, Naval Research Lab., Washington, D.C. 20375, U.S.A.).

525 GREENSTEIN, H., and BATES, C. W., Line-width and tuning effects in resonant excitation, *J. Opt. Soc. Am.*, 1975, **65**, 33. (Dept. of materials Science and Engineering and Electrical Engineering, Stanford University, Stanford, Calif. 94305, U.S.A.).

526 FAIRBANK, W. M., HANSCH, T. W., and SCHAWLOW, A. L., Absolute measurement of very low Na-vapour densities using laser resonance fluorescence, *J. Opt. Soc. Am.*, 1975, **65**, 199. (Dept. of Physics, Stanford University, Stanford, Calif. 94305, U.S.A.).
527 CARLSEN, J. B., Thiol group determination in proteins by flameless AAS of Hg, *Anal. Biochem.*, 1975, **64**, 53. (Dept. of Biochemistry, University of Copenhagen, Copenhagen, Denmark).
528 WEI, P. S. P., TANG, K. T., and HALL, R. B., Forbidden transitions in the emission spectrum of atomic Al, *J. Chem. Phys.*, 1974, **61**, 3593. (Boeing Aerospace Co., Seattle, Wash., U.S.A.).
529 BROWNE, R. C., ELLIS, R. W., and WEIGHTMAN, D., Interlaboratory variation in measurement of blood Pb levels, *Lancet*, 1974, **2**, (7889), 1112. (Medical School, University of Newcastle upon Tyne, Newcastle upon Tyne, England).
530 SANZ BUENO, G., and PALACIOS, M. A., AA and its application to the analysis of biological substances, *Med. Segur. Trab.*, 1974, **22** (86), 17. (Dept. Quim. Anal., Cons. Super. Invest. Cient., Madrid, Spain).
531 ROSENTHAL, A. R., HOPKINS, J. L., APPLETON, B., and ZIMMERMAN, R., Studies on intraocular Cu foreign bodies, AAS, *Arch. Ophthalmol.*, 1974, **92**, 431. (Walter Reed Army Inst. Res., Walter Reed Army Medical Center, Washington, D.C., U.S.A.).
532 STONE, I. C., and PETTY, C. S., Examination of gunshot residues, *J. Forensic Sci.*, 1974, **19**, 784. (Physical Evidence Section, Southwest Institute of Forensic Science, Dallas, Tex., U.S.A.).
533 FIALA, K., STUDENY, M., and FIALOVA, K., Determination of plant-available Mg by AAS, *Arch. Acker-Pflanzenbau Bodenkd.*, 1974, **18**, 861. (Research Institute for Soil Science and Agronomy, Bratislava, Czechoslovakia).
534 LOCKWOOD, T. H., and LIMTIACO, L. P., Determination of Be, Cd and Te in animal tissues using electronically excited oxygen and AAS, *J. Am. Ind. Hyg. Assoc.*, 1975, **36**, 57. (Address as in ref. 44).
535 LINDE, T. R., CODY, T. E., GROHOWSKI, J. A., HORVATH, G. J., and DAVISON, A. L., Portable spectrometric apparatus, U.S. Appl. 343/305, 1973.
536 RASMUSSEN, W., SCHIEDER, R., and WALTHER, H., AF under monochromatic excitation: level crossing experiment on the 6s6p $^1P_1$ level of Ba, *Opt. Commun.*, 1974, **12**, 315. (Physics Institute, University of Koeln, Cologne, West Germany).
537 LERNER, L. A., TIKHOMIROVA, E. I., and PLOTNIKOVA, L. F., AA determination of total Ca, Mg and Mn in a soil decomposed by fusion with a borax and soda mixture, *Pochvovedenie*, 1975, (1), 122. (Pochv. Inst. im. Dokuchaeva, Moscow, U.S.S.R.).
538 ANDERSON, R. L., and SCOTT, W. R., Chemical identification of pressed boards, *AIChE, Symp. Ser.*, 1974, **70**, 106. (Pacific Northwest Division, Reichhold Chem., Inc., Tacoma, Wash., U.S.A.).
539 NOGUCHI, C., *et al*, Determination of microamounts of Cd, Cu, Ni and Mn in fats and oils by AAS, *Yukagaku*, 1975, **24**, 100. (Nippon Oils and Fats Co., Tokyo, Japan).
540 KRISHNAN, S. S., Detection of gunshot residue on the hands by neutron activation and AA analysis, *J. Forensic Sci.*, 1974, **19**, 789. (Chemistry Section, Centre for Forensic Science, Toronto, Ont., Canada).
541 DENKERT, E., Improved method for the determination of water, electrolyte and substrate content of small tissue probes, *Med. Lab.*, 1974, **27**, 241. (Abt. Herzchir, University of Munchen, Munich, West Germany).
542 HUNT, D. C., Screening technique for the presence of Pb and Cd in solid samples by AAS, *Lab. Pract.*, 1975, **24**, 411. (Laboratory of the Government Chemist, Cornwall House, Stamford Street, London, S.E.1, England).
543 KNIPPENBERG, W. F., Inorganic chemical analysis, *Philips Tech. Rev.*, 1974, **34**, 298.
544 BOUMANS, P. W. J. M., Multi-element analysis by optical emission spectrometry — rise or fall of an empire, *Philips Tech. Rev.*, 1974, **34**, 305. (Philips Research Laboratories, NV Philips Gloeilampenfabrieken, Eindhoven, The Netherlands).
545 WITMER, A. W., JANSEN, J. A. J., VAN GOOL, G. H., and BROUWER, G., A system for the automatic analysis of photographically recorded emission spectra, *Philips Tech. Rec.*, 1974, **34**, 322. (Address as in ref. 544).
546 KNEEBONE, B. M., and FREISER, H., Determination of CS_2 in industrial atmo-

spheres by an extraction–AA method, *Anal. Chem.*, 1975, **47**, 942. (Chemistry Dept., University of Arizona, Tucson, Ariz. 85721, U.S.A.).

547 TOMLJANOVIC, M., and GROBENSKI, Z., The analysis of Fe ores by AAS after pressure decomposition with HF in a PTFE autoclave, *At. Absorpt. Newsl.*, 1975, **14**, 52. (Bodenseewerk Perkin-Elmer GmbH, Uberlingen, West Germany).

548 SLAVIN, S., PETENSON, G. E., and LINDAHL, P. C., Determination of heavy metals in meats by AAS,*At. Absorpt. Newsl.*, 1975, **14**, 57. (Address as in ref. 8).

549 SLAVIN S., and LAWRENCE, D. M., An AA bibliography for January–June 1975, *At. Absorpt. News.*, 1975, **14**, 81. (Address as in ref. 8).

550 FULLER, C. W., The determination of trace metals in high-purity Pb silicate glasses by flameless AAS, *At. Absorpt. Newsl.*, 1975, **14**, 73. (Address as in ref. 364).

551 OWENS, J. W., and GLADNEY, E. S., Determination of Be in environmental materials by flameless AAS, *At. Absorpt. Newsl.*, 1975, **14**, 76. (Los Alamos Scientific Laboratory, P.O. Box 1663, Los Alamos, N.Mex. 87545, U.S.A.).

552 WRIGHT, F. C., and RINER, J. C., Determination of Cd in blood and urine with the graphite furnace, *At. Absorpt. Newsl.*, 1975, **14**, 103. (Address as in ref. 1).

553 MANNING, D. C., Aspirating small volume samples in flame AAS, *At. Absorpt. Newsl.*, 1975, **14**, 99. (Perkin-Elmer Corp., Norwalk, Conn. 06856, U.S.A.).

554 PICKFORD, C. J., and ROSSI, G., Determination of some trace elements in NBS (SRM-1577) bovine liver using flameless AA and solid sampling, *At. Absorpt. Newsl.*, 1975, **14**, 78. (Chemistry Division, Euratom, J.R.C., Ispra (Varese), Italy).

555 MINKKINEN, P., A method for the correction of the background absorption in Ag analysis of calcareous samples, *At. Absorpt. Newsl.*, 1975, **14**, 71. (Mineral Exploration in two areas (Phase II), Turkey U.N. Development Programme, PK 407, Ankara, Turkey).

556 TRIVINO, F., Determination of Ca in cement by AA, *At. Absorpt. Newsl.*, 1975, **14**, 70. (Instituto Eduardo Torroja, Avda. Pio XIII, s/n 'Costillares' (Chamartin), Madrid 33, Spain).

557 EDGAR, R. M., An alternative set of conditions for Mo determination by AAS, *At. Absorpt. Newsl.*, 1975, **14**, 68. (Eutectic Corp., Flushing, N.Y. 11358, U.S.A.).

558 MILLER, R. G, and DOERGER, J. U., Determination of Pt and Pd in biological samples, *At. Absorpt. Newsl.*, 1975, **14**, 66. (U.S. Environmental Protection Agency, Environmental Research Center, Environmental Toxicology Research Center, Cincinatti, Ohio 45268, U.S.A.).
Center, Cincinatti, Ohio 45268, U.S.A.).

559 SPERLING, K. R., Heavy metal determination in sea water and in marine organisms with the aid of flameless AAS. Part 2: Determination of Cd in biological material, *At. Absorpt. Newsl.*, 1975, **14**, 60. (Biologische Anstalt Helgoland, Laboratorium Sulldorf, Wustland 2, 2 Hamburg 55, West Germany).

560 BALL J. W., and GOTTSCHALL, W. C., Matrix interferences in the determination of Fe by AAS, *At. Absorpt. Newsl.*, 1975, **14**, 63. (University of Denver, Denver, Colo. 80210, U.S.A.).

561 BAGLIANO, G., BENISCHEK, F., and HUBER, I., Application of graphite furnace AA to the determination of impurities in U oxides without preliminary separation, *At. Absorpt. Newsl.*, 1975, **14**, 45. (International Atomic Energy Laboratory, A-2444, Seibensdorf, Austria).

562 JULSHAMN, K., and BRAEKKAN, O. R., Determination of trace elements in fish tissue by the standard addition method, *At. Absorpt. Newsl.*, 1975, **14**, 49. (Government Vitamin Institute, Directorate of Fisheries, P.O. Box 187, 5001, Bergen, Norway).

563 TREDWELL, C. J., and WEST, M. A., Lasers in physical chemistry, *Lab. Equip. Dig.*, 1975, **13**, 52. (Davy Faraday Research Laboratory, The Royal Institution, London, England).

564 HUSBANDS, A., AAS in the analysis of biological fluids, *Lab. Equip. Dig.*, 1975, **13**, 69. (Pye Unicam Ltd., York Street, Cambridge, England).

565 THOMPSON, K. C., and THOMERSON, D. R., Recent advances in flameless atomization techniques in AAS — Paper presented at the 1975 'LABEX' International Exhibition, London, England. (Address as in ref. 19).

566 FULLER, C. W., Flameless atomization: inorganic applications — Paper presented at the 1975 'LABEX' International Exhibition, London, England. (Address as in ref. 364).

567 BEVAN, D. G., and KIRKBRIGHT, G. F., The use of a demountable hollow

cathode lamp source and a piezo-electrically scanned Fabry–Perot interferometer for investigation of the isotopic composition of Pb ores, *Anal. Chim. Acta,* 1975, **76**, 361. (Address as in ref. 4).

568 SHIGEMATSU, T., MATSUI, M., FUJIMO, O., and KINOSHITA, K., Determination of Mn in natural waters by AAS with a carbon tube atomizer, *Anal. Chim. Acta,* 1975, **76**, 329. (Institute for Chemical Research, Kyoto University, Uji, Kyoto 611, Japan).

569 WIMBERLEY, J. W., The determination of total Hg at the p.p.b. level in soils, ores and organic materials, *Anal. Chim. Acta,* 1975, **76**, 337. (Analytical Research Section, Research and Development Dept., Continental Oil Company, Ponca City, Okla. 74601, U.S.A.).

570 DOGAN, S., and HAERDI, W., Study of the preconcentration and separation of traces of Hg by reduction on metallic Cu and its determination by flameless AAS, *Anal. Chim. Acta,* 1975, **76**, 345. (Dept. de Chimie Minerale et Analytique de l'Universite, Sciences II, 1211-Geneve 4, Switzerland).

571 CAMPBELL, D. E., and PASSMORE, W. O., $BaCO_3$–H_3BO_3, an advantageous flux for analysis of refractory materials by flame spectrometry, *Anal. Chim. Acta,* 1975, **76**, 355. (Sullivan Park Research and Development Laboratories, Corning Glass Works, Corning, N.Y. 14830, U.S.A.).

572 KORKISCH, J., and SORIO, A., Determination of Cu, Cd and Pb in natural waters after anion-exchange separation, *Anal. Chim. Acta,* 1975, **76**, 393. (Address as in ref. 191).

573 MITCHELL, D. G., WARD, A. F., and KAHL, M., Use of a microsampling cup system with a nitrous oxide-acetylene flame for determining less volatile metals, *Anal. Chim. Acta,* 1975, **76**, 456. (Address as in ref. 80).

574 McCULLOUGH, J. D., and DULEY, W. W., Isolation of reactive metals atoms in cyclohexane matrices at 55 to 133 K, *Spectrosc. Lett.,* 1975, **8**, 51. (Centre for Research in Experimental Space Science and Physics Dept., York University, Toronto, Ont. M3J 1P3, Canada).

575 MAY, L. A., and PRESLEY, B. J., Comparison of flameless AAS with neutron activation analysis for V in beach asphalt, *Spectrosc. Lett.,* 1975, **8**, 201. (Address as in ref. 5).

The following papers (*) were presented at the XVIII Colloqium Spectroscopicum Internationale, 15–19 September 1975, Grenoble, France.

576* DECKER, R. J., and KOBUS, H. J., The action of a buffer used in a d.c. arc spectrographic analysis in controlling the conditions in the anode (sample) electrode. (Address as in ref. 211).

577* BARNES, M., and SCHLEICHER, R. C., Computer simulation and experimental verification of an inductively coupled plasma discharge. (Address as in ref. 81).

578* KEIR, M. J., DAWSON, J. B., and ELLIS, D. J., An evaluation of Hadamard-transform spectroscopy as applied to the study of atomic line spectra. (Dept. of Medical Physics, The General Infirmary, Leeds LS1 3EX, England).

579* STRASHEIM, A., OAKES, A. R., SCOTT, R. H., and BLUM, F., A study of the medium-voltage and controlled-waveform spark resources as light sources for the analysis of Al using a sequential spectrometer. (Address as in ref. 304).

580* SCOTT, R. H., STRASHEIM, A., and OAKES, A. R., Application of inductively coupled plasmas to the analysis of Fe/Mn alloys. (Address as in ref. 304).

581* BELTCHEFF, B., and SALTCHEVA, M., Universal instrument for dispersion of dust samples in different gases and their precise concentration in the plasma arc during ES analysis. (Institut 'Mineralproekt' et Institut 'Niprorouda', Sofia, Bulgaria).

582* BARNETT, W. B., KERBER, J. D., and KNOTT, A. R., Optimiziation of instrumental electronics for use with flameless AA sampling devices. (Address as in ref. 220).

583* BIEBER, I. B., and STEJSKALOVA, A., The determination of Mg in the perlitic and ferritic phases of nodular iron by laser microanalysis. (Brno, Krenova 62, Czechoslovakia).

584* THOMPSON, K. C., and GODDEN, R. G., A high-solids nitrous oxide/acetylene burner and its application to the analysis of Al in steels. (Address as in ref. 19).

585* THOMPSON, K. C., and GODDEN, R. G., A method for monitoring excessive levels of Cd and Pb in whole blood using a rapid direct nebulization technique. (Address as in ref. 19).

586* THOMPSON, K. C., and GODDEN, R. G., The AF determination of Sb, As, Se and Te using a hydride generation technique. (Address as in ref. 19).
587* LAKATOS, I. T., The role of the heat of evaporation in solution spectrochemical analysis. (Petroleum Engineering Research Laboratory of the Hungarian Academy of Sciences, 3515 Miskolc-Egyetemvaros, P.O.B.2, Hungary).
588* BERNERON, R., Analysis in depth with the aid of the glow discharge for some metallic elements in steels. (Institut de Recherches de la Siderurgie Francaise, 185 Rue President Roosevelt, 78104 St. Germain en Laye, France).
589* BERNERON, R., and MOREAU, J. P., Study of the erosion conditions during spectral analysis using a glow discharge source. (Address as in ref. 588).
590* FRANK, P., and KRAUSS, L., The problem of appearance of the normal maxima of molecular bands in electric arcs. (DFVLR, Institut fur Reaktionskinetik, Stuttgart 80, West Germany).
591* VAN DER PIEPEN, H., and CLAASE, C., The use of a television camera as simultaneous multichannel photodetector on a spectrometer. (National Physical Research Laboratory, C.S.I.R., P.O. Box 395, Pretoria 0001, South Africa).
592* MAST, F., and PFEILSTICKER, K., The production of a low-voltage spark with a high repetition rate. (Institut fur Plasmaforschung, Stuttgart, West Germany).
593* GILMORE, J., Instruments and analysis in the XUV and VUV part of the spectrum. (Harvard, Mass., U.S.A.).
594* WAGENAAR, H. C., and DE GALAN, L., The influence of spectral line profiles upon analytical curves in AAS. (Address as in ref. 207).
595* KILROE-SMITH, T. A., Optimization of working conditions in flameless AAS. (Address as in ref. 204).
596* SIRE, J.,* COLLIN, J.,* and VOINOVITCH, J. A.,† Direct AAS determination of Cr, Cu, Mn, Ni and V in steels. (*Laboratoire Regional des Ponts et Chaussees, 75 Rue de la Grande Haie, 54510 Tombaline, France; †Laboratoire Central des Ponts et Chaussees, 58 Boulevard Lefebvre, 75732 Paris Cedex 15, France).
597* KORNBLUM, G. R., and DE CALAN, L., Temperature measurement and interference study in an induction coupled r.f. plasma by means of emission and absorption measurements. (Address as in ref. 207).
598* ALDUAN, F. A., CAPREVILA, C., and ROCA, M., Study of processes of vaporization, transport and excitation in the spectrochemical determination of impurities in $NH_4F.HF$. (Junta de Energia Nuclear, Division de Quimica Analitica, Madrid, Spain).
599* ZENTAI, P., Standardized excitation or excited standardization. (Hungarian Geological Survey, Budapest XIV, Nepstadion 14, Hungary).
600* POSMA, F. D., BALKE, J., and MAESSEN, F. J. M. J., Some investigations on analytical signals in flameless AAS. (Address as in ref. 245).
601* POSMA, F. D., Determination of trace metals in blood, plasma and serum by flameless AAS. (Address as in ref. 245).
602* DURRANT, K.,* AMBROSE, A. D.,† and ROBINSON, J. D.,‡ The application of optical ES to the problem of rapid alloy confirmation for production control. (*A.R.L. Ltd., Luton, England; †B.S.C., Tubes Works, Corby, England; ‡High Duty Alloys Ltd., Forgings Div., Redditch, England).
603* MOORE, J. L., and MOSTYN, R. A., The determination of trace metals in Ni-base alloys by flameless AAS. (Materials Quality Assurance Directorate, Royal Arsenal East, London SE18 6TD, England).
604* NICKEL, H., MAZURKIEWICZ, M., and PEUSER, F., Investigations to determine the contamination of reactor graphite by means of laser micro ES analysis. (Institut fur Reaktorwerkstoffe der Kernforschungsanlage Julich GmbH, Julich, West Germany).
605* NICKEL, H.,* MAZURKIEWICZ, M.,* MOLLER, H.,* VUKANOVIC, D. D.,† SIMIC, M. M.,† and VUKANOVIC, V. M.,† Investigations at the d.c. arc in graphite cylinder. (*Address as in ref. 604; †Address as in ref. 213).
606* BOUMANS, P. W. J. M., and DE BOER, F. J., A low-powered inductively coupled h.f. argon plasma for simultaneous multi-element analysis of solutions by ES: a progress report. (Address as in ref. 544).
607* BOUMANS, P. W. J. M.,* DE BOER, F. J.,* DAHMEN, J.,† HOLZEL, H.,† and MEIER, A.,† Unambiguous comparison of some analytical performance characteristics of an inductively coupled h.f. plasma and a capacitively coupled microwave

plasma for solution analysis by ES. (*Address as in ref. 544; † E. Merck, Analytisches Zentrallaboratorium, Darmstadt, West Germany).
608* MAESSEN, F. J. M. J.,* ELGERSMA, J. W.*, and BOUMANS, P. W. J. M.,† A systematic and unbiased statistical approach for establishing the uncertainty of analytical results: application to a comparison of alternative experimental procedures for simultaneous multi-element trace analysis in geological materials using dc. arc ES. (*Address as in ref. 206; †Address as in ref. 544).
609* KO, J. B., and LAQUA, K., Comparison of different excitation methods by the ES analysis of Al alloys. (Institut fur Spektrochemie und Angewandte Spektroskopie, 46 Dortmund, West Germany).
610* LEIS, F., QUENTMEIER, A., HAGENAH, W. D., and LAQUA, K., Two new methods in laser spectral analysis. (Address as in ref. 609).
611* MASSMANN, H., and EL GOHARY, Z., Background measurement in AAS. (Institut fur Spektrochemie und Angewandte Spektroscopie, 4600 Dortmund, West Germany).
612* KLOCKENKAMPER, R., and LAQUA, K., Detection limit for absolute quantities by laser analysis with additional transverse spark excitation. (Address as in ref. 609).
613* NEDDEN, P., Reproducibility with a 500 Hz monoalternance in the analysis of steels: a statistical study. (M.B.L.E., Dept. Spectrographie, Rue des Deux Gares 80, B-1070 Bruxelles, Belgium).
614* RIANDEY. C., and PINTA, M., Direct determination of metallic traces in atmospheric particles in contact with the oceans by flameless AA. (Laboratoire de Spectrographie, O.R.S.T.O.M., 93140 Bondy, France).
615* DAHLQUIST, R. L., and KNOLL, J. W., Solids and liquids analysis using r.f. inductively coupled argon plasma optical ES. (Hasler Research Center, A.R.L., 95 La Patera Lane, Goleta, Calif. 93017, U.S.A.).
616* TOROK, T., BUZASE, A., and ZARAY,G., Experiments with a low temperature hollow cathode. (Institut fur Anorganisch und Analytische Chemie der L. Eotvos Universitat, Budapest, Hungary).
617* ZIMMER, K.,* JARO, M.,* and GEGUS, E.,† Application of spectrochemical methods to the investigation of Cu and bronze findings. (*Zentralinstitut fur Museumswesen, Budapest, Hungary; †Institut fur Analytische Chemie der Universitat fur Chemische Industrie in Veszprem, Hungary).
618* PRICE, W. J., and WHITESIDE, P. J., An investigation of the determination of B by AA and flame ES. (Address as in ref. 172).
619* ABDALLAH, M. H., JAROSZ, J., MERMET, J. M., TRASSY, C., and ROBIN, J., Comparison of h.f. plasma generators used for ES analysis. (Address as in ref. 182).
620* TRASSY, C., and ROBIN, J., Elimination of noise and signal detection in AAS. (Address as in ref. 182).
621* POUEY, M.,* LEPERE, D.,† and FLAMAND, J.,† Recent improvements in vacuum ultraviolet monochromators. (*Address as in ref, 23; †Instruments S.A. Div., Jobin et Yvon, 91160 Longjumeau, France).
622* CALOP, J.,* BADOR, R.,† VANDROUX, J. C.,† ISOARD, P.,‡ and FONTANGES, R.,‡ Determination of trace elements in the Lyon atmosphere. (*Faculte de Medecin et de Pharmicie de Grenoble, Domaine de la Merci, 38700 La Tronche, France; †Laboratoire de Physique, U.E.R. de Pharmacie, 69008 Lyon, France; ‡Centre de Recherches du Service de Sante des Armees, 108 Blvd. Pinel, 69272 Lyon Cedex 1, France).
623* IKONOMOV, N., and PAVLOVIC, B., Evaluation of the heat of the chemical reaction from the arc discharge data. (Faculty of Sciences, Faculty of Technology and Metallurgy and Institute of Chemistry, Belgrade, Yugoslavia).
624* DESQUESNES, W., and URBAIN, H., Instrumental substraction of flame backgrounds with a differential emission device. (Address as in ref. 217).
625* TRIPOVIC, M., and VUKANOVIC, V., Addition of I in spectrochemical analyses of traces in a d.c. arc. (Institute of Physics, Faculty of Sciences of the University of Belgrade, Belgrade, Yugoslavia).
626* FIJALKOWSKI, J., CZAKOW,J., and KUCHARZEWSKI, B., Some remarks concerning the determination of non-metallic elements by ES. (Dept. of Analytical Chemistry, Institute of Nuclear Research, Warsaw, Poland).
627* STRZYEWSKA, B., and FIJALKOWSKI, J., Excitation of high-purity Al_2O_3 in the constant-current arc under different protective gas atmospheres. (Address as in ref. 626).

628* RADMACHER, H. W., Spectrochemical analysis for geochemical surveys. (South African Iron and Steel Industrial Corp., P.O. Box 450, Pretoria, South Africa).
629* PETROVIC, D., IKONOMOV, N., and PAVLOVIC, B., Radial distribution of particles in the arc burning in vertical water-cooled metal tubes. (Adress as in ref. 623).
630* BERNHARD, A. E., and WESLEY KEMP, J., A new spectrochemical data aquisition and readout system. (Labtest Equipment Co., Systron Donner Corp., 11828 La Grange Avenue, Los Angeles, Calif. 90025, U.S.A.).
631* ZLANTANOVIC, M., TODOROVIC, M. S., and GEORGIJEVIC, V. J., The relative influence of electrodes and the neutral surrounding gas on the rotating arc motion. (Institute of Physics, Belgrade, Yugoslavia).
632* CANTLE, J. E., and MULLINS, C. B., Peak area measurement of transient signals in AAS. (Instrumentation Laboratory (U.K.) Ltd., Station House, Stamford New Road, Altrincham, Cheshire, England).
633 THOMPSON, K. C., and GODDEN, R. G., Improvements in the AF determination of Hg by the cold vapour technique, *Analyst,* 1975, **100**, 544. (Address as in ref. 19).
634 NALL, W. R., BRUMHEAD, D., and WITHAM, R., A composite scheme for the analysis of steels by AAS using the air/acetylene flame, *Analyst,* 1975, **100**, 555. (Address as in ref. 384).
635 GREEN, H. C., The effect of valency on the determination of Cr in $HClO_4$ media by AAS, *Analyst,* 1975, **100**, 640. (Metallurgy Section, Auckland Industrial Development Div., D.S.I.R., Auckland, New Zealand).
636 DAMIANI, M., DEL MONTE TAMBA, M. G., and BIANCHI, F., Determination of Al, Ca, Mn and Ti in Fe/Si alloys by AAS, *Analyst,* 1975, **100**, 643. (Centro Sperimentale Metallurgico SpA, Via di Castel Romano, 00129, Italy).
637 MATSUNAGA, K., Concentration of Hg by three species of fish from Japanese rivers, *Nature,* 1975, **257**, 49. (Chemistry Dept., Faculty of Fisheries, Hokkaido University, Hakodate, Japan).
638 HIEFTJE, G. M., and BYSTROFF, R. I., An investigation of noise spectra from sheathed and unsheathed air/acetylene flames, *Spectrochim. Acta, Part B,* 1975, **30***B*, 187. (Address as in ref. 226).
639 HUMAN, H. G. C., and ZEEGERS, P. J. T., Molecular fluorescence of CaOH, SrOH and BaCl in flames, *Spectrochim. Acta, Part B,* 1975, **30***B*, 203. (Fysisch Laboratorium, Ryksuniversitiet, Utrecht, The Netherlands).
640 PIEPMEIER, E. H., and DE GALAN, L., Profiles of the Ca resonance line emitted by a modulated hollow cathode lamp, *Spectrochim. Acta, Part B,* 1975, **30***B*, 211. (Address as in ref. 207).
641 MOENKE BLANKENBURG, L., *et al.*, New aspects of apparatus and method for laser emission microanalysis, *Spectrochim. Acta, Part B,* 1975, **30***B*, 227. (Jenoptik Jena GmbH, Carl Zeiss St. 1, 69 Jena, East Germany).
642 GATZE, J., Intense line source for the long wavelength vacuum ultraviolet region, *Spectrochim. Acta, Part B,* 1975, **30***B*, 235. (Zentralinstitut fur Optik und Spekstroskopie, Akademie der Wissenschaften der DDR, 1199 Berlin, East Germany).
643 VAN CALKER, J., and HOLLENBERG, K., An electronically regulated spark generator for frequencies up to 3000 Hz, *Spectrochim. Acta, Part B,* 1975, **30***B*, 243. (Physikalisches Institut der Universitat Dusseldorf, Dusseldorf, West Germany).
The following papers (*) were presented at the 170th American Chemical Society National Meeting, 24–29 August, 1975, Chicago, Ill., U.S.A.
644* PIEPMEIER, E. H., and DE GALAN, L., Temporal behaviour of line profiles of pulsed hollow cathode lamps. (Address as in ref. 207).
645* HARRISON, W. W., The hollow cathode discharge as an emission and ionization source. (Dept. of Chemistry, University of Virginia, Charlottesville, Va. 22903, U.S.A.).
646* CROUCH, S. R., A miniature spark discharge system for multi-element analysis of solution samples. (Address as in ref. 99).
647* HAAS, W. J., FASSEL, V. A., and KNISELEY, R. N., Inductively coupled plasma-optical ES: simultaneous multi-element determination of trace elements in urine. (Address as in ref. 76).
648* VICKERS, T. J., and RIPPETOE, W. E., A new d.c. plasma jet device for optical ES. (Address as in ref. 90).
649* OLIVARES, D., and HIEFTJE, G. M., Tunable dye lasers in AFS. (Address as in ref. 226).

650* FITZGERALD, J. J., JOHNSON, D. J., and WINEFORDNER, J. D., Multi-element analysis via AFS. (Address as in ref. 487).
651* KELIHER, P. N., Applications of echelle spectrometry to multi-element AS. (Chemistry Dept., Villanova University, Villanova, Pa. 19085, U.S.A.).
652* PARDUE, H. L., COOK, T. E., and FELKEL, H. L., Evaluation of a custom designed vidicon spectrometer for multi-element analyses. (Address as in ref. 471).
653* GANJEI, J. D., ROTH, J. R., HOWELL, N. G., and MORRISON, G. H., Multi-element AS using a computerized vidicon spectrometer. (Address as in ref. 130).
654* CODDING, E. G., A silicon photodiode array detector system for multi-element spectrochemical analysis. (Address as in ref. 89).
655* HINNERS, T. A., FAEDER, E. J., and KING, L. C., AA analysis of un-ashed tissues. (Address as in ref. 7).
656* FUWA, K., and HARAGUCHI, H., Molecular flame absorption spectroscopy and its application to analytical chemistry. (Address as in ref. 242).
657* WOODIS, T. C., HUNTER, G. B., and JOHNSON, F. J., Statistical studies of matrix effects on flameless AAS determination of Cd and Pb. (Div. of Chemical Development, Tennessee Valley Authority, Muscle Shoals, Ala. 35660, U.S.A.).
658* WEI, P. S. P., The ES of atomic Al in laser-produced plasmas. (Address as in ref. 528).
659 VANDEBERG, J. T., SWAFFORD, H. D., and SCOTT, R. W., Determination of low concentrations of Pb in paint by AAS, *J. Paint Technol.*, 1975, **47** (604), 84.
660 HARRISON, R. M., PERRY, R., and SLATER, D. H., An adsorption technique for the determination of organic Pb in street air, *Atmos. Environ.*, 1974, **8**, 1187.
661 ANON., Analysis for trace quantities of Pb — Report issued by Ethyl Corp., U.S.A., (undated); *Lead Abstr.*, 1975, **15** (4: July), abstr. 15-0837. (Ethyl Corp., P.O. Box 341, Baton Rouge, La. 70821, U.S.A.).
662 KNAPP, G., Mechanized system for the performance of wet decomposition methods for organic matrices, *Fresenius' Z. Anal. Chem.*, 1975, **274**, 271. (Institut fur Allgemeine Chemie, Mikro- u. Radiochemie der Technischen Hochschule, Graz, Austria).
663 BEHNE, D., and MATAMBA, P. A., Drying and ashing of biological samples in trace element determinations by neutron activation analysis, *Fresenius' Z. Anal. Chem.*, 1975, **274**, 195. (Hahn-Meitner Institut, Bereich Kernchemie und Reaktor, D-1000 Berlin, 39-Wannsee, West Germany).
664 GRAMPUROHIT, S. V., and KAIMAL, V. N. P., Spectrographic determination of rare earths in high purity Gd_2O_3 using Stallwood jet. *Fresenius' Z. Anal. Chem.*, 1975, **274**, 181. (Spectroscopy Div., Bhabha Atomic Research Centre, Trombay, Bombay-85, India).
665 RADIC-PERIC, J. B., and VUKANOVIC, V. M., Formation of CaF and AlF in the arc plasma and spectrochemical determination of Ca and Al, *Fresenius' Z. Anal. Chem.*, 1975, **274**, 177. (Address as in ref. 213).
666 PERIC, M. N., VUKANOVIC, V. M., and TODOROVIC, P. S., Transport of CaO in a d.c. arc plasma, *Fresenius' Z. Anal. Chem.*, 1975, **274**, 109. (Address as in ref. 213).
667 DEWIT, A., DEWIJN, R., SMEYERS-VERBEKE, J., and MASSART, D. L., Ion-association extraction combined with low--temperature ashing for the determination of Cd in foodstuffs, *Bull. Soc. Chim. Belg.*, 1975, **84**, 91. (Farmaceutisch Institut, Vrije Universieit Brussel, B-1640 Sint-Genesius-Rode, Brussels, Belgium).
668 WATTERSON, J. R., and NEUERBERG, G. J., Analysis for Te in rocks to 5 p.p.b., *J. Res. U.S. Geol. Surv.*, 1975, **3**, 191. (U.S. Geological Survey, Denver, Colo., U.S.A.).
669 TALMI, Y, Applicability of TV-type multichannel detectors to spectroscopy, *Anal. Chem.*, 1975, **47**, 658A. (Address as in ref. 161).
670 TALMI, Y., TV-type multichannel detectors, *Anal. Chem.*, 1975, **47**, 697A. (Address as in ref. 161).
671 KAWAGUCHI, H., and VALLEE, B. L., Microwave excitation ES: determination of pg quantities of metals in metalloenzymes, *Anal. Chem.*, 1975, **47**, 1029. (Address as in ref. 263).
672 ALDOUS, K. M., MITCHELL, D. G., and JACKSON, K. W., Simultaneous determination of seven trace metals in potable water using a vidicon AA spectrometer, *Anal. Chem.*, 1975, **47**, 1034. (Address as in ref. 80).
673 AGEMAIN, H., ASPILA, K. I., and CHAU, A. S. Y., Comparison of the perform-

ance of the single and triple slot air/acetylene burners for AAS, *Anal. Chem.*, 1975, **47**, 1038. (Address as in ref. 185).

674 BEGNOCHE, B. C., and RISBY, T. H., Determination of metals in atmospheric particulates using low-volume sampling and flameless AAS. *Anal. Chem.*, 1975, **47**, 1041. (Dept. of Chemistry, Pennsylvania State University, University Park, Pa. 16802, U.S.A.).

675 LUKASIEWICZ, R. J., BERENS, P. H., and BUELL, B. E., Rapid determination of Pb in gasoline by AAS in the nitrous oxide/hydrogen flame, *Anal. Chem.*, 1975, **47**, 1045. (Research Laboratories, Union Oil Co. of California, Brea, Calif. 92621, U.S.A.).

676 PIERCE, F. D., GORTATOWSKI, M. J., MECHAM, H. D., and FRASER, R. S., Improved automated extraction method for AAS, *Anal. Chem.*, 1975, **47**, 1132. (Utah State Division of Health, Bureau of Laboratories, Salt Lake City, Utah, U.S.A.).

677 SWAIN, H. A., LEE, C., and ROZELLE, R. B., Determination of the solubility of $Mn(OH)_2$ and MnO_2 at 25° C by AAS, *Anal. Chem.*, 1975, **47**, 1135. (Dept. of Chemistry, Wilkes College, Wilkes-Barre, Pa. 18703, U.S.A.).

678 McCORRISTON, L. L., and RITCHIE, R. K., Determination of Pb in gasoline by AAS using a total consumption burner, *Anal. Chem.*, 1975, **47**, 1137. (Gulf Oil Canada Ltd., Research and Development Dept., Sheridan Park, Ont., Canada).

679 SHEINSON, R. S., and WILLIAMS, F. W., Cool flames: use of the term in combustion chemistry and analytical chemistry, *Anal. Chem.*, 1975, **47**, 1197. (Chemical Dynamics Branch, Chemistry Div., Naval Research Laboratory, Washington, D.C. 20375, U.S.A.).

680 ALLKINS, J. R., Tunable lasers in analytical spectroscopy, *Anal. Chem.*, 1975, **47**, 752A. (15485 One Oak Lane, Monte Sereno, Calif. 95030, U.S.A.).

681 HORLICK, G., and YUEN, W. K., AS measurements with a Fourier transform spectrometer, *Anal. Chem.*, 1975, **47**, 775A. (Address as in ref. 89).

682 JOHNSON, D. J., SHARP, B. L., WEST, T. S., and DAGNALL, R. M., Some observations on the vaporization and atomization of samples with a carbon filament atomizer, *Anal. Chem.*, 1975, **47**, 1234. (Address as in ref. 112).

683 STURGEON, R. E., CHAKRABARTI, C. L., MAINES, I. S., and BERTELS, P. C., Atomization in graphite-furnace AAS. Peak height *vs.* integration method of measuring absorbance: CRA 63, *Anal. Chem.*, 1975, **47**, 1240. (Dept. of Chemistry, Carleton University, Ottawa, Ont. K1S 5B6, Canada).

684 STURGEON, R. E., CHAKRABARTI, C. L., and BERTELS, P. C., Atomization in graphite-furnace AAS. Peak height method *vs.* integration method of measuring absorbance: HGA 2100, *Anal. Chem.*, 1975, **47**, 1250. (Address as in ref. 683).

685 RUNNELS, J. H., MERRYFIELD, R., and FISHER, H. B., Analysis of petroleum for trace metals: a method for improving detection limits for some elements with the graphite furnace atomizer, *Anal. Chem.*, 1975, **47**, 1258. (Phillips Petroleum Company, Bartlesville, Okla. 74004, U.S.A.).

686 KNAUER, H. E., and MILLIMAN, G. E., Analysis of petroleum for trace metals: determination of Hg in petroleum and petroleum products, *Anal. Chem.*, 1975, **47**, 1263. (Analytical Section, Mobil Research and Development Corp., Paulsboro, N.J., U.S.A.).

687 ROBBINS, W. K., and WALKER, H. H., Analysis of petroleum for trace metals: determination of trace quantites of Cd in petroleum by AAS, *Anal. Chem.*, 1975, **47**, 1269. (Address as in ref. 139).

688 TREYTL, W. J., MARICH, K. W., and GLICK, D., Spatial differentiation of optical emission in Q-switched laser-induced plasmas and effects on spectral line analytical sensitivity, *Anal. Chem.*, 1975, **47**, 1275. (Div. of Histochemistry, Dept. of Pathology, Stanford University School of Medicine, Stanford, Calif. 94305, U.S.A.).

689 BRINKMAN, D. W., and SACKS, R. D., Exploding wires as an intense ultraviolet continuum excitation source with preliminary application to AFS, *Anal. Chem.*, 1975, **47**, 1279. (Address as in ref. 78).

690 BLOOD, E. R., and GRANT, G. C., Determination of Cd in fish tissue by flameless AA with a tantalum ribbon, *Anal. Chem.*, 1975, **47**, 1438. (Dept. of Chemistry, Virginia Commonwealth University, Richmond, Va. 23284, U.S.A.).

691 BEDARD, M., and KERBYSON, J. D., Determination of trace Bi in Cu by hydride-evaluation AAS, *Anal. Chem.*, 1975, **47**, 1441. (Noranda Research Centre, Pointe Claire, Que., Canada).

692 COX, L. E., Drift-compensating integrator for measurement of transient AA signals,

Anal. Chem., 1975, **47**, 1493. (University of California, Los Alamos Scientific Laboratory, Los Alamos, N. Mex. 87544, U.S.A.).

693 DENTON, H., SHARP, B. L., and WEST, T. S., A stable d.c. capillary arc plasma for solution analysis, *Talanta*, 1975, **22**, 379. (Address as in ref. 112).

694 HARRINGTON, D. E., and BRAMSTEDT, W. R., Determination of Ru and Ir in anode coatings by AAS, *Talanta*, 1975, **22**, 411. (Address as in ref. 168).

695 OHTA, K., and SUZUKI, M., Trace metal analysis of rocks by flameless AAS with a metal micro-tube atomizer, *Talanta*, 1975, **22**, 465. (Dept. of Chemistry, Faculty of Engineering, Mie University, Kamihama-cho, Tsu-chi, Mie-ken, Japan).

696 HARIZANOV, Y., and JORDANOV, N., Spectrographic determination of impurities in ultra-pure W and WO_3, *Talanta*, 1975, **22**, 485. (Institute of General and Inorganic Chemistry, Bulgarian Acamedy of Sciences, Sofia 13, Bulgaria).

697 BROOKS, R. R., and SMYTHE, L. E., The progress of analytical chemistry 1910–1970, *Talanta*, 1975, **22**, 495. (Address as in ref. 159).

698 HIRAIDE, M., and MIZUIKE, A., Flotation of traces of Ag and Cu(II) ions with a methyl cellosolve solution of dithiozone, *Talanta*, 1975, **22**, 539. (Address as in ref. 371).

699 FOGG, A. G., SOLEYMANLOO, S., and THORBURN BURNS, D., Masking of Fe with F in the extractive AAS determination of Cr in steel, *Talanta*, 1975, **22**, 541. (Chemistry Dept., University of Technology, Loughborough, Leicestershire, England).

700 GOLEMBESKI, T., Determination of submicrogram amounts of Se in rocks by AAS, *Talanta*, 1975, **22**, 547. (Dept. of Chemistry, Lowell Technological Institute, Lowell, Mass., U.S.A.).

701 GREENFIELD, S., McGEACHIN, H. McD., and SMITH, P. B., Plasma emission sources in analytical spectroscopy. II, *Talanta*, 1975, **22**, 553. (Address as in ref. 104).

702 ORTNER, H. M., and KANTUSCHER, E., Impregnation of graphite tube with metal salts for improvement of the AAS determination of Si, *Talanta*, 1975, **22**, 581. (Metallwerk Plansee AG and Co., KG, A-6600 Reutte, Austria).

The following papers (*) were presented at the 26th Pittsburgh Conference, 3–7 March, 1975, Cleveland, Ohio, U.S.A.

703* BEGNOCHE, B. C., and RISBY, T. H., Determination of metals in atmospheric particulates using low-volume sampling and flameless AAS. (Address as in ref. 674).

704* McAVOY, R. L.,* and ERDMANN, D. E.,† The precision and accuracy that can be expected when analysing natural water samples. (U.S. Geological Survey, *W.R.D., Central Laboratory, 1745, West 1700 South, Salt Lake City, Utah 84104, U.S.A.; †Analytical Methods Research, Denver Federal Center, Lakewood, Colo. 80225, U.S.A.).

705* JACKSON, K. W., and MITCHELL, D. G., The rapid determination of metals in biological tissues by microsampling cup AAS. (Address as in ref. 80).

706* SLAVIN, S., PETERSON, G. E., and LINDAHL, P. C., Determination of heavy metals in food products by AA. (Address as in ref. 8).

707* BUONO, J. A., MORGENTHALER, L. P., and LUCIANO, V. J., The rapid determination of heavy metals in whole blood and blood sera. (Analytical Instrument Div., Fisher Scientific Co., 590 Lincoln Street, Waltham, Mass. 02154, U.S.A.).

708* ALDOUS, K. M., JACKSON, K. W., and MITCHELL, D. G., Determination of Pb in biological and environmental samples by AAS. (Address as in ref. 80).

709* WOODRUFF, T., and MALMSTADT, H. V., A new arc/flame source in an automated spectrometer for multi-element analyses via AE and AF. (School of Chemical Sciences, University of Illinois at Urbana-Champaign, Urbana, Ill. 61801, U.S.A.).

710* KUO, P. K., and MALMSTADT, H. V., Application of hollow cathode lamps in a programmed high-current mode for fluoresecence and absorption molecular spectrometry. (Address as in ref. 709).

711* BRINKMAN, D. W., and SACKS, R. D., Controlled wire explosions: a new continuum excitation source for AF. (Address as in ref. 78).

712* TOWNE, D. P., CALLICOAT, D. L., and McCLASKEY, T. L., Hollow cathode ES analysis of trace elements in Ni-base alloys. (Huntington Alloy Products Division, International Nickel Co. Inc., Huntington, W.Va. 25720, U.S.A.).

713* NASH, D. L., and WOOD, D. L., Spectrographic analyses for diffused metals in Au plating with a glow discharge source. (Bell Laboratories Murray Hill, N.J. 07974, U.S.A.).

714* GLASS, E., and CROUGH, S. R., Characterization and design of a miniature spark

discharge for solution and gas analysis. (Address as in ref. 99).
715* MARKS, J. Y., WELCHER, G. G., and SCUSSELL, D. J., Determination of traces of Ge, In, Ga, Zn, Sb, Cd, and Tl by the carrier distillation–ES technique in high-temperature alloys. (Materials Engineering and Research Laboratory, Pratt and Witney Aircraft, East Hartford, Conn. 06108, U.S.A.).
716* GOLIGHTLY, D. W., THOMAS, C. P., DORRZAPF, A. F., and ANNELL, C. S., Improved accuracy in computerized ES analysis of geologic materials through corrections for matrix effects. (Address as in ref. 218).
717* BLEVINS, D. R., and O'NEILL, W. R., Spectrographic calculations with computerized emulsion calibration and programmable calculator computations. (Ethyl Corporation Research Laboratories, 1600 West 8 Mile Road, Ferndale, Mich. 48220, U.S.A.).
718* BRINKMAN, D. W., and SACKS, R. D., Computer-aided photographic emulsion calibration for quantitative spectrometry. (Address as in ref. 78).
719* STEINHAUS, D. W., and COX, L. E., Direct calculation of a simple formula for the calibration of a photographic emulsion using exposures through a two-step filter. (Address as in ref. 692).
720* BRIGGS, T. H.,* and KRAFT, E. A.,† Effect of different spectrographs on laser emission microprobe sensitivities. (Western Electric, *555 Union Boulevard, Allentown, Pa. 18103, U.S.A.; †6200 East Broad Street, Columbus, Ohio 43213, U.S.A.).
721* LUEDTKE, N. A., FASCHING, J. L., and HAMMOCK, J. P., Elemental analysis of selected sediments by neutron activation analysis and AAS. (Dept. of Chemistry, University of Rhode Island, Kingston, R.I. 02881, U.S.A.).
722* SANYAL, R. M.,* BANERJEE, B. K.,* and CHAKRABURTTY, A. K.,† Spectrophotometric and AAS assaying of metallic micronutrients in soil with buiret as reagent. (*Physical Research Wing, Planning and Development Div., Fertiliser Corporation of India Ltd., Sindri, Bihar, India; †Dept. of Chemistry, Jadavpur University, Calcutta 32, India).
723* BISHOP, J. N., TAYLOR, L. A., and DIOSADY, P. L., High temperature acid digestion for the determination of Hg in environmental samples. (Ministry of the Environment, Laboratory Services Branch, P.O. Box 213, Rexdale, Ont. M9W 5L1, Canada).
724* DIOSADY, P. L., DUHOLKE, W. K., and WANG, D. T., Identification of methyl-Hg in environmental samples. (Address as in ref. 723)
725* GOODE, S. R., and OTTO, D. C., Optimization of experimental parameters influencing microwave-excited electrodeless discharge lamps. (Address as in ref. 109).
726* VICKERS, T. J., JOHNSON, E. R., and WOLFE, T. C., Automated optimization of pulsed hollow cathode discharge tubes. (Address as in ref. 90).
727* SPILLMAN, R., and MALMSTADT, H. V., New instrumental developments in an automated multi-element AF/AE spectrometer. (Address as in ref. 709).
728* MARTIN, S. J., and MALMSTADT, H. V., Application of Fourier transform techniques to multi-element non-dispersive AF determinations. (Address as in ref. 709).
729* EPSTEIN, M. S., * RAINS, T. C.,* and O'HAVER, T. C.,† The analysis of S in metal alloys by molecular emission on a carbon cup atomizer. (*†Addresses as in refs. 34 and 219).
730* HORLICK, G., YUEN, W. K., and BETTY, K. R., Simultaneous multi-element AES using a Fourier transform spectrometer. (Address as in ref. 89).
731* KELIHER, P. N., and WOHLERS, C. C., Some AA continuum studies at wavelengths below 3000 Å. (Address as in ref. 651).
732* ALLKINS, J. R., The tunable dye laser and spectroscopy: some recent applications. (Address as in ref. 680).
733* MITCHELL, D. G., CANELLI, E., and ALDOUS, K. M., Determination of organic carbon in waters using a microsampling cup/plasma ES technique. (Address as in ref. 80).
734* WOOD, D. L., DARGIS, A. B., and NASH, D. L., A computerized TV spectrometer for emission analysis. (Address as in ref. 713).
735* BRECH, F, and CRAWFORD, R., Trace level multi-element determinations through inductively coupled argon plasma excitation. (Address as in ref. 17).
736* SILVESTER, M. B., and ABERCROMBIE, F. N., Argon r.f. plasma sample introduction. (Barringer Research Ltd., 304 Carlingview Drive, Rexdale, Ont. M9W 5G2, Canada).

737* ABERCROMBIE, F. N., and SILVESTER, M. B., Applications of the argon r.f. plasma to real sample systems. (Address as in ref. 736).
738* MONTASER, A., FASSEL, V. A., and GOLDSTEIN, S. A., AFS with an inductively coupled plasma atomization source: some preliminary observations. (Address as in ref. 98).
739* SOBEL, H. R., KNISELEY, R. N., SUTHERLAND, W. L., and FASSEL, V. A., Simultaneous multi-element determination of trace elements in foods by inductively-coupled plasma excitation. (Address as in ref. 98).
740* DAHLQUIST, R. L., KNOLL, J. W., and HOYT, R. E., Thermal and direct aerosol generation: alternate method for sample presentation to the inductively coupled plasma. (Address as in ref. 615).
741* AJHAR, R. M., DALAGER, P. D., and DAVISON, A. L., Multi-element analysis in the laboratory with an inductively coupled plasma/optical emission system. (Applied Research Laboratories, 9545 Wentworth Street, Sunland, Calif. 91040, U.S.A.).
742* CULVER, B. R., ROWE, C. J., and DELLES, F., Trace metals in biological samples by non-flame AA. (Address as in ref. 248).
743* DENTON, M. B., and ROUTH, M. W., Investigations into flame spectrochemical systems utilizising interactive computer control. (Address as in ref. 244).
744* HORLICK, G., YUEN, W. K., and BETTY, K. R., Simultaneous multi-element AA analysis using a computer-coupled photodiode-array spectrometer. (Address as in ref. 89).
745* FUTRELL, T. L., and MORROW, R. W., An automated AA spectrometer utilizing a programmable desk calculator. (Address as in ref. 460).
746* ISSAQ, H. J., and ZIELINSKI, W. L., A new design of a graphite tube atomizer for AAS. (N.C.I., Frederick Cancer Research Center, P.O. Box B, Frederick, Md. 21701, U.S.A.).
747* POULOS, T. J., The operation and performance of a new digital AA spectrometer. (Jarrell-Ash Division, Fisher Scientific Co., Waltham, Mass., U.S.A.)
748* BARNETT, W. B., BOHLER, W., and KERBER, J. D., Optimization of instrumental electronics for use with flameless AA sampling devices. (Address as in ref. 220).
749* SOTERA, J. J., BANCROFT, M. F., and HWANG, J. Y., Recent development of in situ sample pretreatment techniques and their applications in flameless AA. (Address as in ref. 313).
750* INGLE, J. D., and HAWLEY, J. E., Comparison of cold vapour AA and AF Hg analysis. (Address as in ref. 67).
751* JOHNSON, J. D., Analyses with the demountable hollow cathode. (Spectrogram Corporation, 358 State Street, North Haven, Conn. 06473, U.S.A.).
752* BOSS, C. B., and HIEFTJE, G. M., Studies on the diffusion of atoms and ions from individual solute particles vaporizing in a laminar flame. (Address as in ref. 226).
753* SMITH, S. B., BANCROFT, M. F., and HWANG, J. Y., Peak height or area integration measurements? Theory and practice as applied to nonflame AA techniques. (Address as in ref. 313).
754* FERNANDEZ, F. J., EDIGER, R. D., and KERBER, J. D., Instrumental and chemical methods for the reduction of nonspecific absorption with the graphite furnace. (Address as in ref. 199).
755* INGLE, J. D., and BOWER, N. W., Comparison of experimental and theoretical precision of AA measurements. (Address as in ref. 67).
756* ADAMS, M. J., KIRKBRIGHT, G. F., and WILSON, P. J., The application of AAS with a graphite furnace atomizer to the direct determination of trace amounts of I, S and P. (Address as in ref. 4).
757* SHRADER, D. E., and CULVER, B. R., The effects of H_2 on flameless AA analyses. (Address as in ref. 248).
758* GOLEB, J. A., Determination of Ba and Sb in gunshot residue with flameless AA using various collection techniques. (Address as in ref. 88).
759* MOHAMED, M. M., and SORIANO, P. R., Solvent extraction AA determination of Pd in ores, sweeps and slimes. (AMAX Base Metals Research and Development Dept. Inc., 400 Middlesex Avenue, Carteret, N.J. 07008, U.S.A.).
760* KIRK, M., PERRY, E. G., and ARRITT, J. M., The separation and AA measurement of trace amounts of Pb, Ag, Zn, Bi and Cd in high Ni alloys. (Huntington

Alloy Products Div., International Nickel Co. Inc., Huntington, W.Va. 25720, U.S.A.).
761* MITCHELL, D. G., KAHL, M., and WARD, A. F., An instrumental system for microsampling-up AAS utilizing a nitrous oxide/acetylene flame. (Address as in ref. 80).
762* WALSH, P. R., and FASCHING, J. L., The determination of As using graphite-tube atomization AAS. (Address as in ref. 721).
763* SELLERS, N. G., SCHMITT, E. C., and IKENBERRY, L. C., Analysis of slags, ores, sinters and refractories by $Li_2B_4O_7$ fusion and AAS. (Armco Steel Corp., Research and Technology, Middletown, Ohio 45042, U.S.A.).
764* BRAMSTEDT, W. R., and HARRINGTON, D. E., Analysis of Ru, Sn and Ti in solution by AAS. (Address as in ref. 168).
765* MANNING, D. C., Aspirating small volume samples in flame AAS. (Address as in ref. 553).
766* MITCHELL, D. G., WARD, A. I., WEINBLOOM, R. C., SMITH, R. M., and ALDOUS, K. M, Use of a microsampling-cup nitrous oxide/acetylene atomization system to minimise matrix interferences. (Address as in ref. 80).
767* HOYT, R. E.,* and DRYER, H. T.,† Verification of alloy type with a remote hand held probe. (*Hasler Research Center, Applied Research Laboratories, 95 La Patera Lane, Goleta, Calif. 93017, U.S.A.); †Applied Research Laboratories, 20200 West Outer Drive, Dearborn, Mich. 48124, U.S.A.).
768* McMAHON, M A., and FRICIONI, R. B., Separation and determination of Al in low alloy steels into three soluble fractions. (Allegheny Ludlum Industries, Research Center, Brackenridge, Pa. 15014, U.S.A.).
769* EVENS, F. M., COWLEY, T. G., MONN, D. E., HASSELL, C. L., and COOPER, D. E., Spectrometric applications for an innovative curve-fitting routine. (Continental Oil Co., Ponca City, Okla. 74601, U.S.A.).
770* HADEISHI, T., and McLAUGHLIN, R., Application of IZAA to multiple element analyses. (Lawrence Berkeley Laboratory, Berkeley, Calif. 94720, U.S.A.).
771 HELD, A., and STEPHENS, R., Effect of photon trapping on line intensities in AF and ES, *Can. J. Spectrosc.*, 1975, **20**, 10. (Address as in ref. 216).
772 FLINN, C. G., and STEPHENS, R., Multi-cathode discharge source for AAS, *Can. J. Spectrosc.*, 1975, **20**, 14. (Address as in ref. 216).
773 EPSTEIN, M. S., RAINS, T. C., and MENIS, O., Determination of Cd and Zn in standard reference materials by AFS with automatic scatter correction, *Can. J. Spectrosc.*, 1975, **20**, 22. (Address as in ref. 34).
774 LUZAR, O., and SLIVA, V., Determination of CaO, MgO, SiO_2, Fe, Cu, Zn, Pb, Cd, Na_2O and K_2O in Fe ores and agglomerates, *Hutn. Listy,* 1975, **30**, 55. Research Institute NHKG, Ostrava-Kuncice, Czechoslovakia).
775 MAKHNEV, Y. A., PETROV, B. I., and ZHIVOPISTSEV, V. P., Spectrochemical determination of Pt, *Izv. Vyssh. Uchebn. Zaved., Khim. Khim. Tekhnol.*, 1974, **17**, 172.
776 SYCHRA, V., SVOBODA, V., and RUBESKA, I., AFS — Book published by Van Nostrand–Reinhold, London, 1975.
777 CORNIL, J., and LEDENT, G., Spectrographic determination of some non-metallic elements in water and in biological or geological samples, *Analusis,* 1975, **3**, 11. (Institut de Recherches chimiques, Ministere belge de l'Agriculture, 5 Molenstraat, Tervuren, B-1980 Belgium).
778 RADECKI, A., LAMPARCZYK, H., GRZYBOWSKI, J., and HALKIEWICZ, J., Acid interferences in indirect determination of Si by AAS, *Spectrosc. Lett.*, 1974, **7**, 627. (Dept. of Physical Chemistry, Institute of Chemistry and Analytics, Medical Academy, Gdansk, Poland).
779 ENG, K., Determination of heavy metals in paints by AAS, *Skand. Tidskr. Farg Lack,* 1975, **21**, 7. (Scandinavian Paint and Printing Ink Research Institute, Copenhagen, Denmark).
780 PATENT, Improvements in or relating to flame spectrometry apparatus, British Patent 1,382,254.
781 TEMMA, T., and MIWA, S., Spectrographic determination of small amounts of Sb in synthetic polyester fibres, *Bunseki Kagaku,* 1974, **23**, 1475. (Central Customs Laboratory, Ministry of Finance, Matsudo-shi, Chiba, Japan).
782 GOMEZ COEDO, A., and DORADO, M. T., Determination of As in different

matrices by AAS, *Rev. Metal. (Madrid),* 1974, **10**, 355. (Sec. Anal. Quim. CENIM, Ciudad University, Madrid, Spain).
783 WINDEMANN, H., and MUELLER, U., Determination of Cd in tobacco by AAS, *Mitt. Geb. Lebensmittelunters. Hyg.*, 1975, **66**, 64. (Kanton Laboratory Lebensmittel-u. Trinkwasserkontrolle, Bern, Switzerland).
784 ROUSSELET, F., EL SOLH, N., and GIRARD, M. L., Nature and variations of the interactions occurring in AAS during the determination of Sr in biological media, *Analusis,* 1975, **3**, 44. (Laboratoire de Biochimie Applique, U.E.R. de Biologique Humaine et Experimentale, 4 Avenue de l'Observatoire, F-75270 Paris Cedex 06, France).
785 LE TRUNG TAM, Determination of B in C steel and low-alloy steel by use of a d.c. arc, *Analusis,* 1975, **3**, 23. (Laboratoire de Spectrographie, Laboratoire Nationale d'Essais, F 75015 Paris, France).
786 WARREN, J., and CARTER, D., Determination of trace amounts of Cu, V, Cr, Ni, Co and Ba in silicate rock by flame AAS, *Can. J. Spectrosc.*, 1975, **20**, 1. (Laboratory of the Government Chemist, Cornwall House, Stamford Street, London S.E.1, England).
787 NICHOLSON, N. M., and DAVEY, J., Analysis of small quantities of environmental dust samples, *Br. Steel Corp., Open Rep.*, GS/EX/44/74/C, 1974. (Corporate Development Laboratory, British Steel Corporation, Sheffield, England).
788 KORNILOVA, O. A., Spectrographic analysis of alloys based on V and Zr, *Zh. Anal. Khim.*, 1974, **29**, 1427. (Donetsk Chem.-Metallurg. Processing Plant, U.S.S.R.).
789 GARDNER, R. D., HENICKSMAN, A. L., and ASHLEY, W. H., Determination of selected rare-earth metals in U alloys by AAS, *U.S. At. Energy Com., Rep.*, LA-5539, 1974. (Los Alamos Science Laboratory, Los Alamos, N.Mex., U.S.A.).
790 VOROB'EVA, G. A., *et al.,* Extraction of noble metals for their group concentration and subsequent spectrographic determination, *Zh. Anal. Khim.*, 1974, **29**, 497.
791 JOSEPHSON, M., and DIXON, K., Determination of minor amounts of Sb in ores and concentrates by AAS, *Natl. Inst. Metall. Repub. S. Afr., Rep.* No. 1665, 1974:. (National Institute of Metallurgy, Milner Park, Johannesburg, South Africa).
The following papers (*) were presented at the Fifth International Conference on Atomic Spectroscopy, 25–29 August 1975, Melbourne, Australia.
792* SEGAR, D. A., Flameless AS: a quantum jump in investigating the environment. (National Oceanic and Atmospheric Administration, Atlantic Oceanographic and Meteorological Laboratories, Miami, Fla. 33149, U.S.A.).
793* LAQUA, K., Glow discharges: a means to complete and universal spectrochemical analyses. (Address as in ref. 609).
794* TOLK, N., The emission of optical radiation arising from low-energy ion/atom and ion/surface collisions. (Bell Laboratories, 600 Mountain Avenue, Murray Hill, N.J. 07974, U.S.A.).
795* GAYDON, A. G., Spectroscopic studies on the state of equilibrium in flame gases. (Imperial College, London, England).
796* HIEFTJE, G. M., Atom formation processes in analytical flames. (Address as in ref. 226).
797* OMENETTO, N., The possibility of local sensing of physical parameters in flames. (Institute of Inorganic and General Chemistry, University of Pavia, 27100 Pavia, Italy).
798* HANNAFORD. P., The influence of spectral-line profiles in AAS. (C.S.I.R.O., Div. of Chemical Physics, P.O. Box 160, Clayton, Vic. 3168, Australia).
799* YASUDA, K., Application of the Zeeman effect to AAS. (Address as in ref. 369).
800* WOODRIFF, R., Constant-temperature non-flame atomizers. (Address as in ref. 31).
801* AGGETT, J., Experiences with flameless atomization. (Chemistry Dept., University of Auckland, Auckland, New Zealand).
802* HWANG, J. Y., CORUM, T. L., SOTERA, J. J., BANCROFT, M. F., and EMMEL, R. H., A high temperature tungsten–graphite flameless atomizer. (Address as in ref. 313).
803* WELZ, B., Improvement of precision in flameless AA by atomic sample injection. (Bodenseewerk Perkin-Elmer und Co. GmbH, Uberlingen, West Germany).
804* McCLELLAN, B. E., and STEIN, V. B., Enhancement of AA sensitivity for Ni, Mn and Ag and determination of submicrogram quantities of Cd in environmental samples. (Dept. of Chemistry, Murray State University, Murray, Ky. 42071, U.S.A.).
805* BOAR, P. L., BONE, K. M., and HIBBERT, W. D., The determination of trace

concentrations of metals in brown coal ash and effluents from power stations. (Scientific Div., State Electricity Commission of Victoria, Richmond, Vic. 3121, Australia).

806* RAINS, T. C., The determination of trace metals in water by non-flame AAS. (Address as in ref. 34).

807* SEGAR, D. A., and CANTILLO, A. Y., Analysis of sea water for total Fe, Mn, Cu, Ni and Cd by direct-injection flameless AAS without sample pre-processing. (Address as in ref. 792).

808* LEE, M. L., and BURRELL, D. C., Determination of some soluble heavy metals in marine water by carbon filament AS. (Institute of Marine Science, University of Alaska, Fairbanks, Alaska 99701, U.S.A.).

809* OTTAWAY, J. M., and HOUGH, D. C., Multi-element analysis of dust emissions and atmospheric particulates at a steel works by carbon furnace AAS. (Address as in ref. 394).

810* PINTA, M., and RIANDEY, C., Direct analysis by flameless AA of metallic traces of particles suspended in the atmosphere in contact with the ocean. (Address as in ref. 614).

811* NOLLER, B. N., BLOOM, H., and PARKER, C. R., Non-flame AA in the analysis of trace metals in natural and urban air. (Address as in ref. 432).

812* SUZUKI, K.,* and OYAGI, Y.,† DDTC–xylene extraction as a pre-analysis treatment in AAS. (*National Institute of Nutrition, 1 Toyama-cho Shinjuku-ku, Tokyo, Japan; †Chiba University, 1-33 Yayoi-cho, Chiba, Japan).

813* SCHRAMEL, P., CUMPELIK, O., IYENGAR, V., and PAVLU, J., Matrix influences in flameless AAS. (Gesellschaft fur Strahlen und Umweltforschung mbH, Neuherberg/Munich, West Germany).

814* OTTAWAY, J. M., CAMPBELL, W. C., ROWSTON, W. B., SHAW, F., and STRONG, B., Mechanism of interferences in carbon furnace AAS. (Address as in ref. 394).

815* CZOBIK, E. J., and MATOUSEK, J. P., Atom formation and interferences in non-flame AAS. (School of Chemistry, University of N.S.W., P.O. Box 1, Kensington, N.S.W. 2033, Australia).

816* PARKER, C.,* and SCHRADER, D.,† Effects of reactive and inert gases on performance of a non-flame atomizer. (*Varian Techtron Pty. Ltd., Springvale, Vic., Australia; †Varian Instrument Division, Park Ridge, Ill., U.S.A.).

817* KERBER, J. D., BARNETT, W. B., and KNOTT, A. R., Optimization of instrumental electronics for use with flameless AA sampling devices. (Address as in ref. 220).

818* STURGEON, R. E., CHAKRABARTI, C. L., and BERTELS, P. C., Characteristic features of absorption pulses in graphite furnace AAS. (Address as in ref. 683).

819* STURGEON, R. E., CHAKRABARTI, C. L., and BERTELS, P. C., The peak absorbance *vs.* the integrated absorbance in graphite furnace AAS. (Address as in ref. 683).

820* OTTAWAY, J. M., SHAW, F., HUTTON, R., and HAMILTON, T., Determination of trace elements in metals and alloys by carbon furnace AAS. (Address as in ref 394).

821* ANDREWS, D. G., and HEADRIDGE, J. B., A furnace method for the determination of trace elements in alloys without dissolution. (Dept. of Chemistry, University of Sheffield, Sheffield, England).

822* LANGMYHR, F. J., Direct AAS analysis of solid samples. (Dept. of Chemistry, University of Oslo, Oslo 3, Norway).

823* CHAPMAN, J. F., and DALE, L. S., The application of flameless AA to the direct analysis of solids. (Address as in ref. 79).

824* SLAVIN, S., FERNANDEZ, F. J., and KERBER, J. D., Instrumental and chemical methods for the reduction of nonspecific absorption with the graphite furnace. (Address as in ref. 8).

825* MURRAY, R. W., and ROBERTS, E. D., Application of a dual channel instrument to non-flame AAS. (Sulphide Corporation Pty. Ltd., Boolaroo, N.S.W. 2284, Australia).

826* SANDERS, J. B., and STUX, R., Simultaneous background correction: luxury or necessity? (Address as in ref. 429).

827* SCHMIDER, P., and POLT, D., Unique compensation system for nonspecific energy losses in AAS up to 1·4 A. (Beckman Instruments GmbH, 8000 Munchen 40, Frankfurter Ring 115, West Germany).

828* OTTAWAY, J. M., and SHAW, F., Carbon furnace AES. (Address as in ref. 394).
829* WELZ, B., Influence of sample preparation on accuracy and precision in flameless AA. (Address as in ref. 803).
830* VAN LOON, J. C.,* RADZIUCK, B.,* and SILVESTER, M. D.,† A T-tube furnace for the investigation of metal species by AAS. (*University of Toronto, Toronto, Ont. M5S 1A1, Canada; †Address as in ref. 736).
831* CHAU, Y. K., WONG, P. T. S., and GOULDEN, P. O., A gas chromatography/AAS system for the determination of volatile alkyl Pb and Se compounds. (Address as in ref. 167).
832* KIRKBRIGHT, G. F., and ADAMS, M. J., Some studies of photo-ionization detectors and their use in AAS. (Address as in ref. 4).
833* LARSON, G. F., FASSEL, V. A., WINGE, R. K., and KNISELEY, R. N., Ultra-trace determinations by optical ES: the stray light problem. (Address as in ref. 76).
834* McNEILL, J. J., Sixty years of diffraction grating ruling in Australia. (Div. of Chemical Physics, C.S.I.R.O., P.O. Box 160, Clayton, Vic. 3168, Australia).
835* MATOUSEK, J. P., and ORR, B. J., Use of a pulsed laser in non-flame atomization for AAS. (Address as in ref. 815).
836* QUENTMEIER, A., HAGENAH, W. D., and LAQUA, K., A contribution to analytical AAS on laser-produced vapour plume. (Address as in ref. 609).
837* LEIS, F., HAGENAH, W. D., and LAQUA, K., ES analysis with microwave excitation of lased-produced vapour plume. (Address as in ref. 609).
838* GOUGH, D. S., LARKINS, P. L., and WALSH, A., Sputtered atomic vapours in AAS. (Div. of Chemical Physics, C.S.I.R.O., P.O. Box 160, Clayton, Vic. 3168, Australia).
839* McPHERSON, G. L., and PRICE, J. W., Applications of cathodic sputtering to AAS in studies of metal surface composition. (John Lysaght (Australia) Ltd., Newcastle, N.S.W., Australia).
840* SACKS, R. D., and LING, C. S., Exploding-foil extraction for the analysis of trace metals in aqueous media. (Address as in ref. 78).
841* HUMAN, H. G. C., SCOTT, R. H., and WEST, C. D., The study of a spark as sampling/nebulizing device for solid samples in AS. (Address as in ref. 181).
842* BAUDIN, G., and REMY, B., Use of a glow discharge as an atom reservoir for AAS. (Service d'Etudes Analytiques, Commissariat a l'Energie Atomique, 92 Fontenay-aux Roses, France).
843* BUTLER, L. R. P., and WEST, C. D., The measurement of emission radiation from a Grimm-type glow discharge lamp by means of resonance radiation. (National Physical Research Laboratory, C.S.I.R., P.O. Box 395, Pretoria 0001, South Africa).
844* SULLIVAN, J. V, Demountable atomic spectral lamps. (Div. of Chemical Physics, C.S.I.R.O., P.O. Box 160, Clayton, Vic. 3168, Australia):
845* HUMAN, H. G. C., Characteristics of boosted-output hollow cathode lamps. (Address as in ref. 639).
846* LOWE, R. M., A glow discharge source for spectrochemical analysis. (Div. of Chemical Physics, C.S.I.R.O., P.O. Box 160, Clayton, Vic. 3168, Australia).
847* SACKS, R. D., BRINKMAN, D. W., and THOMAS, P. L., A high-intensity excitation source for analytical AF. (Address as in ref. 78)
848* SACKS, R. D., and LAMPERT, J. K., Radiative and electrical properties of exploding silver wires. (Address as in ref. 78).
849* BAUDIN, G., and PICHET, R., Use of a glow discharge in the measurement of diffusion profiles. (Address as in ref. 842).
850* TSONG, I. S. T., and McLAREN, A. C., Detection and determination of H_2 in solids using an ion-beam spectrochemical analyser. (Dept. of Physics, Monash University, Clayton, Vic. 3168, Australia).
851* ARNOLD, J. M., and SMYTHE, L. E., Design and evaluation of a dual channel AF spectrometer. (Address as in ref. 159).
852* LARKINS, P. L., and WILLIS, J. B., Non-dispersive flame fluorescence as a practical analytical technique. (Address as in ref. 838).
853* URBAIN, H., and DESQUESNES, W., Application of a differential technique for multi-element analysis by AES in the nitrous oxide/acetylene flame. (Address as in ref. 217).
854* DAHLQUIST, R. L., and KNOLL, J. W., Solids and liquids analysis using r.f. inductively coupled argon plasma optical ES. (Address as in ref. 615).
855* KOIRTYOHANN, S. R., and LICHTE, F. E., The effect of acid concentration on

the emission intensity from the inductively coupled plasma. (Address as in ref. 203).
856* SCOTT, R. H., and HUMAN, H. G. C., The shapes of spectral lines emitted by an inductively coupled plasma. (Address as in ref. 181).
857* KIRKBRIGHT, G. F., and BEVAN, D. G., Application of a piezoelectrically-scanned Fabry–Perot interferometer and demountable hollow cathode lamp to isotopic studies and atomic line profile measurements. (Address as in ref. 4).
858* HANNAFORD, P., and LOWE, R. M., The isotopic analysis of B by AA. (Address as in ref. 798).
859* SANDLE, W. J., ROBERTS, G. J., and WARNINGTON, D. M., A technique for studying lifetimes of molecular states: transients in the fluorescence following a sudden magnetic field pulse. (Physics Dept., University of Otago, New Zealand).
860* HANNAFORD, P., The Hanle effect in atomic vapours produced by cathodic sputtering. (Address as in ref. 798).
861* PIPER, J. A., LITTLEWOOD. I. M., and WEBB, C. E., Production of excited levels of As(II) in thermal energy collisions. (Clarendon Laboratory, Oxford, England).
862* PERRY, K., AAS in the process control chemistry of a precious metal refinery. (Lonrho Refinery Ltd., Brakpan, South Africa).
863* MICHAELSON, A. S., Pt group metals detection using the atomic analysis system. (M.P. International, P.O. Box 1285, Nogales, Ariz. 85621, U.S.A.).
864* WATTS, J. C., The determination of the rare-earth elements in naturally occurring materials by flame spectroscopy. (Australian Mineral Development Laboratories, Adelaide, South Australia).
865* VOINOVITCH, I., LEGRAND, G., and LOUVRIER, J., Determination of Ca in cement by FES with internal standard. (Address as in ref. 596).
866* SUTTON, M. M., and LOWE, M. D., Interactions of HCl, HF, $HClO_4$, H_2SO_4, H_3BO_3, H_3PO_4 with B, Zr, Ti, Mo, Al, V and lateral diffusion effects in unshielded and argon-shielded nitrous oxide/acetylene flames. (Ruakura Agricultural Research Centre, Hamilton, New Zealand).
867* TAYLOR, R. D., Atomization interference mechanisms in analytical flame spectroscopy. (Agriculture Div., Government Chemical Laboratories, Perth, Western Australia).
868* JOHNSON, G. M., Factors affecting the Al atom concentration in flames. (C.S.I.R.O., Div. of Mineral Chemistry, North Ryde, N.S.W., Australia).
869* BANERJEE, B. K., and SINGHAL, K. C., The effect of CO_3^{--} and HCO_3^- on Ca absorption in AAS. (Address as in ref. 722).
870* SINHA, R. C. P., and BANERJEE, B. K., An indirect method for the estimation of low content of $Ca(NO_3)_2$ in $CaNH_4(NO_3)_3$ fertilizer by AAS. (Address as in ref. 722).
871* HERRMANN, R., F trace determination by flame and non-flame emission and AA methods. (Address as in ref. 20).
872* KNOWLES, D. J.,* ABACHI, M. Q.,† BELCHER, R.,† BOGDANSKI, S. L.,† TOWNSHEND, A.,† The determination of mixtures of inorganic S compounds by MECA. (*Preston Institute of Technology, Melbourne, Australia; †Address as in ref. 35).
873* FRANK, P., and KRAUSS, L., A new contribution to the origin of the green and orange bands of CaO. (Address as in ref. 590).
874* STEVENS, B. J., Determination of metals in small samples of biological material by spike-height method. (Dept. of Applied Biology, Royal Melbourne Institute of Technology, Melbourne, Australia).
875* CERNIK, A. A., and SAYERS, M. H. P., Blood Cd: an analytical method to facilitate its application to industry. (Address as in ref. 195).
876* GLOVER, J. W., Heavy metals in Victorian marine species. (Div. of Agricultural Chemistry, Dept. of Agriculture, 5 Parliament Place, Melbourne, Australia).
877* DAVID, D. J., and WILLIAMS, C. H., AA investigation of the effects on plants and soils of extended application of raw sewage at the Melbourne Metropolitan Board of Works Farm, Werribee. (C.S.I.R.O., Div. of Plant Industry, Box 1600, Canberra City, A.C.T. 2601, Australia).
878* AGGETT, J., and ASPELL, A. C., Determination of As(III) and As(V) by AAS. (Address as in ref. 801).
879* CLINTON, O. E., A routine method for the determination of Se in blood and plant material by hydride generation. (Ruakura Agricultural Research Centre, Hamilton, New Zealand).

880* FLEMING, H. D., and IDE, R. G., Application of the hydride evolution technique to the AAS analysis of volatile hydride-forming metals in steel. (Quality Control Laboratories, Chemical Div., Broken Hill Pty. Co. Ltd., Iron and Steel Works, Newcastle, N.S.W., Australia).
881* TAYLOR, R. D., The use of dual nebulizer in analytical flame spectroscopy. (Address as in ref. 867).
882* ALKEMADE, C. T. J., Fluctuation correlation considerations in analytical spectroscopy. (Address as in ref. 123).
883* LIDDELL, P. R., Noise at detection limit levels in AA flame spectrometry. (Address as in ref. 425).
884* SCHMIDER, P., AUBERG, W., and POLT, D., On-line recalculation of concentration units from non-linear calibration curves in AAS. (Address as in ref. 827).
885* VILLANOVA, R. A., and AZAMBUJA, D. S., A direct procedure for the determination of Nb in steel by AAS. (Acos Finos Piratini SA, Rua Cancio Gomes 127, 90.000-Porto Alegre-RS, Brazil).
886* KIDANI, Y., INAGAKI, K., UNO, E., NAKAMURA, K., and NOJI, M., Indirect determination of drugs by AAS: metal chelate formation of Schiff base and solvent extraction as a ternary complex. (Address as in ref. 379).
887* WILLIS, J. B., AAS analysis by direct introduction of powders into the flame. (Div. of Chemical Physics, C.S.I.R.O., PO. Box 160, Clayton, Vic. 3168, Australia).
888* WIDMER, D. S., BAKER, S. J., and MILLS, K. J., Optimisation of spray chamber and burner design for AA instruments. (Pye Unicam Ltd., York Street, Cambridge, England).
889* UMEBAYASHI, M., and KITAGISHI, K., Direct attachment of AAS to a liquid chromatograph for the identification, estimation and continuous monitoring of metal ions and metal chelates. (Faculty of Agriculture, Mie University, Tsu, Mie 514, Japan).

The following papers (†) were presented at the 169th American Chemical Society National Meeting, 6–11 April, 1975, Philadelphia, Pa., U.S.A.

890† LUKASIEWICZ, R. J., and BUELL, B. E., Utility of the nitrous oxide/hydrogen flame in AA and ES of non-aqueous systems. (Address as in ref. 675).
891† NIEMCZYK, T. M., A new demountable hollow-cathode lamp for use in AES. (Dept. of Chemistry, University of New Mexico, Albuquerque, N.Mex. 87131, U.S.A.)
892† MURTON, R., SIEVERS, R. E., and EISENTRAUT, K. J., Analysis for Ti in aircraft lubricating oil by AAS. (Aerospace Research Laboratories, ARL/LJ, Wright-Patterson Air Force Base, Ohio 45433, U.S.A.).
893† RAINS, T. C., Trace metal analysis of environmental and biological materials by AA and ES. (Address as in ref. 34).
894† GREY, P., Analysis of petroleum and petroleum products for trace quantities of Ni and V. (Mobil Research and Development Corp., Paulsboro, N.J. 08066, U.S.A.).
895† MERRYFIELD, R., Analysis of petroleum and petroleum products for trace quantities of As (Address as in ref. 685).
896† WALKER, H. H., Analysis of petroleum and petroleum products for trace quantities of Se. (Mobil Research and Development Corp., Paulsboro, N.J. 08066, U.S.A.).
897† ROBBINS, W. K., Analysis of petroleum and petroleum products for trace quantities of Cr and Mn. (Address as in ref. 139).
898† GREY, P., Analysis of petroleum and petroleum products for trace quantities of Co and Mo. (Address as in ref. 894).
899† NOWAK, A. V., Analysis of petroleum and petroleum products for trace quantities of Cd, Sb and Pb. (Atlantic Richfield Co., Harvey, Ill. 60426, U.S.A.).
900† RUNNELS, J. H., Analysis of petroleum and petroleum products for Be and the atomization mechanism for low-temperature graphite atomizer. (Address as in ref. 685).
901† MILLIMAN, G. E., Analysis of petroleum and petroleum products for trace quantities of Hg. (Address as in ref. 686).
902† LARSON, J. O., Application of neutron activation and ES to the analysis of trace metals in petroleum. (Chevron Research Company, Richmond, Calif. 94802, U.S.A.).
903† ABU--ELGHEIT, M., Direct evaluation of Cd in petroleum by AA analysis. (Chemistry Dept., Faculty of Science, Alexandria University, Alexandria, Egypt).
904† WILKINSON, D. R., and RUSANOWSKY, P., The determination of Pb in blood by flameless AAS using a preliminary chemical oxidation step. (Box 64, Dept. of

Chemistry, Delaware State College, Dover, Del. 19901, U.S.A.).
905† BALL J. W., and JENNE, E. A., Determination of Hg in solid and water samples by flameless AAS. (Address as in ref. 560).
The following papers (*) were presented at the 58th Chemical Conference and Exhibition, 25–28 May, 1975, Toronto, Ont., Canada.
906* FASSEL, V. A., Trace and ultratrace analysis by inductively coupled plasma optical ES: present status and prospects. (Address as in ref. 98).
907* HELD, A., LAU, C., and STEPHENS, R., Sensitivity enhancements to flame AA using a flame atom trap. (Address as in ref. 216).
908* STURGEON, R. E., CHAKRABARTI, C. L., and BERTELS, P. C., Characterization of absorption pulses in AAS with various graphite furnaces. (Address as in ref. 683).
909* DICK, J. G., and FELDMAN, A., The interfering effect of P in the determination of As by the AA/carbon furnace technique. (Concordia University, 1455 de Maisonneuve Blvd., Montreal, Canada).
910* STURGEON, R. E., CHAKRABARTI, C. L., and BERTELS, P. C., The peak mode *vs.* the integration mode of measurement of absorbance in graphite furnace AAS. (Address as in ref. 683).
911* SIMOVIC, D., PAGE, J., and VANLOON, G., The determination of low levels of Hg. (Dept. of Chemistry, Queen's University, Kingston, Ont., Canada).
912* KOOP, D. J., and VAN LOON, J. C., Thermal volatilization of Pb and Pb compounds in solid samples, followed by AA measurements. (Address as in ref. 830).
913* SEN GUPTA, J. G., The determination of lanthanides and Y in rocks and minerals by AA and FES. (Geological Survey of Canada, Ottawa, Ont. K1A 0E8, Canada).
914* JOHNSON, W. M., RALPH, P. F., CHAUDHRY, M. A., and BHAGWANANI, B., A consideration of measurement precision and sampling statistics in fire assay and AA techniques of Au analysis. (Dept. of Mines and Petroleum Resources, 541 Superior St., Victoria, B.C., Canada).
915 CHEN, C. T., and WINEFORDNER, J. D., Interference on Mg by trace concomitants in flame AAS, *Can. J. Spectrosc.*, 1975, **20**, 87. (Address as in ref. 487).
916 CHAKRABARTI, C. L., and McNEIL, D. P. D., An evaluation of the flame parameters in the determination of V by AAS with a nitrous oxide acetylene flame, *Can. J. Spectrosc.*, 1975, **20**, 90. (Address as in ref. 683).
917 GUY, R. D, CHAKRABARTI, C. L., and SCHRAMM, L. L., The application of a simple chemical model of natural waters of metal fixation in particulate matter, *Can. J. Chem.*, 1975, **53**, 661. (Address as in ref. 683).
918 BAILY, P., and KILROE-SMITH, T A., Effect of sample preparation on blood Pb values, *Anal. Chim. Acta*, 1975, **77**, 29. (Address as in ref. 204).
919 MARUTA, T., and SUDOH, G., AsH_3 generation and determination of trace amounts of As by AAS, *Anal. Chim. Acta*, 1975, **77**, 37. (Central Laboratory of Research and Development, Chichibu Cement Co. Ltd., Ohmiya, Chichibu-shi, Saitama-ken, Japan).
920 FRECH, W., Rapid determination of Pb in steel by flameless AAS, *Anal. Chim. Acta*, 1975, **77**, 43. (Dept. of Analytical Chemistry, University of Umea, 901 87 Umea, Sweden).
921 BELCHER, R., BOGDANSKI, S. L., KNOWLES, D. J., and TOWNSHEND, A., MECA: a new flame analytical technique. Part V: The determination of some S anions, *Anal. Chim. Acta*, 1975, **77**, 53. (Address as in ref. 35).
922 NAKASHIMA, R., SASAKI, S., and SHIBATA, S., Determination of Ag in biological materials by h.f. plasma torch, *Anal. Chim. Acta*, 1975, **77**, 65. (Government Industrial Research Institute, Nagoya, Hirate-machi, Kita-ku, Nagoya, Japan).
923 HAWLEY, J. E., and INGLE, J. D., Improvements in the non-flame AF determination of Hg, *Anal. Chim. Acta*, 1975, **77**, 71. (Address as in ref. 67).
924 OHTA, K., and SUZUKI, M., Determination of Se in metallurgical samples by flameless AAS, *Anal. Chim. Acta*, 1975, **77**, 288. (Address as in ref. 695).
925 KNAPP, G.,* SCHREIBER, B.,† and FREI, R. W.,† A simple concentration procedure for trace metals for x-ray fluorescence and AAS, *Anal. Chim. Acta*, 1975, **77**, 293. (*Address as in ref. 662; †Analytical Research and Development, Pharmaceutical Dept., Sandoz Ltd., Basle, Switzerland).
926 FIEDLER. R., and PROKSCH, G., The determination of ^{15}N by emission and mass spectrometry in biochemical analysis: a review, *Anal. Chim. Acta*, 1975, **78**, 1. (I.A.E.A. Laboratory, A-2444 Seibersdorf, Austria).

927 SEBOR, G., LANG, I., VAVRECKA, P, SYCHRA, V., and WEISSER, O., The determination of metals in petroleum samples by AAS. Part I: Determination of V, *Anal. Chim. Acta,* 1975, **78**, 99. (Address as in ref. 41).
928 SUGIMAE, A., ES determination of trace elements in airborne particulate matter collected on glass fibre filter, *Anal. Chim. Acta,* 1975, **78**, 107. (Environmetal Pollution Control Centre, Osaka Prefecture, 1-chome, Nakamichi, Higashinari-ku, Osaka, Japan).
929 ROBINSON, J. W., WOLCOTT, D. K., and RHODES, L., Direct analysis of blood, urine, sea water, filter paper and polyethylene by AAS with the hollow-T atomizer, *Anal. Chim. Acta,* 1975, **78**, 285. (Address as in ref. 141).
930 COBB, W. D., FOSTER, W. W., and HARRISON, T. S., The determination of Ti in Fe and steel by AAS, *Anal. Chim. Acta,* 1975, **78**, 293. (Address as in ref. 53).
931 GUEST, R. J., and MacPHERSON, D. R., The use of flame procedures in metallurgical analysis. Part II: Determination of Al in sulphide and silicate materials and in ores and slags, *Anal. Chim. Acta,* 1975, **78**, 299. (Chemical Analysis Section, Extraction Metallurgy Div., Mines Branch, Department of Energy, Mines and Resources, Ottawa, Ont., Canada).
932 LEE, A. P., and BOLTZ, D. F., An AAS study of the distribution ratio of Mo 1-pyrrolidine carbodithioate for the MIBK/water system, *Anal. Chim. Acta,* 1975, **78**, 466. (Dept. of Chemistry, Wayne State University, Detroit, Mich. 48202, U.S.A.).
933 ROBINSON, J. W., RHODES, L., and WOLCOTT, D. K., The determination and identification of molecular Pb pollutants in the atmosphere, *Anal. Chim. Acta,* 1975, **78**, 474. (Address as in ref. 141).
934 CAREL, A. B., The incremental evolution, collection and determination of Hg in soils at the p.p.b. level as a function of temperature, *Anal. Chim. Acta,* 1975, **78**, 479. (Continental Oil Co., Ponca City, Okla. 74601, U.S.A.).
935 CRESSER, M. S., A course on analytical techniques — Book published by Ministry of Agriculture and Natural Resources, Soil Institute of Iran, Publ. No. 412, April 1975. (Soil Sci. Dept., Aberdeen University, Aberdeen, Scotland).
936 RAMIREZ-MUNOZ, J., Experimental concentration ranges for cement sample analysis by AAS, *Flame Notes,* 1974, **6**, 29. (Address as in ref. 319).
937 MUDGETT, P. S., RICHARDS, L. W., and ROEHRIG, J. R., New technique to measure H_2SO_4 in the atmosphere, pp. 85–105 *in* Stevens, R. K., and Herget, W. F., *Editors,* Analytical methods applied to air pollution measurements — Book published by Ann Arbor Science Publishers, Inc., Ann Arbor, 1974. (Billerica Research Center, Cabot Corp., Billerica, Mass., U.S.A.)
938 BAUMGARDNER, R. E., CLARK, T. A., and STEVENS, R. K., Increased specificity in the measurement of S compounds with the flame photometric detector, *Anal. Chem.,* 1975, **47**, 563. (National Environmental Research Center, Environmental Protection Agency, Research Triangle Park, N.C., U.S.A.).
939 ABJEAN, R., LERICHE, M, and JOHANNIN-GILLES, A., Measurement of oscillator strengths by absorption in an atomic beam, *J. Quant. Spectrosc. Radiat. Transfer,* 1975, **15**, 15. (Faculty of Science, Brest, France).
940 SCHRAMM, G., and KOCH, B., K in vegetable laxatives and intestinal regulators: flame photometric determinations of K in some vegetable laxatives, intestinal regulators and intestinal dietetic agents in a biopharmaceutical model experiment, *Schweiz. Apoth-Ztg.,* 1974, **112**, 416.
941 FAZAKAS, J., GERMAN, A., BAIULESCU, G., and MULLINS, C., Determination of microelements in CO_2-containing mineral water by AA and ES. II: Determination of Cu and Cr by flameless AAS, *Rev. Chim. (Bucharest),* 1974, **25**, 917.
942 VAN DER HURK, J., HOLLANDER, T., and ALKEMADE, C. T. J., Excitation energies of BaO bands measured in flames, *J. Quant. Spectrosc. Radiat. Transfer,* 1975, **15**, 113. (Address as in ref. 123).
943 MONTASER, A., Fundamental investigation of nonflame and flame atomization with computer-controlled spectrometric systems — Thesis, 1974; *Diss. Abstr. Int. B,* 1974, **35**, 2600. (Address as in ref. 99).
944 ABJEAN, R., and JOHANNIN-GILLES, A., Measurement of oscillator strengths by absorption in an atomic beam. II: Oscillator strength of the $^1P_1$ $^1S_0$ transition of Zn, *J. Quant. Spectrosc. Radiat. Transfer,* 1975, **15**, 25. (Address as in ref. 939).
945 SUBBARAM, K. V., VASUDEV, R., and JONES, W. E., Atomic and molecular emission from microwave discharge through $BeCl_2$, *J. Opt. Soc. Am.,* 1975, **65**, 318. (Dept. of Chemistry, Dalhousie University, Halifax, N.S., Canada).

946 BROWN, A., and HUSAIN, D., Collisional quenching of electronically excited Sn atoms by time-resolved AAS, *Int. J. Chem. Kinet.*, 1975, **7**, 77. (Address as in ref. 149).

947 PUECHBERTY, D., and COTTEREAU, M. J., Absorption spectroscopy study of the concentration profiles of the hydroxyl radical in propane oxygen flames under reduced pressure, *C.R. Hebd. Seances Acad. Sci., Ser. C,* 1974, **279**, 537. (Institute of Science Haute-Normandie, Fac. Sci. Tech. Rouen, Mont-Saint-Aignan, France).

948 KOROLEFF, F., HAAPALA, K., and DALE, T., Intercalibration of methods for the determination of Fe, Mn, Cu and Zn in natural waters, *NORDFORSK, Miljoevardssekr., Publ.*, 1974, No. 10.

949 CLYNE, M. A. A., and TOWNSEND, L. W., Atomic oscillator strengths using resonance absorption with a Doppler line source; transitions of Br and I, *J. Chem. Soc., Faraday Trans. 2,* 1974, **70**, 1863. (Dept. of Chemistry, Queen Mary College, London, England).

950 CHUPAKHIN, M. S., Atomization of solid substances, pp. 44–46 *in* Zolotov, Y. A., and Petrikova, M. N., *Editors,* Uspekhi Analiticheskoi Khimii — Book published by "Nauka", Moscow, 1974. (Address as in ref. 44).

951 IORDANOV, N., DASKALOVA, K., KHAVEZOV, I., and TONCHEVA, V., AA determination of Se after its preliminary isolation in the form of an organo-Se compound, pp. 54–58 *in* Zolotov, Y. A., and Petrikova, M. N., *Editors,* Uspekhi Analiticheskoi Khimii — Book published by "Nauka", Moscow, 1974. (Institute of General and Inorganic Chemistry, Sofia, Bulgaria).

952 OHNISHI, Y., OHYA, K., and KOBAYASHI, T., Enhancement of the AA of Sn by organic solvents, *Aichi-Ken Kogyo Shidosho Hokoku,* 1974, No. 10, 80. (Chemistry Div., Aichiken Kogyo Shidosho, Aichiken, Japan).

953 BELCHER, R., BOGDANSKI, S. L., GHONAIM, S., and TOWNSHEND, A., Molecular emission analysis using a hollow flame-heated rod, pp. 353–358 *in* Zolotov, Y. A., and Petrikova, M. N., *Editors,* Uspekhi Analiticheskoi Khimii — Book published by "Nauka", Moscow, 1974. (Address as in ref. 35).

954 EVERSON, R. J., Modification in the official methods for the determination of metals in feeds and fertilizers by AAS, *J. Assoc. Off. Anal. Chem.*, 1975, **58**, 158. (Dept. of Biochemistry, Purdue University, West Lafayette, Ind., U.S.A.).

955 HASTIE, J. W., Sampling reactive species from flames by mass spectrometry, *Int. J. Mass Spectrom. Ion Phys.*, 1975, **16**, 89. (National Bureau of Standards, Washington, D.C., U.S.A.).

956 PUNGOR, E., POLOS, L., BEZUR, L., and HARSANYI, E. G., Methods of modern metal analysis, *Banyasz. Kohasz. Lapok, Ontode,* 1974, **25**, 134. (Address as in ref. 275).

957 SANG, S. L., CHENG, W. C., SHIUE, H. I., and CHENG, H. T, Direct determination of trace metals in cane juice, sugar and molasses by AAS, *Int. Sugar J.*, 1975, **77**, 71. (Taiwan Sugar Research Institute, Tainan, Taiwan).

958 LANIEPCE, B., Population transfers between (6s, 6d) configuration levels of atomic Hg effected by collisions with molecular N, *J. Phys. (Paris),* 1974, **35**, 953. (Laboratoire de Spectroscopie Atomique, University of Caen, Caen, France).

959 TSUJINO, R., and IKEDA, M., Sample decomposition technique for AAS, Japan. Kokai 74 91,696.

960 BURAKOV, V. S., MISAKOV, P. Y., NECHAEV, S. V., and YANKOVSKII, A. A., Determination of trace amounts of a substance by a laser AA method, *Zh. Prikl. Spektrosk.*, 1974, **21**, 979.

961 TARASEVICH, N. I., CHEBOTAREV, V. E., and BOCHKAREVA, I. I., Effect of the matrix on the flame photometric determination of alkali elements in W, Mo and their compounds, *Optich. Metody Kontrolya Khim. Sostava Materialov,* 1974, 8.

962 MASHIREVA, L. G., SOROKINA, S. B., and KOROVIN, V. A., Calculation of petroleum product viscosities during flame photometric analyses, *Khim. Tekhnol. Topl. Masel,* 1975, (1), 55.

963 BACHURINA, L. G., PERMINOVA, V. M., and SAVOSTIN, S. A., Use of a superhigh frequency discharge for AA determination of Ba, Sr and Ca in sputtered deposits, *Zavod. Lab.*, 1974, **40**, 1348.

964 TSUJINO, R., KISHIMOTO, T., IKEDA, M., HIROSHIMA, H., and MORITA, K., Standard addition method for a carbon tube flameless atomizer, *Bunseki Kagaku,* 1974, **23**, 1535. (Nippon Jarrell-Ash Co. Ltd., Kyoto, Japan).

965 GUILE, A. E., and HITCHCOCK, A. H., Oxide films on arc cathodes and their

emission and erosion, *J. Phys. D,* 1975, **8**, 663. (Dept. of Electrical and Electronic Engineering, University of Leeds, Leeds, England).

966 OGURO, H., Interference of perchloric acid in AAS, *Bunseki Kagaku,* 1974, **23**, 1362. (Central Research Laboratory, Matsushita Electrical Industry Co. Ltd., Kadoma, Japan).

967 BAUER, H. J., and BOGARDUS, E. H., Controlled r.f. sputter etching using AAS, *J. Vac. Sci. Technol.,* 1974, **11**, 1144. (Systems Product Division, I.B.M., Hopewell Junction, N.Y., U.S.A.).

968 MURAKAMI, T., KIDA, A., and NAKAI, M., Determination of Cr in bottom sediments by AAS. II: Pretreatment of sediments, *Hiroshima-ken Eisei Kenkyusho To Kohai Kenyusho Kenkyu Hokoku,* 1974, **21**, 17. (Hiroshima Prefecture Institute of Environmental Science, Hiroshima, Japan).

969 GROSS, R. M., Absorption spectra of high temperature solid propellant flames — Thesis, 1974; *Diss. Abstr. Int. B,* 1975, **35**, 3302. (University of Utah, Salt Lake City, Utah, U.S.A.).

970 PATENT, Flame spectrometer, British Patent 1,382,254. (Pye Ltd.).

971 DONELLY, T. H., FERGUSON, J., and ECCLESTON, A. J., High-speed method of continuous background correction in AAS. III: Direct determination of trace metals in sea water using the Varian Techtron carbon rod atomizer model 63, *Appl. Spectrosc.,* 1975, **29**, 158. (Div. of Mineralogy, C.S.I.R.O., Canberra, Australia).

972 ROUSSELET, F., COURTOIS, V., and GIRARD, M. L., Applications of AAS to the analysis of metallic elements in medicinal preparations, *Analusis,* 1975, **3**, 132. (Address as in ref. 784).

973 BIRO, J., FEHER, I., OPAUSZKY, I., SZIVOS, K., POLOS, L., and PUNGOR, E., Emission measurements of aerosol pollutants, *KFKI (Rep.),* 1974, KFKI-74-79. (Address as in ref. 275).

974 VOINOVITCH, I., LEGRAND, G., and LOUVRIER, J., Influence of an internal standard on the precision of the determination of Ca in cement by AAS, *Analusis,* 1975, **3**, 123. (Address as in ref. 596).

975 BARTSCHMID, B. R., Flameless nondispersive AFS — Thesis, 1974; *Diss. Abstr. Int. B,* 1975, **35**, 3191. (University of Houston, Houston, Tex., U.S.A.).

976 WINEFORDNER, J. D, AFS: past, present and future, *Chem. Technol.,* **1975, 5,** 123. (Address as in ref. 487).

977 FULLER, W. C., Application of Fourier transform techniques to multi-element AF determinations — Thesis, 1974; *Diss. Abstr. Int. B,* 1975, **35**, 3193. (University of Illinois, Urbana, Ill., U.S.A.).

978 KOKOT, M. L., Review of the methods for the determination of Hg in geological samples by flameless AAS, *Miner. Sci. Eng.,* 1974, **6**, 236. (Address as in ref. 181).

979 NAKAMURA, Y., and MORIKI, H., Flameless AAS for the determination of inorganic Hg. I: Determination by a reduction/aeration single path method, *Eisei Kagaku,* 1974, **20**, 300. (Fukuoka Environmental Research Centre, Fukuoka, Japan).

980 BELYAEV, Y. I., and KOVESHNIKOVA, T. A., AF method for determining Hg in rocks and lunar regolith, pp. 47–53 *in* Zolotov, Y. A., and Petrikova, M. N., *Editors,* Uspekhi Analiticheskoi Khimii — Book published by "Nauka", Moscow, 1974. (Address as in ref. 334).

981 VITKUN, R. A., ZELYUKOVA, Y. V., and POLUEKTOV, N. S., Flameless AA determination of Hg in Se and Te preparations using formaldehyde as a reducing agent, *Ukr. Khim. Zh. (Russ. Ed.),* 1974, **40**, 1304. (Address as in ref. 514).

982 TITOVA, I. N., NOVIKOV, Y. V., and YUDINA, T. V., Determination of Cu in air and biological materials using AAS analysis, *Gig. Sanit.,* 1975, (1), 57. (Mosk. Nauchno-Issled. Inst. Gig. im. Erismans, Moscow, U.S.S.R.):

983 KJELLSTROM, T., LIND, B., LINNMAN, L., and ELINDER, C. G., Variation of Cd in Swedish wheat and barley, *Arch. Environ. Health,* 1975, **30**, 321. (Dept. of Environmental Hygiene, Karolinska Institute and the National Environmental Protection Board, Stockholm, Sweden).

984 CARR, G., and WILKINSON, A. W., Zn and Cu urinary excretions in children with burns and scalds, *Clin. Chim. Acta,* 1975, **61**, 199. (Dept. of Paediatric Surgery, Institute of Child Health, 30 Guildford Street, London, W.C.1, England).

985 BAAK, J. M.,* HECK, Y. S. L.,* and VAN DER SLIK, W.,† A comparative study on the estimation of Ca in serum, *Clin. Chim. Acta,* 1975, **62**, 125. (*Dept. of Clinical Chemistry, St. Luke's Hospital, Amsterdam, The Netherlands; †Dept. of Clinical Chemistry, State University, Leiden, The Netherlands).

986 BLYTH, A., and STEWART, M. J., Serum Mg levels in patients treated with phenyloin, *Clin. Chim. Acta,* 1975, **62**, 305. (Dept. of Clinical Chemistry, Ninwell's Hospital and Medical School, Dundee DD4 1UD, Scotland).

987 DENTON, M.,* DELVES, H. T.,† and ARNSTEIN, V. M.,* Release of Fe from stroma during erythroblast maturation, *Biochem. Biophys. Res. Commun.*, 1974, **11**, 926. (*Dept. of Biochemistry, Kings College, University of London, London, England; †Institute of Child Health, 30 Guilford Street, London W.C.1, England).

988 WANG, J., and PIERSON, R. N., Distribution of Zn in skeletal muscle and liver tissue in normal and dietary controlled alcoholic rats, *J. Lab. Clin. Med.*, 1975, **85**, 50. (Body Composition Unit, Dept. of Medicine, St. Luke's Center, Columbia University, New York, N.Y., U.S.A.).

989 DELVES, H. T., Recent developments in analytical AS: flameless atomization, organic applications — Paper presented at the 1975 'LABEX' International Exhibition, London, England. (Address as in ref. 987).

990 STEPHENS, R., and RYAN, D. E., An application of the Zeeman effect to analytical AS. I: The construction of magnetically stable spectral sources, *Talanta,* 1975, **22**, 655. (Address as in ref. 216).

991 STEPHENS, R., and RYAN, D. E., An application of the Zeeman effect to analytical AS. II: Background correction, *Talanta,* 1975, **22**, 659. (Address as in ref. 216).

992 KORKISCH, J., GODL, L., and GROSS, H., Application of ion-exchangers to the determination of trace elements in natural waters. VIII: Mo, *Talanta,* 1975, **22**, 669. (Address as in ref. 191).

993 MUZZARELLI, R. A. A., and ROCCHETTI, R., AA determination of Mn, Co and Cu in whole blood and serum, with a graphite atomizer, *Talanta,* 1975, **22**, 683. (Institute of Biochemistry, Faculty of Medicine, University of Ancona, 60100 Ancona, Italy).

994 PICKERING, W. F.,* and THOMAS, P. E.,† Equilibrium effects in the determination of Ta by AAS, *Talanta,* 1975, **22**, 691. (*Dept. of Chemistry, University of Newcastle, N.S.W. 2308, Australia; †Varian Techtron, North Springvale, Vic. 3171, Australia).

995 CAMPBELL, W. C., and OTTAWAY, J. M., Determination of Pb in carbonate rocks by carbon furnace AAS after dissolution in nitric acid, *Talanta,* 1975, **22**, 729. (Address as in ref. 394).

996 KIM, C. H., ALEXANDER, P. W., and SMYTHE, L. E., The use of long chain alkylamines for preconcentration of traces of Mo, W and Re in their determination by AAS. I: General studies, *Talanta,* 1975, **22**, 739. (Address as in ref. 159).

997 VASILYEVA, A. A., YUDELEVICH, I. G., GINDIN, L.M., LANBINA, T. V., SCHULMAN, R. S., KOTLAREVSKY, I. L., and ANDRIEVSKY, V. N., Extractive concentration of Pt group elements and their determination by AAS, *Talanta,* 1975, **22**, 745. (Address as in ref. 287).

998 BURYAK, Z. I., NOVAK, V. P., MALTSEV, V. F., and KOBUS, L. F., Effect of composition of the sample on results of the spectrographic determination of Ce, *Zavod. Lab.*, 1975, **41**, 182. (All-Union Scientific Research and Technological Design Institute Tube Industries, Dnepropetrovsk, U.S.S.R.).

999 FOWLER, B. W., and SUNG, C. C., Doppler and collision-broadening effects in the profile of spectral lines, *J. Opt. Soc. Am.*, 1975, **65**, 949. (Advanced Systems Concepts Office, Army Missile Research, Development and Engineering Laboratory, Redstone Arsenal, Ala. 35807, U.S.A.).

1000 COOLEN, F. C. M., and HAGENDOORN, H. L., Detection of ^{20}Na atoms and measurement of Na vapour densities by means of atomic-resonance fluorescence, *J. Opt. Soc. Am.*, 1975, **65**, 952. (Eindhoven University of Technology, Eindhoven, The Netherlands).

1001 PINTA, M., and RIANDEY, C., Physicochemical study of the mechanism of thermoelectric atomization and its perturbations: application to AAS in a graphite furnace, *Analusis,* 1975, **3**, 86. (Address as in ref. 614).

1002 FULLER, C. W., Problems in the determination of trace elements using spark source mass spectroscopy and flameless AAS, *Proc. Soc. Anal. Chem.*, 1974, **11**, 176. (Address as in ref. 364).

1003 RAMIREZ-MUNOZ, J., Auxiliary conversion tables for sensitivity performance tests of AA instruments, *Microchem. J.*, 1975, **20**, 56. (Address as in ref. 319).

1004 KOIZUMI, H., and UCHINO, K., AA analysis for Hg using the Zeeman effect, *Hitachi Hyoron,* 1974, **56**, 1037. (Address as in ref. 369).

1005 SASANO, H., OSHIDA, H., TOCHIMOTO, H., and KURODA, S., Detecting trace metals by AAS using a graphite furnace atomizer, *Tokyo Toritsu Eisei Kenkyusho Kenkyu Nempo,* 1974, **25**, 709. (Tokyo Metropolitan Research Laboratory for Public Health, Tokyo, Japan).

1006 KUZOVKIN, B. I., MAIOROV, I. A., NEDLER, V. V., and SMIRNOV, A. Y., Use of ion-bombardment for layer--by-layer analysis of materials by spectroscopic methods, *Zh. Anal. Khim.*, 1975, **30**, 33. (State Science Research Des. Int. Rare Metal Industries, Moscow, U.S.S.R.).

1007 LOPEZ HERRERA, M. T., ESTRADA, P. J., MERE DE MELROSE, R. M., and CHACON, D. G., AA determination of W in Al_2O_3 supported catalysts for dehydrogenation of paraffinic hydrocarbons, *Rev. Inst. Mex. Pet.*, 1974, **6**, 56. (Institute of Mexican Petroleum, Mexico City, Mexico).

1008 BRISKA, M., Rapid determination of Cu, Ag and Pd in Pb/Sn tinning baths by AAS, *Fresenius' Z. Anal. Chem.*, 1975, **273**, 283. (I.B.M. Deutsch GmbH, Boblingen, West Germany).

1009 MAEDA, M., ISHITSUKA, F., and MIYAZOE, Y., Dye laser amplified AA flame spectroscopy, *Opt. Commun.*, 1975, **13**, 314. (Dept. of Electrical Engineering, Kyushu University, Fukuoka, Japan).

1010 DONNELLY, T. H., and ECCLESTON, A. J., High speed method of continuous background correction in AAS. II: Assessment of method, *Appl. Spectrosc.*, 1975, **29**, 154. (Address as in ref. 971).

1011 SIEMER, D. D., Development of techniques for nonflame spectroscopic determination of trace metals — Thesis, 1974; *Diss. Abstr. Int. B*, 1975, **35**, 4355. (Address as in ref. 31).

1012 DITMAN, L. S., GAMMON, R. W., and WILKERSON, T. D., High-resolution emission spectra of laser-excited I_2 as the excitation frequency is tuned through and away from resonance, *Opt. Commun.*, 1975, **13**, 154. (Dept. of Electrical Engineering, John S. Hopkins University, Baltimore, Md., U.S.A.).

1013 SUGIYAMA, Y., Determination of Hg in various materials, Japan. Kokai 74 107,795.

1014 PADLEY, P. J., Flame spectrophotometry: interferences arising from catalytic effects, *Proc. Soc. Anal. Chem.*, 1974, **11**, 318. (Dept. of Chemistry, University College of Swansea, Swansea, Wales).

1015 WATLING, R. J., Identification and analysis of lattice-held Hg in sphalerite from Keel prospect, County Longford, Ireland, *Inst. Min. Metall., Trans., Sect. B*, 1974, **83** (Aug), B88. (Address as in ref. 183).

1016 VUL'FSON, E. K., KARYAKIN, A. V., and SHIDLOVSKII, A. I., Possibilities and limitations of the total absorption method during AA measurements in a laser jet, *Zh. Prikl, Spektrosk.*, 1975, **22**, 14. (Address as in ref. 418).

1017 TAKEUCHI, T., AAS: powerful analytical technique, *Mizu Shori Gijutsu*, 1974, **15**, 1189. (Nagoya University, Nagoya, Japan).

1018 GREEN, J. H. S., AS techniques: principles, scope and limitations, *Proc. Soc. Anal. Chem.*, 1974, **11**, 49. (Div. of Chemical Standards, National Physical Laboratory, Teddington, Middlesex, England).

1019 MOLNAR, C. J., Vitreous carbon tube furnace for AFS — Thesis, 1974; *Diss. Abstr. Int. B*, 1975, **35**, 3775. (University of Florida, Gainesville, Fla., U.S.A.).

1020 SVERDLINA, O. A., KUZOVLEV, I. A., SOLOMATIN, V. S., YURKOVA, V. Y., KOVYKOVA, N. V., NISHANOV, D., and SHATALINA, L. G., AA method of layer analysis of GaAs expitaxial structures with chromatographic separation of the matrix, *Zavod. Lab.*, 1975, **41**, 172.

1021 KARPEL, N. G., and FEDORCHUK, O. K., Accuracy of spectrochemical analysis of expitaxial structures, *Zavod. Lab.*, 1975, **41**, 169.

1022 GORBUNOVA, L. B., KUTEINIKOV, A. F., AVDEENKO, M. A., and MURASHKINA, V. N., Spectrographic determination of traces of impurities in high-purity C, *Zavod, Lab.*, 1975, **41**, 178.

1023 PRISENKO, V. S., and SHVARTS, D. M., AA determination of low concentrations of impurities in high-purity Ni, *Zavod. Lab.*, 1975, **41**, 175. (State Scientific Research and Development Institute for the Nickel Industry, Leningrad, U.S.S.R.).

1024 YAKOVLEVA, A. F., and CHUPAKHIN, M. S., AA determination of alkaline earth metals in their mixed niobates, *Zavod. Lab.*, 1975, **41**, 185. (Address as in ref. 44).

1025 EROSHEVICH, T. A., and MAKAROV, D. F., AA analysis of alloys containing Ni, Cu, Fe and Co, *Zavod. Lab.*, 1975, **41**, 186. (Address as in ref. 306).

1026 EROSHEVICH, T. A., and MAKAROV, D. F., AA determination of Ni, Cu, Fe and Mn as impurities in materials of Co production, *Zavod. Lab.*, 1975, **41**, 187. (Address as in ref. 306).
1027 ZAKHAROVA, T. I., *et al.*, Determination of impurities in Al isopropoxide, *Zavod. Lab.*, 1975, **41**, 181.
1028 MAKULOV, N. A., ZHINKIN, D. Y., GRADSKOVA, N. A., and TROKHACHENKOVA, O. P., Spectrographic determination of Si in organo-Si compounds, *Zavod. Lab.*, 1975, **41**, 180.
1029 WATSON, A. E., and RUSSELL, G. M., Analysis of geological samples for trace elements by direct-reading ES, *Natl. Inst. Metall., Repub. S. Afr., Rep.*, No. 1656, 1974. (National Institute of Metallurgy, Milner Park, Johannesburg, South Africa).
1030 HOFTON, M. E., and BAINES, S., Dissoultion of oxide materials in H_3PO_4, *Br. Steel Corp., Open Rep.*, GS/EX/43/72/C, 1975, (British Steel Corporation, Hoyle Street, Sheffield, England).
1031 FAKHRY, A. A., Semiquantitative method of estimation of As, Sb and Bi in siliceous rock samples, *Indian J. Technol.*, 1975, **13**, 46. (Spectroscopy Unit, National Research Centre, Dokki, Cairo, Egypt).
1032 KHAN, S. A., and STONE, M., Determination of Li in silicate rocks and minerals using a simple flame photometer, *Rev. Roum. Chim.*, 1974, **19**, 1669. (C.S.I.R. Laboratory, Karachi, Pakistan)
1033 ALVAREZ-ARENAS, E. A., and SAMPEDRO, A., Spectrochemical determination of Ti in stainless steel, *Rev. Metal. (Madrid)*, 1975, **11**, 13.
1034 HOFTON, M. E., Determination of Ag in Fe and steels by AA, *Br. Steel Corp., Open Rep.*, GS/TECH/558/1/74/C, 1974. (Address as in ref. 1030).
1035 HURLBUT, J. A., Direct determination of Be in urine by flameless AAS, *U.S. At. Energy Comm., Rep.*, RFP-2151, 1974. (Address as in ref. 256).
1036 SAMSONI, Z., and NAGY, Z., Objective method for measuring the length of wedge-shaped spectrum lines, *Acta Chim. Hung.*, 1975, **84**, 1. (Institute for Nuclear Research, Hungarian Academy of Science, Debrecen, Hungary).
1037 MACK, D., Determination of Pb in wine and juices by flameless AA, *Dtsch. Lebensm.-Rundsch.*, 1975, **71**, 71. (Chemische Landesuntersuchungsanst., Stuttgart, West Germany).
1038 HENDERSON, R. W., and ANDREWS, D., Pb extraction from Al, *Bull. Environ. Contam. Toxicol.*, 1975, **13**, 330. (Dept. of Chemistry and Physics, Francis Marion College, Florence, S.C. 29501, U.S.A.).
1039 WALTER, H., Survey of algorithms for concentration determination in computerised spectroscopic analysis, *Arch. Eisenhuettenwes.*, 1975, **46**, 209.
1040 PARKER, C., and PEARL, A., AS, British Patent 1,385,791. (Address as in ref. 816).
1041 FLORIAN, K., FLORIANOVA, K., and MATHERNY, M., Dependence of the performance of spectrochemical methods on the resolving power of the spectrograph. I,*Acta Chim. Hung.*, 1974, **82**, 403. (Dept. of Analytical Chemistry, Technical University, Kosice, Czechoslovakia).
1042 MOENKE-BLANKENBUG, L., *et al.*, LMA-10; a new laser microspectral analyser, *Jena Rev.*, 1975, **20**, 107. (Address as in ref. 641).
1043 SCHRON, W., Systematic investigation of a number of minerals and of steel using the LMA-10 laser microspectral analyser, *Jena Rev.*, 1975, **20**, 112. (Mining Academy, Freiberg, West Germany).
1044 CARTER, D., REGAN, J. G. T., and WARREN, J., AA determination of Sr in silicate rocks: a study of major element interferences in the nitrous/oxide acetylene flame, *Analyst*, 1975, **100**, 721. (Address as in ref. 786).
1045 ANON., The determination of small amounts of Cd in organic matter. II: Submicrogram levels, *Analyst*, 1975, **100**, 761. (The Chemical Society, Burlington House, London W1V 0BN, England).
1046 WALSH, A., The separated flame as a resonance detector, *Analyst*, 1975, **100**, 764. (Address as in ref. 838).
1047 BALLAL, D. R., and LEFEBURE, A. H., The structure and propagation of turbulent flames, *Proc. R. Soc. London, Ser. A*, 1975, **344**, 217. (School of Mechanical Engineering, Cranfield Institute of Technology, Cranfield, Bedfordshire, England).
1048 HUBER, M. C. E., SANDEMAN, R. J., and TUBBS, E. F., The spectrum of Cr(I) between 179·8 and 200 nm, wavelengths, absorption cross-sections and oscillator strengths, *Proc. R. Soc. London, Sec. A*, 1975, **342**, 431. (Center for Astrophysics,

Havard College Observatory and Smithsonian Observatory, Cambridge, Mass. 02138, U.S.A.).

1049 AVERDUNK, R., OSTAPOWICZ, B., and GUNTHER, T., The role of cyclic AMP and Ca in the altered permeability of tumour cells grown with a deficiency of Mg, *Z. Klin. Chem. Klin. Biochem.*, 1975, **13**, 361. (Klinikum Steglitz, D-1000 Berlin 45, Hindenburgdamm 30, West Germany).

1050 GLICK, D., and MARICH, K. W., Potential for clinical use of the analytical laser microprobe for element measurement, *Clin. Chem.*, 1975, **21**, 1238. (Address as in ref. 688).

1051 SHVANGHIRADZE, R. R., and VYSOKOVA, I. L., The influence of argon atmosphere on excitation and spectral lines intensity in arc discharge, *Zh. Prikl. Spektrosk.*, 1975, **22**, 618.

1052 ALEKSEEV, M. A., and KAPITONOV, A. N., Study of electron concentration and of temperature in a.c. discharge: influence of current phase, *Zh. Prikl. Spektrosk.*, 1975, **22**, 595.

1053 NIKITINA, O. I., SHAPAPOV, I. S., ANTIPENKO, L. L., and RYABEKA, V. P., The possibility of application of time-resolved spectra for steel and slag analysis with a high- voltage spark excitation, *Zh. Prikl. Spektrosk.*, 1975, **22**, 599.

1054 DROBYSHEV, A. I., and TURKIN, Y. I., The spectral analysis of pulverised samples in cooled hollow cathode discharge, *Zh. Prikl. Spektrosk.*, 1975, **22**, 755.

1055 RICHTOL, H. H., REEVES, R. R., and NELSON, D. L., Converting a television system into a quantitative AE spectrograph, *J. Chem. Educ.*, 1975, **52**, 198. (Rensselaer Polytechnic Institute, Troy, N.Y. 12181, U.S.A.).

1056 DRESSER, R. D., MOONEY, R. A., HEITHMAR, E. M., and PLANKEY, F. W., AS atomization systems, *J. Chem. Educ.*, 1975, **52**, A403. (Dept. of Chemistry, University of Pittsburgh, Pittsburgh, Pa. 15260, U.S.A.).

1057 THOMERSON, D. R., and THOMPSON, K. C., Sample atomization in AAS, *Chem. Br.*, 1975, **11**, 316. (Address as in ref. 19).

1058 WOLLEY, J. F., Some recent applications of AAS to inorganic analysis, *Scan*, 1975, **6**, 6. (Standard Telecommunication Laboratories Ltd., Harlow, Essex, England).

1059 FAITHFULL, N. T., The automatic determination of Fe in herbage Kjeldahl digests by AAS, *Lab. Pract.*, 1975, **24**, 658. (Address as in ref. 295).

1060 ADAMS, M. J., KIRKBRIGHT, G. F., and RIENVATANA, P., Molecular absorption spectra of some simple inorganic salts in the heated graphite atomizer, *At. Absorpt. Newsl.*, 1975, **14**, 105. (Address as in ref. 4).

1061 MARTIN, T. D.,* KOPP, J. F.,* and EDIGER, R. D.,† Determining Se in water, wastewater, sediment and sludge by flameless AAS, *At. Absorpt. Newsl.*, 1975, **14**, 109. (*Environmental Protection Agency, Support Laboratory, Cincinnati, Ohio 45268, U.S.A.; †Perkin-Elmer Corp., Lombard, Ill. 60148, U.S.A.).

1062 RANTALA, R. T. T., and LORING, D. H., Multi-element analysis of silicate rocks and marine sediments by AAS, *At. Absorpt. Newsl.*, 1975, **14**, 117. (Marine Ecology Laboratory, Bedford Institute of Oceanography, Dartmouth, N.S., Canada).

1063 JACKSON, K. W., FULLER, T. D., MITCHELL, D. G., and ALDOUS, K. M., A rapid microsampling cup AA procedure for the determination of Pb in urine, *At. Absorpt. Newsl.*, 1975, **14**, 121. (Address as in ref. 80).

1064 BETTGER, R. J., FICKLIN, A. C., and REES, T. F., Determination of Be in air samples using the graphite furnace, *At. Absorpt. Newsl.*, 1975, **14**, 124. (Dow Chemical Co., Rocky Flats Div., Golden, Colo. 80401, U.S.A.).

1065 KNOTT, A. R., Investigation of a high-purity water system, *At. Absorpt. Newsl.*, 1975, **14**, 126. (Perkin-Elmer Corp., Norwalk, Conn. 06856, U.S.A.).

1066 EDIGER, R. D., AA analysis with the graphite furnace using matrix modification, *At. Absorpt. Newsl.*, 1975, **14**, 127. (Address as in ref. 1061).

1067 IBRAHIM, R. J., and SABBAH, S., The effect of viscosity differences between standards and samples on Fe measurements in crude oil, *At. Absorpt., Newsl.*, 1975, **14**, 131. (Central Laboratory, Iraq National Oil Co., Basrah, Iraq).

1068 ANON., Anchoring the drift in silicon ultraviolet photo-detectors, *Opt. Spectra.*, 1975, **9** (9), 31.

1069 OLSTAD, R. A., and OLANDER, D. R., Evaporation of solids by laser pulses. I: Fe, *J. Appl. Phys.*, 1975, **46**, 1499. (Inorganic Materials Research Div., College of Engineering, University of California, Berkeley, Calif. 94720, U.S.A.).

1070 OLSTAD, R. A., and OLANDER, D. R., Evaporation of solids by laser pulses. II: ZrH_4, *J. Appl. Phys.*, 1975, **46**, 1509. (Address as in ref. 1069).

1071 CHENG, T. K., and CASPERSON, L. W., Plasma diagnosis by laser beam scanning, *J. Appl. Phys.*, 1975, **46**, 1961. (School of Engineering and Applied Science, University of California, Los Angeles, Calif. 90024, U.S.A.).

1072 MARODE, E., The mechanism of spark breakdown in air at atmospheric pressure between a positive point and a plane. I: Experimental; nature of the streamer track, *J. Appl. Phys.*, 1975, **46**, 2005. (Laboratoire de Physique des Decharges (CNRS)-ESE, 10 Ave. Pierre Larousse, 92240 Malakoff, France).

1073 MARODE, E., The mechanism of spark breakdown in air at atmospheric pressure between a positive point and plane. II: Theoretical; computer simulation of the streamer track, *J. Appl. Phys.*, 1975, **46**, 2016. (Address as in ref. 1072).

1074 GREENE, J. E., SEQUEDA-OSORIO, F., and NATARAJAN, B. R., Glow discharge optical spectroscopy for microvolume elemental analysis, *J. Appl. Phys.*, 1975, **46**, 2701. (Dept. of Metallurgy, University of Illinois, Urbana, Ill. 61801, U.S.A.).

The following papers (*) were presented at a meeting of the Dutch Atomic Spectroscopy Working Group, 23 May, 1975, Amsterdam, The Netherlands.

1075* HOLLANDER, T., Properties of alkaline-earth electrodeless discharge lamps. (Address as in ref. 123).

1076* WITMER, A. W., A system for automatic analysis of photographically recorded emission spectra. (Philips' Research Laboratories, Eindhoven, The Netherlands).

1077* VAN BRASSEM, P., Some fundamental aspects of the microwave-induced plasma. (Address as in ref. 206).

1078* HOUPT, P. M., ES detection of elements in gas chromatographic fractions. (Central Laboratory T.N.O., Schoemakerstraat 97, Delft, The Netherlands).

1079 LEBEDEV, V. P., Signal-to-noise ratio in a scanning spectrometer with a photomultiplier, *Opt. Spectrosc. (USSR)*, 1975, **38**, 217.

1080 BLOKH, M. A., and VORONOV, G. S., Measurement of the degree of ionization of a highly ionized plasma from the relative intensities of spectral lines, *Opt. Spectrosc. (USSR)*, 1975, **38**, 218.

1081 FRY, P. W., Silicon photodiode arrays, *J. Phys. E*, 1975, **8**, 337. (Integrated Photomatrix Ltd., Grove Trading Estate, Dorchester, Dorset, England).

1082 FAULKNER, E. A., The principles of impedance optimization and noise matching, *J. Phys. E*, 1975, **8**, 533. (J. J. Thomson Physical Laboratory, Whiteknights, Reading RG6 2AF, England).

1083 BLAIR, D. P., and SYDENHAM, P. H., Phase-sensitive detection as a means to recover signals, *J. Phys. E*, 1975, **8**, 621. (Dept. of Geophysics, University of New England, N.S.W., Australia).

1084 SIMPSON, S. W., A fast gas valve for plasma sampling, *J. Phys. E*, 1975, **8**, 739. (School of Physics, University of Sydney, Sydney, N.S.W., Australia).

1085 USHER, M. J., Noise and bandwidth, *J. Phys. E*, 1974, **7**, 962. (Dept. of Engineering and Cybernetics, Whiteknights, Reading RG6 2AL, England).

1086 SCHAGEN, P., Image converters and intensifiers, *J. Phys. E*, 1975, **8**, 153. (Mullard Research Laboratories, Redhill, Surrey, England).

1087 SCHMAUCH, G. E., Methods of analysis for the noble gases, *C.R.C. Crit. Rev. Anal. Chem.*, 1974, **4**, 107. (Cryogenic Systems Div., Air Products and Chemicals, Inc., Allentown, Pa., U.S.A.).

1088 KRUGERS, J., Instrument exhibition 'Het Instrument,' Amsterdam, 1975, *Chem. Weekbl.*, 1975, **71**, (38) 15; (39) 15; (42) 15. (I.B.M., Johan Huizingalaan 257, Amsterdam, The Netherlands).

1089 KRUGERS, J., Laboratory instruments, *Chem. Weekbl.*, 1975, **71**, (10) 13, (14) 13; (18) 13; (22) 19; (26) 17; (30) 15. (Address as in ref. 1088).

The following papers (*) were presented at the Federation of Analytical Chemistry and Spectroscopy Societies' Second National Meeting, 6–10 October 1975, Indianapolis, Ind., U.S.A.

1090* WALTERS, J. P., Interchangeable apparatus for research in optical ES. (Dept. of Chemistry, University of Wisconsin, Madison, Wis. 53706, U.S.A.).

1091* WALTERS, J. P., and BEATY, J. S., An interactive computer-assisted method for the determination of wavelength in optical ES. (Address as in ref. 1090).

1092* SCHEELINE, A., and WALTERS, J. P., A new look at a traditional approach to spectroscopic data acquisition and interpretation using the Abel inversion. (Address as in ref. 1090).

1093* COLEMAN, D. M., and WALTERS, J. P., Electronic control and application of an adjustable-waveform spark source. (Address as in ref. 1090).

1094* RENTNER, J., WALTERS, J. P., and UCHIDA, T., Design and application of a 323 MHz quarter-wave spark source for production of positionally stable discharges at high repetition rate. (Address as in ref. 1090).
1095* KLUEPPEL, R. J., WALTERS, J. P., EATON, W. S., GOLDSTEIN, S. A., and SACKS, R. D., Design and performance of mechanically time-gated, radially resolved spectrometers for research on repetitive spark discharges. (Address as in ref. 1090).
1096* WATTERS, R. L., WALTERS, J. P., and HOSCH, J. W., Time-gated Schlieren study of a stabilized spark discharge. (Address as in ref. 1090).
1097* FRICKE, F. L., ROSE, O., and CARUSO, J. A., Microwave-induced plasma coupled to a Harrick detector for simultaneous multi-element trace analysis: carbon cup sample introduction. (Food and Drugs Administration, 1141 Central Parkway, Cincinnati, Ohio 45202, U.S.A.).
1098* DALAGER, P. D., DAVISON, A. L., and AJHAR, R. M., The inductively coupled plasma: the answer to many unsolved spectrochemical problems. (Address as in ref. 741).
1099* SCHLEICHER, R. G., and BARNES, R. M., Theoretical and experimental study of temperature, velocity and particle decomposition in the inductively coupled plasma spectrochemical source. (Address as in ref. 81).
1100* DAHLQUIST, R. L., KNOLL, J. W., WILCOX, A. A., CHEN, R. S., and IRONS, R. D., Simultaneous multi-element analysis of biological fluids and tissue: r.f. inductively coupled plasma optical ES. (Address as in ref. 615).
1101* LARSON, G. F., FASSEL, V. A., WINGE, R. K., and KNISELEY, R. N., Ultra-trace determinations by optical ES: the stray light problem. (Address as in ref. 76).
1102* WALTERS, P. E., CHESTER, T. L., and WINEFORDNER, J. D., Diagnostics of 144 MHz plasmas. (Address as in ref. 487).
1103* KOIRTOYOHANN, S. R., and LICHTE, F. E., The current status of flame spectro-photometric methods for trace analysis. (Address as in ref. 203).
1104* FASSEL, V. A., Trace and ultratrace characterizaton by optical ES. (Address as in ref. 98).
1105* NATUSCH, D. F. S., and THORPE, T. M., Gas evolution analysis: a potentially powerful tool for trace metal speciation. (Address as in ref. 33).
1106* WINEFORDNER, J. D., Characterization via atomic and molecular luminescence spectrometry. (Address as in ref. 487).
1107* PARIAN, R. W., SOLOMONS, E. T., and RUTHERFORD, C. P., The determina-tion of Sb, Ba, Pb and As in samples of interest to the forensic scientist using argon plasma ES. (Georgia Crime Laboratory, P.O. Box 1456, Atlanta, Ga. 30301, U.S.A.).
1108* RIPPETOE, W. E., VICKERS, T. J., and KEIRS, C. D., Recent developments in ES with the d.c. plasma arc. (Address as in ref. 90).
1109* LERMAN, S., and ECK, E. H., Computerised calculations of quantitative ES analysis results. (Con Edison, 21 St. and 20 Ave., Building 138, Astoria, N.Y. 11105, U.S.A.).
1110* KARLINSKI, T., and MISSIO, D., A computer-controlled system for plasma optical ES. (Spectrametrics, Inc., 204 Andover Street, Andover, Md. 01810, U.S.A.).
1111* THOMAS, C. P., A minicomputer-based ES analysis system. (Address as in ref. 218).
1112* SEQUEDA OSORIO, F., and GREEN, J. E., Glow discharge optical spectroscopy: technique for micro-volume elemental analysis and depth profiling. (Address as in ref. 1074).
1113* SKOGERBOE, R. K., COLEMAN, G. N., LAMOTHE, P. J., FREELAND, S. J., and VARCOE, F. T., Optical emission analysis with a spectrometer designed for automatic background correction. (Address as in ref. 85).
1114* SKOGERBOE, R. K., and URASA, I., Applications of a plasma emission/echelle spectrometer system to trace analysis of solutions. (Address as in ref. 85).
1115* MARRS, J. M., Real-time analytical spectrometry. (Tektronix, Inc., Beaverton, Ore. 97077, U.S.A.).
1116* COOK, T. E., and PARDUE, H. L., Derivative spectroscopy with a vidicon spectro-meter. (Address as in ref. 471).
1117* HUNTER, T. W., and HIEFTJE, G. M., Adaptive computer optimization of an isolated-droplet flame spectrometer. (Address as in ref. 226).
1118* BOSS, C. B., and HIEFTJE, G. M., Diffusion of atoms from individual solute particles vaporizing in analytical flames: matrix effects. (Address as in ref. 226).
1119* SATURDAY, K. A., and HIEFTJE, G. M., Production and application of a stable helium/oxygen/acetylene laminar flame for use in AS. (Address as in ref. 226).

1120* COCHRAN, R. L., and HIEFTJE, G. M., A new method for continuous source AA using selective resonance line modulation. (Address as in ref. 226).
1121* WOODRIFF, R., New developments in constant-temperature furnace AA. (Address as in ref. 31).
1122* GOODE, S. R., and OTTO, D. C., Multi-element microwave-excited electrodeless discharge lamps for AA and AFS. (Address as in ref. 109).
1123* HOWELL, J. A., and PIZARRO, G., AS using a chlorine/hydrogen flame. (Western Michigan University, Dept. of Chemistry, Kalamazoo, Mich. 49008, U.S.A.).
1124* AMORE, F., Direct microblood Pb analysis by AA using double-beam background correction. (Illinois Department of Public Health, 134 North Ninth St., Springfield, Ill. 62701, U.S.A.).
1125* BARNETT, W. B., Application of an AA instrument containing a microprocessor. (Perkin-Elmer Corp., Norwalk, Conn. 06856, U.S.A.).
1126* PIEPMEIER, E. H., and DE GALAN, L., Temporal behaviour of line profiles of pulsed sources and cells for AS. (Address as in ref. 207).
1127* EVONIUK, C., and WOODRIFF, R., Investigation of the contained d.c. arc column for spectrochemical analysis. (Address as in ref. 31).
1128* ALDOUS, K. M., GARDEN, J., and MITCHELL, D. G., The application of an image dissector multichannel spectrometer to multi-element AAS. (Address as in ref. 80).
1129* BATH, D. A., and WOODRIFF, R., Sequential hollow cathodes for elemental analysis and background correction in AAS. (Address as in ref. 31).
1130* MARKS, J. Y., SPELLMAN, R. J., and WYSOCKI, B., Effect of non-analyte light attenuation on accuracy in trace element analysis in complex alloys. (Address as in ref. 715).
1131* BANCROFT, M. F., SCHLEICHER, R. G., and EMMEL, R. H., Using a programmable calculator in AA. (Instrumentation Laboratory, Inc., Jonspin Road, Wilmington, Mass. 01887, U.S.A.).
1132* BENTLEY, G. E., and PARSONS, M. L., The use of covalent hydrides in the production of electrodeless discharge lamps. (Address as in ref. 137).
1133* RISBY, T. H., Ultratrace metal analysis in environmental and biological systems. (Address as in ref. 674).
1134* JOHNSON, J. D., Optical emission analysis of water wastes with a demountable hollow cathode source. (Address as in ref. 751).
1135* CANELLI, E., ALDOUS, K.M., and MITCHELLL, D. G., The determination of total organic C in natural waters by plasma ES. (Address as in ref. 80).
1136* SANDS, M. D., and COX, W. G., Determination of trace metal concentrations in environmental samples using d.c. argon plasma. (Oceanographic and Environmental Services, Raytheon Co., Portsmouth, R.I. 02871, U.S.A.).
1137* FRANKLIN, M. L., and PICKETT, E. E., Effects of blood collection devices on trace metal analysis.(Address as in ref. 231).
1138* MANNING, D. C., Determination of trace metals in foods by graphite furnace AAS. (Address as in ref. 553).
1139* SOHLER, A., WOLCOTT, P., and PFEIFFER, C. C., Determination of Zn in fingernails by non-flame AAS. (Brain Bio Center, 1225 State Road, Princeton, N.J. 08540, U.S.A.).
1140* SOTERA, J. J., BANCROFT, M. F., and HWANG, J. Y., Determination of refractory elements Be, Mo, Ni, Si, Sn, Ti and V in 'real world' samples by a new high-temperature flameless atomizer. (Address as in ref. 313).
1141* BURDO, R. A., and WISE, W. M., Determination of Si in glasses and minerals by AAS. (Corning Glass Works, Sullivan Research Park, Corning, N.Y. 14830, U.S.A.).
1142* EPSTEIN, M. S., RAINS, T. C., and BARNES, I. L., Some considerations for precision and accuracy in the analysis of Hg in standard reference materials by cold-vapour AA. (Address as in ref. 34).
1143* CORUM, T. L., SCHLEICHER, R. G., EMMEL, R. H., and HWANG, J. Y., An innovation in the determination of V, Ni, Fe and Cu in crude oils by flameless atomizer. (Address as in ref. 313).
1144* YOUNG, C. M., and BALDWIN, J. M., Determination of Co in highly complexed radioactive samples by flameless AAS. (Allied Chemical Corporation, 550 Second Street, Idaho Falls, Idaho 93401, U.S.A.).

1145* AMOS, M. D., and BENNETT, P. A., Some analytical possibilities of cathodic sputtering in AAS. (Address as in ref. 429).
1146* MITCHELL, D. G., JACKSON, K. W., KAHL, M., and ALDOUS, K. M., Application of a microsampling cup/nitrous oxide/acetylene flame AAS. (Address as in ref. 80).
1147* OLSEN, R. A., MURBARAK, A., and WOODRIFF, R., Nutrients in soil solution analysis using constant temperature furnace AA. (Address as in ref. 31).
1148* LUNDQUIST, J. A., Oil analysis: a proven method to reduce operating costs. (Quaker State Oil Refining Corp., Research Center, P.O. Box 989, Oil City, Pa. 16301, U.S.A.).
1149* VARNES, A. W., VIGLER, M. S., and ESKAMAN, A., Trace metal determination in petroleum and petroleum products by AA and induction coupled plasma ES. (Standard Oil of Ohio, Research Laboratory, 4440 Warrensville Road, Cleveland, Ohio 44128, U.S.A.).
1150* WESTWOOD, L. C., AAS in the automobile industry. (Scientific Research Staff, Ford Motor Co., P.O. Box 2053, Dearborn, Mich. 48121, U.S.A.).
1151 GUSARSKII, V. V., On the system of standard samples for spectral analysis of Al and Al alloys, *Zavod. Lab.*, 1974, **40**, 1343.
1152 KARPEL, N. G., and FEDORCHUK, O. K., The precision of chemical spectrographic analysis of epitaxial structures, *Zavod. Lab.*, 1975, **41**, 169.
1153 SVERDLINA, O. A., KUZOVLEV, I. A., SOLOMATIN, V. S., YURKOVA, V. Y., KOUYKOVA, N. V., NISHANOV, D., and SHATALINA, L. G., AA method for analysis of layers of epitaxial structures of GaAs with chromatographic seperation of the matrix, *Zavod. Lab.*, 1975, **41**, 172.
1154 PRISENKO, V. A., and SHVARTS, D. M., AA determination of low concentrations of impurities in high-purity Ni, *Zavod. Lab.*, 1975, **41**, 175. (Address as in ref 1023).
1155 GORBUNOVA, L. B., KUTEINIKOV, A. F., AVDEENKO, M. A,. and MURASHKINA, V. N., Determination of microimpurities in high-purity carbon materials, *Zavod. Lab.*, 1975, **41**, 178.
1156 FLORIAN, K., LAVRIN, A., and MATHERNY, M., The dependence of the course of calibration curves on the evaluation procedure in ES, *Chem. Zvesti*, 1974, **28**, 760. (Address as in ref. 1041).
1157 DANILOVA, F. I., OROBINSKAYA, V. A., PARFENOVA, V. S., NAZARENKO, R. M., KHITROV, C. G., and BELOUSOV, G. E., Chemical-spectrographic determination of Pt metals and Au in Cu/Ni alloys obtained by smelting Cu/NL ores, *Zh. Anal. Khim.*, 1974, **29**, 2142. (Siberian State Design and Scientific Research Institute of Non-ferrous Metallurgy, U.S.S.R.).
1158 TARASOVA, I. I., DUDENKOVA, L. S., KHITROV, V. G., and BELOUSOV, G. E., Chemical-spectrographic determination of Pt group metals and Au in products of sulphide Cu/Ni ore processing, *Zh. Anal. Khim.*, 1974, **29**, 2147. (Norilsk Mining and Metallurgical Group of Enterprises and Institute of Geology of Ore Deposits, Petrography, Mineralogy and Geochemistry, U.S.S.R. Academy of Sciences, Moscow, U.S.S.R.).
1159 DANILOVA, F. I., OROBINSKAYA, V. A., PARFENOVA, V. S., PROPITSOVA, R. F., and SAVVIN, S. B., A new scheme for the determination of Pt metals in Cu based standards for spectrographic analysis, *Zh. Anal. Khim.*, 1974, **29**, 2150. (Address as in ref. 1157).
1160 APOLITSKII, V. N., Increase of the sensitivity and accuracy of ES determination of the noble metals, *Zh. Anal. Khim.*, 1974, **29**, 2273. (Address as in ref. 411).
1161 DANILOVA, F. I., OROBINSKAYA, V. A., KHUDOLEI, G. N., and DMITREVA, G. A., Determination of Os in Cu/Ni collectors after smelting Cu/Ni sulphide ores, *Zh. Anal. Khim.*, 1974, **29**, 2276. (Address as in ref. 1157).
1162 KARPOVA, A. F., LITVINSKAYA, V. I., KASHLINSKAYA, S. E., KHARDINA, V. K., and KUNILOVA, N. M., Determination of microamounts of noble metals in technological solutions, *Zh. Anal. Khim.*, 1974, **29**, 2279. (Norilsk Mining and Metallurgical Group of Enterprises, Norilsk, U.S.S.R.).
1163 BROVKO, A. I., NAZAROV, S. N., and RISH, M. A., AA determination of Zn, Cd, Co, Cu and Ni after their extraction concentration in the diphenylcarbazone/pyridine/toluene system, *Zh. Anal. Khim.*, 1974, **29**, 2387. (Samarkand State University, Samarkand, U.S.S.R.).
1164 BELYAEV, Y. I., ORESHKIN, V. N., and VNUKOVSKAYA, G. L., AA deter -mination of trace elements in rocks using pulse thermal atomization of solid

samples, *Zh. Anal. Khim.*, 1975, **30**, 503. (Address as in ref. 334).
1165 BEGAK, O. Y., Determination of Mo by AAS in acetylene/air flame, *Zh. Anal. Khim.*, 1975, **30**, 619.
1166 L'VOV, B. V., KRUGLIKOVA, L. P., POLZIK, L. K., and KATSKOV, D. A., Theory of AA analysis. III: Transverse diffusion of atoms in flames supported on slot burners, *Zh. Anal. Khim.*, 1975, **30**, 839. (Address as in ref. 348).
1167 L'VOV, B. V., KRUGLIKOVA, L. K., and KATSKOV, D. A., Theory of AA analysis. IV: Calculated temperature values and equilibrium composition of acetylene/nitrous oxide and acetylene/air flames, *Zh. Anal. Khim.*, 1975, **30**, 846. (Address as in ref. 348).
1168 SHAPKINA, Y. S., and PRUDNIKOV, E. D., Ultramicro method of flame-spectrometric determination of Rb and Cs traces, *Zh. Anal. Khim.*, 1975, **30**, 906. (Address as in ref. 330).
1169 KRASILSCHIK, V. Z., and MANOVA, T. G., Spectrographic analysis of boric acid using preliminary electrochemical concention of impurities, *Zh. Anal. Khim.*, 1975, **30**, 971. (Address as in ref. 43).
1170 ANISIMOVA, Z. A., and PETRIKEEVA, N. A., Electric discharge method for determining C in powdered ZrC, *Zh. Anal. Khim.*, 1975, **30**, 1005. (Voroshilovgrad Machine Building Institute, Voroshilovgrad, U.S.S.R.).
1171 MORGENTHALER, L., A primer for flameless atomization, *Am. Lab.*, 1975, (April), 41. (Address as in ref. 707).
1172 HARRINGTON, D. E., and EVANS, T. R., AA data processing with remote computer terminal, *Am. Lab.*, 1975, (Aug.), 15. (Address as in ref. 168).
1173 ROBBINS, W. K., Microchemical changes in heated vaporization AA, *Am. Lab.*, 1975, (Sept.), 23. (Address as in ref. 139).
1174 ROBBINS, W. K., Microchemical changes in heated vaporization AA, *Am. Lab.*, 1975, (Sept.), 38. (Address as in ref. 139).
1175 BARNETT, W. B., and KERBER, J. D., Instrumental electronics for use with a flameless AA sampling device, *Am. Lab.*, 1975, (Oct.), 43. (Address as in ref. 220).
1176 HADEISHI, T., and McLAUGHLIN, R. D., Isotope Zeeman AA: a new approach to chemical analysis, *Am. Lab.*, 1975, (Oct.), 57. (Address as in ref. 770).
1177 SLAVIN, S., AAS for the analysis of foods, *Perkin-Elmer Instrum. News*, 1974, **24** (2), 4. (Address as in ref. 8).
1178 SLAVIN, S., The new model 460 AAS with microcomputer, *Perkin-Elmer Instrum. News*, 1975, **25**, (2), 3. (Address as in ref. 8).
1179 FRANCO, V., and HOLAK, W., Collaborative study of the determination of H_3BO_3 in caviar by ES, *J. Assoc. Off. Anal. Chem.*, 1975, **58**, 293. (Address as in ref. 50).
1180 CARY, E. E., and OLSON, O. E., AA determination of Cr, *J. Assoc. Off. Anal. Chem.*, 1975, **58**, 433. (Plant, Soil and Nutrition Laboratory, U.S. Department of Agriculture, Ithaca, N.Y. 14853, U.S.A.).
1181 ISAAC, R. A., and JOHNSON, W. C., Collaborative study of wet and dry ashing techniques for the elemental analysis of plant tissue by AAS, *J. Assoc. Off. Anal. Chem.*, 1975, **58**, 436. (Soil Testing and Plant Analysis Laboratory, University of Georgia, Athens, Ga., U.S.A.).
1182 JONES, J. B., Collaborative study of the elemental analysis of plant material, *J. Assoc. Off. Anal. Chem.*, 1975, **58**, 764. (Dept. of Horticulture, University of Georgia, Athens, Ga. 30602, U.S.A.).
1183 NEUNER, T. E., PICKETT, E. E., and GEHRKE, C. W., Spectrophotometric determination of B in boronated fertilizers, *J. Assoc. Off. Anal. Chem.*, 1975, **58**, 920. (Address as in ref. 231).
1184 NEUNER, T. E., PICKETT, E. E., GEHRKE, C. W., and WIPPLER, J. F., Automated flame photometric determination of K_2O in fertilizers, *J. Assoc. Off. Anal. Chem.*, 1975, **58**, 923. (Address as in ref. 231).
1185 FRAHM, L. J., ALBRECHT, M. E., and McDONNELL, J. P., AAS determination of 4-hydroxy-3-nitrobenzenearsonic acid in premixes, *J. Assoc. Off. Anal. Chem.*, 1975, **58**, 945. (Salisbury Laboratories, Charles City, Iowa 50616, U.S.A.).
1186 NG, S., and McSHARRY, W., AAS determination of Se in animal feed premix using the vapour generation technique, *J. Assoc. Off. Anal. Chem.*, 1975, **58**, 987. (Address as in ref. 454).
1187 HAGEMAN, L., TORMA, L., and GINTHER, B. E., Analysis of feed grains and forages for traces of Co by flameless AAS, *J. Assoc. Off. Anal. Chem.*, 1975, **58**,

990. (Montana Department of Agriculture, Montana State University, Bozeman, Mont. 59715, U.S.A.).
1188 NG, S., MUNROE, M., and McSHARRY, W., Determination of Se in animal feed premix by AAS, *J. Assoc. Off. Anal. Chem.*, 1974, **57**, 1260. (Address as in ref. 454).
1189 GLASER, E., The determination of Cu in the p.p.b. range in the water of power stations with the aid of the AAS 1, *Jena Rev.*, 1975, **20**, 244. (VEB Kernkraftwerke Greifswald-Rheinsberg, East Germany).
1190 KIRKBRIGHT, G. F., Optical density: its rewards and penalties, *Proc. Anal. Div. Chem. Soc.*, 1975, **12**, 8. (Address as in ref. 4).
1191 PRICE, W. J., Teaching AS: the instrument manufacturer's viewpoint, *Proc. Anal. Div. Chem. Soc.*, 1975, **12**, 119. (Address as in ref. 172).
1192 BROWNER, R. F., Sources for AFS in multi-element analysis, *Proc. Anal. Div. Chem. Soc.*, 1975, **12**, 138. (Department of Industry, Laboratory of the Government Chemist, Cornwall House, London SE1 9NQ, England).
1193 BOUMANS, P. W. J. M., and DE BOER, F. J., An assessment of the inductively coupled plasma for simultaneous multi-element analysis, *Proc. Anal. Div. Chem. Soc.*, 1975, **12**, 140. (Address as in ref. 544).
1194 HARRISON, T. S., Further applications of AAS in the steel industry, *Proc. Anal. Div. Chem. Soc.*, 1975, **12**, 152. (Address as in ref. 53).
1195 ROBINSON, J. W., Carbon atomizers in AAS, *Proc. Anal. Div. Chem. Soc.*, 1975, **12**, 275. (Address as in ref. 141).
1196 KANTOR, T., ES, *chapter in* Wilson, C. L., and Wilson, D. W., *Editors,* Comprehensive analytical chemistry, Vol. V—Book published by Elsevier, Amsterdam and New York, 1975.
1197 HURLBUT, J. A., History, uses, occurrences, analytical chemistry and biochemistry of Be: a review, *U.S. At. Energy Comm., Rep.*, RFP-2152, 1974. (Address as in ref. 256).
1198 CHANDOLA, L. C., and MAHAJAN, V., Spectrographic determination of noble and other metals in C, *Bhabha At. Res. Cent., Rep.*, BARC-738, 1974. (Address as in ref. 291).
1199 GOMEZ-COEDO, A., and DORADO LOPEZ, M. T., Determination of low contents of Pb in Al and its alloys, *Rev. Metal. (Madrid)*, 1975, **11**, 61. (Address as in ref. 782).
1200 AZIZ, A., and SOLGAARD, P., Spectrochemical determination of trace impurities in U_3O_8 by carrier-distillation method using the d.c. arc cathode-central region, *Dan. At. Energy Comm., Riso, Rep.*, M-1735, 1974. (Chemistry Dept., Danish Atomic Energy Commission Research Establishment, Riso, Roskilde, Denmark).
1201 IGARASHI, S., Determination of Cd in dust, *Bunseki Kagaku*, 1975, **24**, 270. (Pollution Control Section, City Office, Yokkaichi, Mie, Japan).
1202 TUTAKINA, N. M., and MARSHAKOV, I. K., Use of a rotating disc electrode with an annular ring for determining the chemical composition of alloys, *Zavod. Lab.*, 1975, **41**, 259. (Voronezh State University, U.S.S.R.).
1203 MOROSHKINA, T. M., and PEREZ SANFIEL, F., Separation of Nb and Ta by paper chromatography and their subsequent spectrographic determination, *Vestn. Leningr. Univ., Fiz., Khim.*, 1975 (2), 129. (University of Leningrad, Leningrad, U.S.S.R.).
1204 HOCQUELLET, P., Determination of Co in animal feeding stuffs by flame and flameless AAS, *Ann. Falsif. Expert. Chim.*, 1974, **67**, 495. (Inst. Munic. Rech. sur l'Alimentation Humaine et Anim. Lab. Munic., Bordeaux, France).
1205 LUECKE, W., ESCHERMANN, F., LENNARTZ, U., and PAPASTAMATAKI, A. J., Interference problems in flameless AAS; spectrometric analysis of geochemical samples, *Neues Jahrb. Mineral., Abh.*, 1974, **120**, 178.
1206 CULVER, B. R., LECH, J. F., and PRADHAN, N. K., Trace metal analysis of foods by non-flame AAS, *Food Technol.*, 1975, **29**, 16. (Address as in ref. 31).
1207 KING, H. G., and MORROW, R. W., Determination of As and Se in surface water by AA to support environmental monitoring programmes, *U.S. At. Energy Comm., Rep.*, Y-1956, 1974. (Address as in ref. 288).
1208 HODKINSON, A., Method for the spectrographic analysis of small amounts of non-metallic steelworks materials, *Br. Steel Corp., Open Rep.*, SM/589/A, 1975. (British Steel Corporation, Hoyle Street, Sheffield, England).
1209 NEWSTEAD, R. A., Improvements in and relating to atomic spectrometers, British Patent 1,397,286.

1210 WOOLEY, J. F., Graphite tube furnace, British Patent 1,399,050.
1211 MAY, R. W., and PORTER, J., Evaluation of common methods of paint analysis, *J. Forensic Sci.*, 1975, **15**, 137. (Home Office Central Research Establishment, Aldermaston, Berkshire, England).
1212 PATENT, Improvements in or relating to AF spectrometers, British Patent 1,396,806.
1213 MAINES, I. S., ALDOUS, K. M., and MITCHELL, D. G., Determination of Pb in potable waters using Delves cup AAS with signal integration, *Environ. Sci. Technol.*, 1975, **9**, 549. (Address as in ref. 80).
1214 ANON., Methods for the analysis of Pb alloys. Part I: Sb in Pb alloys — Australian Standard, 1671, Pt I: 1974.
1215 ANON., Toothpastes — British Standard, BS 5136: 1974.
1216 FAITHFULL, N. T., Spectrum of the acetylene/air flame used in the determination of Fe by AAS, *Lab. Pract.*, 1975, **24**, 737. (Address as in ref. 295).
1217 DUNCAN, L., and PARKER, C. R., Applications of $NaBH_4$ for AA determination of volatile hydrides, *Varian Tech. Topics*, 1974. (Address as in ref. 429).
1218 PARKER, C. R., SCOBBIE, R., and DUNCAN, L., The analysis of NBS standard reference materials using the carbon rod atomizer, *Varian Tech. Topics*, 1974, (Aug.). (Address as in ref. 429).
1219 VINCIGUERRA, G., and THOMPSON, B. G., Forensic applications of the carbon rod atomizer in the investigation of shooting cases, *Varian Tech. Topics*, 1974, (Sept.). (Address as in ref. 429).
1220 VINCIGUERRA, G., Determination of Sb, Ba and Pb in gunshot residues by means of the carbon rod atomizer, *Varian Tech. Topics*, 1974, (Sept.). (Address as in ref. 429).
1221 PARKER, C. R., and STUX, R. L., Background correction in AA, *Varian Tech. Topics*, 1974, (Dec.). (Address as in ref. 429).
The following papers (*) were presented at the Eastern Analytical Symposium, 21 November, 1975, New York, N.Y., U.S.A.
1222* MARKS, J. Y., WELCHER, G. G., and SPELLMAN, R. J., Analysis of metals and alloys for major constituents and trace elements by AA. (Address as in ref. 715).
1223* MITCHELL, D. G., Difficult trace metal analysis in clinical chemistry. (Address as in ref. 80).
1224* WISE, W. M., and WILLIAMS, J. P., The analysis of glass, refractories, ceramics and related materials by AS. (Address as in ref. 1141).
1225* RAINS, T. C., Application of AAS to the analysis of heavy metals in water and air particulates. (Address as in ref. 34).
1226 MOSIER, E. L., ANTWEILER, J. C., and NISHI, J. M., Spectrochemical determination of trace elements in galena, *J. Res. U.S. Geol. Surv.*, 1975, **3**, 625. (Federal Center, U.S. Geological Survey, Denver, Colo., U.S.A.).
1227 TALMI, Y., and NORVELL, V. E., Determination of As and Sb in environmental samples using gas chromatography with a microwave ES system, *Anal. Chem.*, 1975, **47**, 1510. (Address as in ref. 161).
1228 WARD, A. F., MITCHELL, D. G., and ALDOUS, K. M., Use of a nitrous oxide/acetylene flame to minimise interferences in microsampling-cup AAS, *Anal. Chem.*, 1975, **47**, 1656. (Address as in ref. 80).
1229 ISSAQ, H. J., and MORGENTHALER, L. P., Utilization of ultrasonic nebulization in AAS: a study of parameters, *Anal. Chem.*, 1975, **47**, 1661. (Address as in ref. 707).
1230 ISSAQ, H. J., and MORGENTHALER, L. P., Utilization of ultrasonic nebulization in AAS: trace metal analysis in aqueous solutions, *Anal. Chem.*, 1975, **47**, 1668. (Address as in ref. 707).
1231 FUJIWARA, K., HARAGUCHI, H., and FUWA, K., Response surface and atomization mechanism in air-acetylene flames: acid interference in AA of Cu and In, *Anal. Chem.*, 1975, **47**, 1670. (Address as in ref. 242).
1232 LUKASIEWICZ, R. J., and BUELL, B. E., Study of Zn determination in lubricating oils and additives by AAS, *Anal. Chem.*, 1975, **47**, 1673. (Address as in ref. 675).
1233 STEPHENS, B. G., and FELKEL, H. L., Extraction of Cu (II) from aqueous thiocyanate solutions into proplyene carbonate and subsequent AAS determination, *Anal. Chem.*, 1975, **47**, 1676. (Dept. of Chemistry, Wofford College, Spartanburg, S.C. 29301, U.S.A.).
1234 ISSAQ, H. J., and ZIELINSKI, W. L., Modification of a graphite tube atomizer for flameless AAS, *Anal. Chem.*, 1975, **47**, 2281. (Address as in ref. 746).

1235 KOIZUMI, H., and YASUDA, K., New Zeeman method for AAS, *Anal. Chem.*, 1975, **47**, 1679. (Address as in ref 369).
1236 GREIG, R. A., Comparison of AA and neutron activation analyses for the determination of Ag, Cr and Zn in various marine organisms, *Anal. Chem.*, 1975, **47**, 1682. (Milford Laboratory, Middle Atlantic Coastal Fisheries Center, National Marine Fisheries Service, Milford, Conn. 06460, U.S.A.).
1237 JOHNSON, D. J., PLANKEY, F. W., and WINEFORDNER, J. D., Multi-element analysis via computer-controlled rapid-scan AFS with a continuum source, *Anal. Chem.*, 1975, **47**, 1739. (Address as in ref. 487).
1238 ISSAQ, H. J., and MORGENTHALER, L. P., Utilization of ultrasonic nebulization in AAS: trace metal analysis in samples of high salt content, *Anal. Chem.*, 1975, **47**, 1748. (Address as in ref. 707).
1239 WILLIS, J. B., AAS analysis by direct introduction of powders into the flame, *Anal. Chem.*, 1975, **47**, 1752. (Address as in ref. 887).
1240 SIMON, S. J., and BOLTZ, D. F., AAS study of the extractability of selected molybdoheteropoly acids, *Anal. Chem.*, 1975, **47**, 1758. (Address as in ref. 932).
1241 SUGIMAE, A., Sensitive ES method for the analysis of airborne particulate matter, *Anal. Chem.*, 1975, **47**, 1840. (Address as in ref 928).
1242 NEWTON, M. P., and DAVIS, D. G., Flameless AAS employing a wire loop atomizer, *Anal. Chem.*, 1975, **47**, 2003. (Dept. of Chemistry, University of New Orleans, New Orleans, La. 70122, U.S.A.).
1243 SIMMONS, W. J., Determination of low concentrations of Co in small samples of plant material by flameless AAS, *Anal. Chem.*, 1975, **47**, 2015. (Address as in ref. 236).
1244 FRICKE, F. L., ROSE, O., and CARUSO, J. A., Simultaneous multi-element determination of trace metals by microwave induced plasma coupled to vidicon detector: carbon cup sample introduction, *Anal. Chem.*, 1975, **47**, 2018. (Address as in ref. 1097).
1245 LING, C. S., and SACKS, R. D., AE determination of selected trace metals in microsamples using exploding foil excitation, *Anal. Chem.*, 1975, **47**, 2074. (Address as in ref. 78).
1246 RIPPETOE, W. E., and VICKERS, T. J., Rotating arc plasma jet for ES, *Anal. Chem.*, 1975, **47**, 2082. (Address as in ref. 90).
1247 POSMA, F. D., SMIT, H. C., and ROOZE, A. F., Optimization of instrumental parameters in flameless AAS, *Anal. Chem.*, 1975, **47**, 2087. (Address as in ref. 245).
1248 TOFFALETTI, J., and SAVORY, J., Use of $NaBH_4$ for determination of total Hg in urine by AAS, *Anal. Chem.*, 1975, **47**, 2091. (Depts. of Medicine, Biochemistry, Pathology and Hospital Laboratories, University of North Carolina, Chapel Hill, N.C. 27514, U.S.A.).
1249 ROBBINS, W. K.,* RUNNELS, J. H.,† and MERRYFIELD, R.,† Analysis of petroleum for trace metals; determination of trace quantities of Be in petroleum and petroleum products by heated vaporization AA, *Anal. Chem.*, 1975, **47**, 2095. (*Address as in ref. 139; †Address as in ref. 685).
1250 SERRAVALLO, F. A., and RISBY, T. H., Effect of doping gases on microwave-induced ES detectors for gas chromatography, *Anal. Chem.*, 1975., **47**, 2141. (Address as in ref. 674).
1251 TALMI, Y., and BOSTICK, D. T., Determination of alkyl-As acids in pesticide and environmental samples by gas chromatography with a microwave ES detection system, *Anal. Chem.*, 1975, **47**, 2145. (Address as in ref. 161).
1252 CHAU, Y. K., WONG, P. T. S., and GOULDEN, P. D., Gas chromatography/AA method for the determination of dimethyl selenide and dimethyl diselenide, *Anal. Chem.*, 1975, **47**, 2279. (Address as in ref. 167).
1253 ESPINASSOU, E., LE RESTE, C., and GUERNET, M., Periodimetry in non-aqueous medium, *Analusis*, 1975, **3**, 19. (Laboratoire de Chimie Analytique, Centre d'Etudes Pharmaceutiques de Chatenay-Malabry, Universite Paris-Sud, Chatenay-Malabry, France).
1254 CONDYLIS, A., and MEJEAN, B., Recent applications of AA to metallurgical analysis: determination of low concentrations of the elements Al, Ba, Ti, Sn, Sb and Ta, *Analusis*, 1975, **3**, 94. (Creusot-Loire, Centre de Recherches Unieux, F-42701, Firminy, France).
1255 URBAIN, H., and CARRET, G., Spectral interferences of V in the determination of Si by AAS, *Analyusis*, 1975, **3**, 110. (Address as in ref. 217).

1256 GOVINDARAJU, K., Ion exchange dissolution of silicates for optical emission and x-ray spectrochemical analysis, *Analusis,* 1975, **3**, 116. (Centre de Recherches Petrographiques et Geochimiques, C.N.R.S., F-54500, Vandoeuvre-les-Nancy, France).
1257 COUETTE, S., COUREAU, C., KUNTZINGER, H., ANTONETTI, A., and AMIEL, C., Ultramicrodetermination of Ca by AA, *Analusis,* 1975, **3**, 126. (I.N.S.E.R.M., 64, Hopital Tenon, F-75020, Paris, France).
1258 GABRIELS, R., Analysis of horticultural products, soil mixes and irrigation waters by FE and AAS, *Analusis,* 1975, **3**, 139. (Dept. de Chimie de la Station des Plantes Ornamentales, Caritasstraat, 21, B-9230 Melle, Belgium).
1259 ECREMONT, F., and BURELLI, F. P., Determination of Li in waters by AAS: use of Li as a tracer in hydrology, *Analusis,* 1975, **3**, 146. (Laboratoire de la Societe pour la Mise en Valeur Agricule de la Corse, B.P. 226, F-20298 Bastia, France).
1260 BERMEJO-MARTINEZ, F., BALUJA-SANTOS, C., and RAVINA-PEREIRO, J. A., Determination of Fe in wines by AA, *Analusis,* 1975, **3**, 157. (Dept. de Chimie Analytique de la Faculte des Sciences, Section de Chimie Analytique du Conseil de Recherches Scientifiques, Santiago de Compostela, Spain).
1261 GOVINDARAJU, K., Iron screw rod powder technique using flame for direct AA determination of rare alkali elements (Li, Rb, Cs) in silicate rock samples, *Analusis,* 1975, **3**, 164. (Address as in ref. 1256).
1262 FARHAN, F. M., and PAZANDEH, H., Determination of traces of metals in petroleum crudes by arc spectrography, *Analusis,* 1975, **3**, 201. (Laboratory of Spectrography and Laboratory of Physical Chemistry, Faculty of Engineering, University of Tehran, Tehran, Iran).
1263 LOPEZ DE AZCONA, J. M., and ALVAREZ-ARENAS, E. A., Effect of the microstructure of Ti and its alloys on spectral emission, *Analusis,* 1975, **3**, 291. (Instituto Nacional de Tecnica Aerospacial, Torrejon de Ardoz, Madrid, Spain).
1264 RIANDEY, C., LINHARES, P., and PINTA, M., Application of flameless AA to the determination of trace elements in natural minerals, *Analusis,* 1975, **3**, 303. (Address as in ref. 614).
1265 SLICKERS, K., and SCHMITT, J. C., Spectrometric determination of Al and its alloys, *Analusis,* 1975, **3**, 317. (A.R.L. France, B.P. 3, F-78320, Le Mesnil-Saint-Denis, France).
1266 BEGUIN, F., and SETTON, R., Determination of K in the presence of $Ag_2Cr_2O_7$ in H_2SO_4 medium using AAS: application to the determination of C-K insertion compounds, *Analusis,* 1975, **3**, 336. (Laboratoire de Chimie IV, U.E.R. de Sciences Fondamentales et Appliquees, F-45045, Orleans Cedex, France).
1267 PINTA, M., Plant standards for the analysis of leaves, *Analusis,* 1975, **3**, 345. (Address as in ref. 614).
1268 PARDHAN, S. I., and OTTAWAY, J. M., Determination of trace elements in soaps and phosphate materials by carbon furnace AAS, *Proc. Anal. Div. Chem. Soc.*, 1975, **12**, 291. (Address as in ref. 394).
1269 ROZENBLUM, V., Successive determination of pg amounts of P and As in pure water by indirect flameless AAS, *Anal. Lett.*, 1975, **8**, 549. (Dept. of Chemistry, Technion-Israel, Institute of Technology, Haifa, Israel).
1270 FLOYD, M., and SOMMERS, L. E., Determination of alkyl-Hg compounds in lake sediments by steam distillation flameless AA, *Anal. Lett.*, 1975., **8**, 525. (Dept. of Agronomy, Purdue University, West Lafayette, Ind. 47907, U.S.A.).
1271 SIEMER, D. D., and HAGEMANN, L., An improved hydride generation AA apparatus for Se determination, *Anal. Lett.*, 1975, **8**, 323. (Address as in ref. 31).
1272 MANTEL, M., and ALADJEM, A., Determination of Fe in Zr by electrolytic dissolution and AAS, *Anal. Lett.*, 1975, **8**, 415. (Nuclear Chemistry Dept., Soreq Nuclear Research Centre, Yavne, Israel).
1273 JONES, D. R., and MANAHAN, S. E., Measurement of chelating agent levels in media with high concentrations of interfering metal ions, *Anal. Lett.*, 1975, **8**, 421. (Dept. of Chemistry, University of Missouri, Columbia, Mo. 65201, U.S.A.).
1274 MESMAN, B. B., and THOMAS, T. C., A study of two AA methods for the determination of sub-microgram amounts of As and Se, *Anal. Lett.*, 1975, **8**, 449. (U.S.A.F. Environmental Health Laboratory, McClellan A.F.B., Calif. 95652, U.S.A.).
1275 LE BIHAN, A., and COURTOT-COUPEZ, J., Determination of Hg in natural waters after extraction by the pyrrolidine dithiocarbamate/propylene carbonate system, *Anal. Lett.*, 1975, **8**, 269. (Address as in ref. 258).
1276 *Delted.*

1277 HARAGUCHI, H.,* TODA, S.,* HIRABAYASHI, N.,† and FUWA, K.,* Background correction in two-flames automatic recording spectrophotometer for flame absorption spectra, *Bunseki Kagaku,* 1975, **24**, 392. (*Address as in ref. 242; †Oyama Plant, Scientific Instrument Div., Daini Seiko-sha Co., Ltd., Oyama-machi, Shizuoka, Japan).

1278 MUSHA, S.,* and TAKAHASHI, Y.,† Enrichment of trace amounts of Au in water utilizing the coagulation of soybean protein and its determination by AAS and ES, *Bunseki Kagaku,* 1975, **24**, 395. (*Address as in ref. 186; † Industrial Research Laboratory, Kao Soap Co. Ltd., 1334, Minato-Yakushubata, Wakayama-shi, Wakayama, Japan).

1279 KONO, T., and NEMORI, A., Extraction of Cu, Cd, Pb, Ag and Bi with iodid/MIBK and application to AAS, *Bunseki Kagaku,* 1975, **24**, 419. (Industrial Research Institute of Iwate, Tonan-mura, Shiwa-gun, Iwate, Japan).

1280 TANAKA, I., SATO, K., and MATSUMOTO, R., Influence of Mn on ES analysis of Nb in steel: interference of Nb(II) 319·4977 A. *Bunseki Kagaku,* 1975, **24**, 423. (Fundamental Research Laboratories, Nippon Steel Corp., 1618, Ida, Nakhara-ku, Kawasaki-shi, Kanagawa, Japan).

1281 MISHIMA, M., Determination of trace metals in a human blood by AAS, *Bunseki Kagaku,* 1975, **24**, 433. (Institute of Public Health, 4-6-1, Shirokane-dai, Minato-ku, Tokyo, Japan).

1282 MATSUSAKI, K., AAS determination of microamounts of Be in Al and Cu using solvent extraction with acetylacetone, *Bunseki Kagaku,* 1975, **24**, 442. (Technical College, Yamaguchi University, Tokiwadai, Ube-shi, Yamaguchi, Japan).

1283 AIHARA, M., and KIBOKU, M., AAS of Cd and Cu by using solvent extraction with K ethylxanthate/MIBK, *Bunseki Kagaku,* 1975, **24**, 447. (Dept. of Applied Chemistry, Faculty of Engineering, Kinki University, 1000, Hiro-machi, Kure-shi, Hiroshima, Japan).

1284 SAKAMOTO, T., KAWAGUCHI, H., and MIZUIKE, A., Microwave ES determination of As at the p.p.b. level in Al, *Bunseki Kagaku,* 1975, **24**, 457. (Address as in ref. 371).

1285 KIDANI, Y., SAOTOME, T., INAGAKI, K., and KOIKE, H., Indirect determination of *p*-aminobenzoic acid by AAS, *Bunseki Kagaku,* 1975, **24**, 463. (Address as in ref. 379).

1286 AIHARA, M., and KIBOKU, M., AAS of Co and Ni using solvent extraction with K ethylxanthate/MIBK, *Bunseki Kagaku,* 1975, **24**, 501, (Address as in ref. 1283).

1287 GOTO, T., AAS analysis of Sb: iodide complex extraction with high molecular amine, *Bunseki Kagaku,* 1975, **24**, 520. (Faculty of Engineering, Nihon University, Tokusada, Tamura-machi, Kouriyama-shi, Fukushima, Japan).

1288 MUSHA, S., * and TAKAHASHI, Y.,† Enrichment of trace Hg by soybean portein for analysis by the flameless AA method, *Bunseki Kagaku,* 1975, **24**, 535. (*Address as in ref. 186; †Address as in ref. 1278).

1289 MUSHA, S.,* and TAKAHASHI, Y., † Enrichment of trace Cd by soybean protein for analysis by the AA method, *Bunseki Kagaku,* 1975, **24**, 540. (*Address as in ref. 186; †Address as in ref. 1278).

1290 MIYAZAKI, A., and UMEZAKI, Y., Determination of submicrogram amounts of Hg in water by d.c. plasma arc, *Bunseki Kagaku,* 1975, **24**, 562. (Address as in ref. 376).

1291 MITSUI, T.,* and FUJIMURA, Y.,† Indirect determination of barbituric acid derivatives by AAS, *Bunseki Kagaku,* 1975, **24**, 575. (*Criminology Laboratory, Aichi Prefecture, 2–1–1, Sanuomaru, Naka-ku, Nagoya-shi, Aichi, Japan; †Dept. of Industrial Chemistry, Chubu Institute of Technology, 1200, Matsumoto-cho, Kasugai-shi, Aichi, Japan).

1292 SANEMASA, I., NODA, H., HISANAGA, A., DEGUCHI, T., and NAGAI, H., FES determination of La, Eu and Yb in an air/acetylene flame, *Bunseki Kagaku,* 1975, **24**, 579. (Dept. of Chemistry, Faculty of Science, Kumamoto University, 2–39–1, Kurokami, Kumamoto-shi, Kumamoto, Japan).

1293 SATO, A., OIKAWA, T., and SAITOH, N., Resin treatment for the analysis of trace heavy metals in estuarine water by flameless AAS, *Bunseki Kagaku,* 1975, **24**, 584. (Iwate Prefectural Institute of Public Health, Uchimanu, Morioka-shi, Iwate, Japan).

1294 IDE, Y., YANAGISAWA, M., KITAGAWA, K., and TAKEUCHI, T., Temperature measurement by maximum absorbance comparison method: a resonance line and doublet lines of Pb, *Bunko Kenkyu,* 1975, **24**, 143. (Address as in ref. 1017).

1295 FUKUSHIMA, H., and NAKAJIMA, T., Isotopic analysis of U by optical spectral methods. I: Determination of $^{235}U/^{238}U$ ratios using a hollow cathode discharge source, *Bunko Kenkyu,* 1975, **24**, 148. (Japan Atomic Energy Research Institute, Tokai-mura, Naka-gun, Ibaraki-ken, Japan).

1296 KUBOTA, M., and ISHIDA, R., Choice of spectral lines and published Ag lines in the measurement of excitation temperature of spectroscopic light sources, *Bunko Kenkyu,* 1975, **24**, 272. (Address as in ref. 372).

1297 MUSHA, S.,* and TAKAHASHI, Y.,† Enrichment of trace metals in water utilizing the coagulation of soybean protein, *Bunseki Kagaku,* 1975, **24**, 365. (*Address as in ref. 186; †Address as in ref. 1278).

1298 NISHIMURA, M., MATSUNAGA, K., and KONISHI, S., Determination of Hg in natural waters, *Bunseki Kagaku,* 1975, **24**, 655. (Address as in ref. 637).

1299 SHIGEMATSU, T.,* MATSUI, M.,* FUJINO, O.,* MITSUNO, S.,† and NAGAHIRO, T.,† Determination of Cu in sea water and shell fish by AAS with a carbon tube atomizer, *Nippon Kagaku Kaishi,* 1975, 1328. (*Address as in ref. 568; †Himeji Institute of Technology, Himeji-shi, Hyogo, Japan).

1300 DEAN, J. A., and RAINS, T. C., *Editors,* Flame emission and AAS. Volume 3: Elements and matrices — Book published by Marcel Dekker, Inc., New York, 1975.

The following papers (*) were presented at Euroanalysis II, 25–30 August 1975, Budapest, Hungary.

1301* LAKATOS, I. J., Some problems of spectrochemical standards applied in organic media. (Address as in ref. 587).

1302* KISS, L., SZIVOS, K., and DUNGOR, E., Optical investigation of the expansion and temperature field of flames in the presence of organic solvent. (Address as in ref. 275).

1303* OHLS, K., Spectrographic analysis by the liquid-layer-on-solid-sample spark technique. (Estel Hoesch Huttenewerk AG, 4600 Dortmund, Postfach 902, West Germany).

1304* MORITZ, P., GEGUS, E., and REPAS, P., Computer methods for the determination of the concentration of homogeneity of standard samples for spectrometric analysis. (Address as in ref. 617).

1305* BARNES, R. M., SCHLEICHER, R. G., and NIKDEL, S., Computer simulations of r.f. inductively coupled plasma discharges in argon and nitrogen. (Address as in ref. 81).

1306* PAVLOVIC, B., and MIHAILIDI, T., Influence of external rotating magnetic field on spectral line intensities. (Address as in ref. 623).

1307* KANTOR, T., POLOS, L., FODOR, P., and PUNGOR, E., Flame emission and AAS of arc and laser nebulized samples. (Address as in ref. 110).

1308* POSGAY, M., and BORSODY-KOVACS, M., Determination of Mg in high purity Al by AAS. (Femipari Kutato Intezet, Budapest, Hunagry).

1309* MAGYAR, B., and WECHSLER, P., Use of AAS with graphite furnace atomizer in the study of extraction and solubility equilibria. (Laboratorium fur Anorganische Chemie, Eidg. Technische Hochschule, Universitatstrasse 6, 8006 Zurich, Switzerland).

1310* ROTH, E., Comparison of methods in trace analysis. (Dept. de Recherche et Analyse, Centre d'Etudes Nucleaires de Saclay, 91190 Gif-sur-Yvette, France).

1311* BELTCHEV, S., Investigation on a d.c. arc in an alternating magnetic field. (Faculty of Chemistry, University of Sofia, Sofia 26, Bulgaria).

1312* TOROK, T., ZARAY, G., and BUZASI, A., Investigation of the low temperature hollow cathode excitation of He and Ar carrier gases. (Address as in ref. 616).

1313* ZADGORSKA, Z., and KRASNOBAEVA, N., Effect of the anionic composition of the additive on the fundamental parameters of a d.c. arc plasma. (Address as in ref. 37).

1314* PAKSY, L., Some spectrochemical properties of superintensive radiation. (3532 Miskolc, Hungary).

1315* SZABO, E., A spectrographic method for determination of trace impurities in quartz. (United Incandescent Lamps and Electrical Co. Ltd., Research Institute, 1138 Budapest, Hungary).

1316* WOIDICH, H., and PFANNHAUSER, W., AA analysis of As in food. (Address as in ref. 72).

1317* DITTRICH, K., and WENNRICH, R., Determination of light vaporizable elements with flameless AF. (Address as in ref. 107).
1318* DELIJSKA, A., and PRAVCHEVA, C., Spectrochemical determination of As, Sb, Pb, Sn, Bi, Cd and Zn in ferromolybdenum. (Address as in ref. 36).
1319* HARSANYI, E. G., BEZUR, L., POLOS, L., and PUNGOR, E., New high-sensitivity AA methods for Hg determination. (Address as in ref. 275).
1320* POSTA, J.,* PAPP, L.,* and POLOS, L.,† The problems of the interference of Al with alkaline-earth metals in AAS. (*Institute of Inorganic and Analytical Chemistry, Kossuth Lajos University, 4010 Debrecen, Hungary; †Address as in ref. 275).
1321* ZIMMER, K., Computer program for determination of the radial distribution of the temperature and electron pressure of the carbon arc plasma. (Address as in ref 617).
1322* WELZ, B., and GROBENSKI, Z., Improvements in convenience, reproducibility and accuracy in flameless AA applying automatic sample dispensing. (Address as in ref. 547).
1323* GROBENSKI, Z., Possibilities of determination of different elements in food stuff with AAS after decomposition under pressure in PTFE autoclave, with low temperature plama ashing and 'wet' acid digestion. (Address as in ref. 547).
1324 TAI, M. H., HARWIT, M., and SLOANE, N. J. A., Errors in Hadamard spectroscopy or imaging caused by imperfect masks, *Appl. Opt.*, 1975, **14**, 2678. (Cornell University, Center for Radiophysics and Space Research, Ithaca, N.Y. 14853, U.S.A.).
1325 NEISWANDER, R. S., and PLEWS, G. S., Low noise extended frequency response with cooled silicon photodiodes, *Appl. Oct.*, 1975, **14**, 2720. (T.R.W. Systems, Redondo Beach, Calif. 90278, U.S.A.).
1326 ANON., Determination of airborne particulate V by AAS, *Pure Appl. Chem.*, 1974, **40**, 38.
1327 ANON., Determination of airborne particulate Pb by AAS, *Pure Appl. Chem.*, 1974, **40**, 35.
1328 ANON., Determination of airborne particulate Cd by AAS, *Pure Appl. Chem.*, 1974, **40**, 37.
1329 HOULE, C. R., AA digital data acquisition by a programmable calculator, *Chem. Instrum.*, 1975, **6**, 143. (Pacific Utilization Research Centre, National Marine Fisheries Service, National Oceanic and Atmospheric Administration, Seattle, Wash., U.S.A.).
1330 BUTTERWORTH, A., The laser microspectral analyser, *J. Forensic Sci. Soc.*, 1974, **14**, 123. (Home Office Central Research Establishment, Aldermaston, Reading, Berkshire RG7 4PN, England).
1331 NORSTROM, R. J., and REYNOLDS, L. M., Environmental Hg analysis: application to the Magos method to the rapid determination of total Hg and methyl-Hg — Paper presented at the First Chemical Congress of the North American Continent, 1–5 December, 1975, Mexico City, Mexico. (Canadian Wildlife Service, Ottawa, Ont. K1A 0H3, Canada).
1332 BRENDT, H., and JACKWERTH, E., AAS determination of small amounts of substances and analysis of trace element concentrates by the injection method, *Spectrochim. Acta, Part B*, 1975, **30*B***, 169. (Address as in ref. 58).
1333 MOLNAR, C. J., CHUANG, F. S., and WINEFORDNER, J. D., Emission from atomic vapour produced in a tubular vitreous carbon furnace, *Spectrochim. Acta, Part B*, 1975, **30*B***, 183. (Address as in ref. 487).
1334 BUTTGEREIT, G., AA, FE and AFS, pp. 710–725 *in* Korte, F., *Editor,* Methodicum chemicum. Vol. 1: Analytical methods; Part B: Micromethods, biological methods, quality control, automatization — Book published by Academic Press, New York, 1974. (Organische-analytisches Laboratorium, Farbenfabriken Bayer AG, Leverkusen, West Germany).
1335 REIF, I., FASSEL, V. A., and KNISELEY, R. N., Spectroscopic flame temperature measurements and their physical significance. III: Existence of isothermal zones in laboratory flames. *Spectrochim. Acta, Part B,* 1975, **30*B***, 163. (Address as in ref. 76).
1336 PETERSON, E. A., Application of AAS to the analysis of Mo in high alloy steels, *U.S. NTIS, AD-A Rep.* No. 007136, 1974. (General Thomas J. Rodman Laboratory, Rock Island Arsenal, Ill., U.S.A.).
1337 COGHE, W., and KASTELIJN, H., Determination of trace Co, Ni and Sn in drinking water by flameless AAS, *Farm. Tijdschr. Belg.*, 1975, **52**, 97. (Lab. Bromatol. Rijksuniv. Ghent, Belgium).
1338 HURLBUT, J. A., and BOKOWSKI, D. L., Determination of submicrogram amounts

of Be in air filter samples by nonflame AAS, *U.S. At. Energy Comm., Rep.*, RFP 2287, 1974. (Address as in ref. 256).

1339 GARCIA DE JALON, A., and FRIAS PANTOJA, J. E., Determination of Hg in wine by flameless AAS, *An. Estac. Exp. Aula Dei*, 1974, **12**, 176. (Lab. Agrar. Reg. Ebro, Ministry of Agriculture, Spain).

1340 L'VOV, B. V., POLZIK, L. K., KATSKOV, D. A., and KRUGLIKOVA, L. P., Interferometric measurement of line shifts in flames in relation to the interpretation of the line absorption method in AAS, *Zh. Prikl. Spektrosk.*, 1975, **22**, 787. (Address as in ref. 348).

1341 YAMASAKI, S., YOSHINO, A., and KISHITA, A., Determination of submicrogram amounts of elements in soil solution by flameless AAS with a heated graphite atomizer, *Soil Sci. Plant Nutr.*, 1975, **21**, 63. (Hokkaido National Agricultural Experimental Station, Sapporo, Japan).

1342 ECKSTEIN, E. W., COBURN, J. W., and KAY, E., Diagnostics of an r.f. sputtering glow discharge correlation between AA and mass spectrometry, *Int. J. Mass Spectrom. Ion Phys.*, 1975, **17**, 129. (Research Laboratory, I.B.M., San Jose, Calif. U.S.A.).

1343 NOLLER, B. N., and BLOOM, H., Determination of atmospheric particulate Pb using low volume sampling and nonflame AA, *Atmos. Environ.*, 1975, **9**, 505. (Address as in ref 432).

1344 VOLKHONOVITCH, O. P., Determination of K, Na and Ca levels in solated nerve cells using an integrating flame microspectrophotometer, *Fiziol. Zh.*, 1975, **21**, 413. (Institute Fiziol. im. Bogomol'tseva, Kiev, U.S.S.R.).

1345 MARINKOVIC, M., JANJIC, J., and JANKOVIC, D., ES determination of Al, Cu and Fe in solutions with a gas stabilized arc, *Hem. Ind.*, 1975, **29**, 171. (Boris Kidric Institute, Vinca, Belgrade, Yugoslavia).

1346 MANOLIU, C., Comparative study of the spectrophotometric determination of Pd in catalysts by using 2-nitroso-1-naphthol with the AA determination, *Rev. Chim. (Bucharest)*, 1974, **25**, 671.

1347 ZYNGER, J., and CROUCH, S. R., Miniature low-energy spark discharge system for ES analysis of solutions, *Appl. Spectrosc.*, 1975, **29**, 244. (Address as in ref. 99).

1348 TANIGUCHI, B., and MIYAJI, T., Estimation of total and inorganic Hg by flameless AA analysis, *Rinsho Byori*, 1975, **23**, 142. (School of Medicine, Yamaguchi University, Ube, Japan).

1349 COX, L. E., Flameless AA apparatus for analysing alpha radioactive materials, *U.S. At. Energy Comm., Rep.*, LA 5791, 1974. (Address as in ref. 692).

1350 MUSHA, S., Determination of trace inorganic elements in table and industrial-grade salt by AAS. I: Determination of microgram traces of As in salt by the AA method, *Nippon Kaisui Gakkai-Shi*, 1974, **27**, 255. (Address as in ref. 186).

1351 OSTROUMENKO, P. P., and EREMENKO, A. M., Anion effect in the AAS of Ni, *Zh. Prikl. Spektrosk.*, 1975, **22**, 387.

1352 PANNEKOEK, W. J., KELSALL, J. P., and BURTON, R., Methods of analysing feathers for elemental content, *Tech. Rep., Fish. Mar. Serv. (Can.)*, No. 498, 1974. (Dept. of Chemistry, University of Saskatchewan, Saskatoon, Sask., Canada).

1353 SUZUKI, T., MORINAGA, H., and SASAKI, A., Indirect determination of P in Fe and steel by AAS, *Tetsu To Hagane*, 1975, **61**, 1063. (Research Centre, Japan Steel Works Ltd., Muroran, Japan).

1354 TSUKUI, H., and TOGASHI, Y., Indirect determination of F in a coating on low-hydrogen type arc-welding electrodes and raw materials by separating precipitated CaF_2 and AAS, *Tetsu To Hagane*, 1975, **61**, 388. (Tsuruya Works Co. Ltd., Tokyo, Japan).

1355 MATSUSHITA, K., KOHNO, K., KOSHIRO, J., KODAMA, Y., and MATSUMOTO, K., Direct determination of serum Li using an AAS, *Rinsho Kensa*, 1975, **19**, 636. (Faculty of Medicine, Kagoshima University, Kagoshima, Japan).

1356 MOROZOV, N. P., and DEMINA, L. L., Extraction and AAS in determining heavy metals in sea waters, *Tr. Vses. Nauchno Isled. Ints. Mork. Rybn. Khoz. Okeanogr.*, 1974, **100**, 23.

1357 ANON., AA method for determining the content of Zn bacitracin, *Fed. Regist.*, 1975, **40**, 15088. (Food and Drugs Administration, Washington, D.C., U.S.A.).

1358 OSHIMA, T., Decomposition of samples for use in AA analysis, *Jitsumu Hyomen Gijutsu*, 1975, No. 253, 87. (Osaka City Industrial Research Laboratory, Osaka, Japan).

1359 PATENT, Glow discharge tube hollow cathode for AA analysis, British Patent 1,365,139.

1360 DEUBERT, K. H., and GRAY, R., Determination of Zn in individual terrestrial nematodes, *Nematologica,* 1974, **20**, 365. (Laboratory of Experimental Biology, University of Massachusetts, East Wareham, Mass., U.S.A.).

1361 CHOJNICKA-BRZOZOWSKA, B., and SOKOLOWSKA, R., Rejected methods of determination of certain metals in food products. I: Pb determination, *Rocz. Panstw. Zakl. Hig.*, 1975, **26**, 65. (Zahl. Badania Zywn. Przedmiotow Uztku, Panstw. Zakl. Hig. Warsaw, Poland).

1362 KORKISCH, J., and SORIO, A., Determination of seven trace elements in natural waters after separation by solvent extraction and anion-exchange chromatography, *Anal. Chim. Acta,* 1975, **79**, 207. (Address as in ref. 191).

1363 OOGHE, W., and VERBEEK, F., Spectral line interference in the AAS of lanthanides, *Anal. Chim. Acta,* 1975, **79**, 285. (Address as in ref. 1337).

1364 BELCHER, R., BOGDANSKI, S. L., KNOWLES, D. J., and TOWNSHEND, A., MECA: a new flame analytical technique. Part VI: The simultaneous determination of SO, SO_3 or S_2O_3, *Anal. Chim. Acta,* 1975, **79**, 292. (Address as in ref. 35).

1365 MILLER, J. E., HASSETT, J. J., and KOEPPE, D. E., The effect of soil properties and extractable Pb levels on Pb uptake by soybeans, *Commun. Soil Sci. Plant Anal.*, 1975, **6**, 339. (Dept. of Agronomy, University of Illinois, Urbana, Ill. 61801, U.S.A.).

1366 MILLER, J. E., HASSETT, J. J., and KOEPPE, D. E., The effect of soil Pb sorption capacity on the uptake of Pb by corn, *Commun. Soil Sci. Plant Anal.*, 1975, **6**, 349. (Address as in ref. 1365).

1367 DONNELLY, T. H., ECCLESTON, A. J., and GULLY, R. L., A high speed method of continuous background correction in AAS. I: Instrumental, *Appl. Spectrosc.*, 1975, **29**, 149. (Address as in ref. 971).

1368 HODGES, R. J., and BELCHER, C. B., Alkali analysis using light reflected from the grating of an optical emission spectrometer, *Appl. Spectrosc.*, 1975, **29**, 163. (Broken Hill Pty. Co. Ltd., Central Research Laboratories, Shortland, N.S.W. 2307, Australia).

1369 HORLICK, G., and CODDING, E. G., Simultaneous multi-element and multi-line AA analysis using a computer-coupled photodiode array spectrometer, *Appl. Spectrosc.*, 1975, **29**, 167. (Address as in ref. 89).

1370 KELIHER, P. N., and WOHLERS, C. C., Spectral line profile measurements from Ca, Ag, and Al hollow cathode lamps, *Appl. Spectrosc.*, 1975, **29**, 198. (Address as in ref. 651).

1371 SOLDANO, B. A., and KWAN, P. W. O., An empirical light intensity relationship for elemental Hg emission, *Appl. Spectrosc.*, 1975, **29**, 271. (Dept. of Physics, Furman University, Greenville, S.C. 29613, U.S.A.).

1372 GOLIGHTLY, D. W., and HARRIS, J. L., Spectrographic analysis of geological materials with an argon plasma jet, *Appl. Spectrosc.*, 1975, **29**, 233. (U.S. Geological Survey, Reston, Va. 22092, U.S.A.).

1373 SIEMER, D. D.,* and STONE, R. W.,† Analytical potential of non-resonance line flameless AAS, *Appl. Spectrosc.*, 1975, **29**, 240. (*Address as in ref. 31; †Dept. of Chemistry, Montana State University, Bozeman, Mont. 59715, U.S.A.).

1374 MORRIS, W. F., and WORDEN, E. F., Study of the excitation mechanism in the low-current carbon arc in argon and helium, *Appl. Spectrosc.*, 1975, **29**, 255. (Lawrence Livermore Laboratory, University of California, Livermore, Calif. 94550, U.S.A.).

1375 BERNICE, G., DA SILVA, C. T. C. B., and MELLO, V. N. A., A method of interpreting geochemical results obtained by a six-step semi-quantitative spectrographic procedure, *Appl. Spectrosc.*, 1975, **29**, 269. (Companhia de Pesquisa de Recursos Minerais, Rio de Janeiro, 2000, Brazil).

1376 CURRY, K. J., COOLEY, E. F., and DIETRICH, J. A., An automatic filter positioner device for ES, *Appl. Spectrosc.*, 1975, **29**, 274. (U.S. Geological Survey, Denver, Colo. 80225, U.S.A.).

1377 WOOD, D. L., DARGIS, A. B., and NASH, D. L., A computerized television spectrometer for emission analysis, *Appl. Spectrosc.*, 1975, **29**, 310. (Address as in ref. 713).

1378 NAKAHARA, T., and MUSHA, S., Chemical interference effects in the AAS determination of Pb with premixed inert gas (entrained air)/hydrogen flames, *Appl. Spectrosc.*, 1975, **29**, 352. (Address as in ref. 186).

1379 WINEFORDNER, J. D., FITZGERALD, J. J., and OMENETTO, N., Review of multi-element AS methods, *Appl. Spectrosc.*, 1975, **29**, 369. (Address as in ref. 487).

1380 FOLSOM, T. R., HANSEN, N., WEITZ, W. E., and PARKS, G. J., Determination of Na and K in blood of oceanic fish by FES using an exponential extrapolation with multiple standard additions, *Appl. Spectrosc.*, 1975, **29**, 404. (Mount Soledad Laboratory for Marine Radioactivity Studies, Scripps Institution for Oceanography, La Jolla, Calif. 92037, U.S.A.).

1381 McDERMOTT, W. E.,* and NASH, C. P.,† The measurement of oscillator strengths for Cu lines involving auto-ionizing levels, *Appl. Spectrosc.*, 1975, **29**, 408. (*Frank J. Seiler Research Laboratory, (AFSC), U.S. Air Force Academy, Colo. 80840, U.S.A.); †Dept. of Chemistry, University of California at Davis, Davis, Calif. 95616, U.S.A.).

1382 CHUANG, F. S., and WINEFORDNER, J. D., AFS with an Eimac continuum excitation source and graphite filament atomizer, *Appl. Spectrosc.*, 1975, **29**, 412. (Address as in ref. 487).

1383 PARKER, L. R., MORGAN, S. L., and DEMING, S. N., Simplex optimization of experimental factors in AAS, *Appl. Spectrosc.*, 1975, **29**, 429. (Dept. of Chemistry, Emory University, Atlanta, Ga. 30322, U.S.A.).

1384 VOGEL, R. S., A semi-automated device for controlled atmospheres in optical ES, *Appl. Spectrosc.*, 1975, **29**, 436. (Institute for Environmental Studies, University of Illinois, Urbana, Ill. 61801, U.S.A.).

1385 HETZ, G. R., HUGGETT, R. J., and HILL, J. M., Behaviour of Mn, Fe, Cu, Zn, Cd and Pb discharged from a wastewater treatment plant into an estuarine environment, *Water Res.*, 1975, **9**, 631. (Dept. of Chemistry, University of Maryland, College Park, Md. 20742, U.S.A.).

1386 PITA, F. W., and HYNE, N. J., The depositional environmental of Zn, Pb and Cd in reservoir sediments, *Water Res.*, 1975, **9**, 701. (Dept. of Earth Sciences, University of Tulsa, Tulsa, Okla. 74104, U.S.A.).

1387 KRAVOVSKA, E., and MATHERNY, M., Background correction in ES analysis. I: Spectral line and background correlation studies, *Chem. Zvesti,* 1975, **29**, 177. (Address as in ref. 1041).

1388 BEGAK, O. Y., GALKIN, M. A., and FEDOROV, V. V., The determination of N and H in metallurgical materials, *Zav. Lab.*, 1975, **41**, 950.

1389 NEMETS, V. M., PETROV, A. A., and YAKIMOVA, V. A., Spectroscopic analysis of gases using isotope addition and chromatography at low pressure, *Zav. Lab.*, 1975, **41**, 952. (Zhdanov State University, Leningrad, U.S.S.R.).

1390 UFIMTSEVA, R. N., OSLOPVSKIKH, V. N., and MARKOVA, E. A., Quantitative determination of isomorphous Fe in sphalerites using a laser microanalyser, *Zav. Lab.*, 1975, **41**, 954. (Scientific Research and Design Institute of Ore Treatment, Sverdlovsk, U.S.S.R.).

1391 L'VOV, B. V., KRUGLIKOVA, L. P., POLZIK, L. K., and KATSKOV, D. A., Theory of flame AA analysis. 5: Study of behaviour of Sn in the H_2/air flame on the basis of thermodynamic calculation of temperature and composition of combustion products, *Zh. Anal. Khim.*, 1975, **30**, 1045. (Address as in ref. 348).

1392 VIGANT, G. T., SEMENENKO, K. A., and TARASEVICH, N. I., Effect of matrix material on the mean residence time of V, Nb and Ta atoms in the d.c. arc plasma, *Zh. Anal. Khim.*, 1975, **30**, 1223. (Lomonosov Moscow State University, Moscow, U.S.S.R.).

1393 BELYAEV, Y. I., KOVESHNIKOVA, T. A., and ORESHKIN, V. N., Comparative estimation of sensitivity, limits of detection and precision of AA and AF analysis with a nonflame powder atomizer, *Zh. Anal. Khim.*, 1975, **30**, 1267. (Address as in ref. 334).

1394 OGNEV, V. R., OGNEVA, E. Y., and ZAIROVA, G. M., Improvements in accuracy of spectrographic analysis by cooling the electrodes with air flow, *Zh. Anal. Khim.*, 1975, **30**, 1280. (Zhadov Irkutsk State University, Irkutsk, U.S.S.R.).

1395 SHELPAKOVA, I. R., YUDELEVICH, I. G., CHANYSHEVA, T. A., SHCHERBAKOVA, O. I., and USOVA, V. A., Layer by layer chemical-spectrographic determination of B in Si structures, *Zh. Anal. Khim.*, 1975, **30**, 1310. (Address as in ref. 287).

1396 KARPENKO, L. I., FADEEVA, L. A., and SHEVCHENKO, L. D., Determination of rare earths in some fluorides, *Zh. Anal. Khim.*, 1975, **30**, 1330. (Address as in ref. 45).

1397 ZAKHAROV, E. A., MYASOEDOV, B. F., and LEBEDEV, I. A., Determination of Am in Cm using a hollow cathode discharge, *Zh. Anal. Khim.*, 1975, **30**, 1344. (Vernadskii Institute of Geochemistry and Analytical Chemistry, U.S.S.R. Academy of Sciences, Moscow, U.S.S.R.).

1398 STAIKOV, A. I., and ZAKHARIYA, N. F., Use of distillation separation of elements for increasing the sensitivity of spectrographic analysis, *Zh. Anal. Khim.*, 1975, **30**, 1375. (Institute of General and Inorganic Chemistry, Ukrainian Academy of Sciences, Odessa Laboratory, U.S.S.R.).

1399 TROKHACHENKOVA, O. P., GRADSKOVA, N. A., ZHINKIN, D. Y., and MAKULOV, N. A., Selection of optimum conditions for spectrographic determination of Cr in polymethylsiloxane liquids, *Zh. Anal. Khim.*, 1975, **30**, 1380. (Address as in ref. 43).

1400 BURMISTROV, M. P., and MOREISKAYA, L. V., On a way to reduce the number of standards for spectrographic analysis, *Zh. Anal. Khim.*, 1975, **30**, 1477. (State Scientific and Design Institute of Alloys and Non-ferrous Metal Working, Moscow, U.S.S.R.).

1401 TIPSOVA-YAKOVLEVA, V. G., DVORTSAN, A. G., and SEMENOVA, I. B., Chemical-spectrographic methods of analysis of semi-conductor Te and tellurides of some second-group elements, *Zh. Anal. Khim.*, 1975, **30**, 1577. (Moscow Institute of Steel and Alloys, U.S.S.R.).

1402 LARINA, L. K., and MAKARTSEVA, V. N., Spectrographic determination of Hg in metals and alloys, *Zh. Anal. Khim.*, 1975, **30**, 1623. (All-Union Scientific Mining and Metallurgy Institute of Nonferrous Metals, Ust'-Kamenogorsk, U.S.S.R.).

1403 KLEINMANN, I., and SVOBODA, V., Plasma jets for spectrometric analysis, *Chem. Listy,* 1975, **69**, 833. (Institute for Research, Production and Application of Radioisotopes, Pristavni 24, 17000 Prague 7, Czechoslovakia).

1404 KOLIHOVA, D., DUDOVA, N., JANOUSKOVA, J., and SYCHRA, V., Analysis of ilmenite and inorganic pigments on TiO_2 base by AAS. I: Determination of Mn and Cr, *Chem. Listy,* 1975, **69**, 613. (Address as in ref. 41).

1405 SYCHRA, V., JANOUSKAVO, J., KOLIHOVA, D., DUDOVA, N., and MAREK, S., Analysis of ilmenite and TiO_2 based pigments by AAS. II: Determination of Cu, *Chem. Listy,* 1975, **69**, 623. (Address as in ref. 41).

1406 FITCHETT, A. W., DAUGHTREY, E. H., and MUSHAK, P., Quantitative measurement of inorganic and organic As by flameless AAS, *Anal. Chim. Acta,* 1975, **79**, 93. (Dept. of Pathology, University of North Carolina at Chapel Hill, Chapel Hill, N.C. 27514, U.S.A.).

1407 GARCIA, W. J., BLESSIN, C. W., INGLETT, G. E., and CARLSON, R. O., Physical-chemical characterisation and heavy metal content of corn grown on sludge-treated strip-mine soil, *J. Agric. Food Chem.*, 1974, **22**, 810. (Address as in ref. 477).

1408 FURR, A. K.,* MERTENS, D. R.,† GUTENMANN, W. H.,‡ BACHE, C. A.,‡ and LISK, D. J.,‡ Fate of polychlorinated biphenyls, metals and other elements in papers fed to lactating cows, *J. Agric. Food Chem.*, 1974, **22**, 954. (*Nuclear Reactor Laboratory, Virginia Polytechnic Institute, Blacksburg, Va. 24061, U.S.A.; †Animal Science Dept., Iowa State University, Ames, Iowa 50010, U.S.A.; ‡Pesticide Residue Laboratory, Dept. of Food Science, New York State College of Agriculture, Cornell University, Ithaca, N.Y. 14853, U.S.A.).

1409 SCHUTH, C. K., ISENSEE, A. R., WOOLSON, E. A., and KEARNEY, P. C., Distribution of ^{14}C and As derived from (^{14}C) cacodylic acid in an aquatic ecosystem, *J. Agric. Food Chem.*, 1974, **22**, 99. (Agricultural Environmental Quality Institute, Agriculture Research Center, U.S. Department of Agriculture, Beltsville, Md. 20705, U.S.A.).

1410 BAETZ, R. A.,* and KENNER, C. T.,† Determination of trace metals in foods using chelating ion exchange concentration, *J. Agric. Food Chem.*, 1975, **23**, 41. (*Food and Drug Administration, Dallas District, Dallas, Texas 75204, U.S.A.; †Dept. of Chemistry, Southern Methodist University, Dallas, Texas 75275, U.S.A.).

1411 WELCH, R. M., and CARY, E. E., Concentrations of Cr, Ni and V in plant materials, *J. Agric. Food Chem.*, 1975, **23**, 479. (Address as in ref. 1180).

1412 TEENY, F. M., Rapid method for the determination of Hg in fish tissue by AAS, *J. Agric. Food Chem.*, 1975, **23**, 668. (U.S. Department of Commerce, National Oceanic and Atmospheric Administration, National Marine Fisheries Service, Pacific Utilization Research Center, Seattle, Wash. 98112, U.S.A.).

1413 DASSANI, S. D., McCLELLAN, B. E., and GORDON, M., Submicrogram level determination of Hg in seeds, grains and food products by cold-vapour AAS, *J. Agric. Food Chem.*, 1975, **23**, 671. (Address as in ref. 804).

1414 FURR, A. K., KOSIKOWSKI, F. V., BACHE, C. A., and LISK, D. J., Elemental analysis of protein-containing food materials from various sources, *J. Food Sci.*, 1974, **39**, 887. (Address as in ref. 1408).

1415 ODLAND, D., and EHEART, M. S., Ascorbic acid, mineral and quality retention in frozen broccoli blanched in water, steam and ammonia/steam, *J. Food Sci.*, 1975, **40**, 1004. (College of Human Ecology and Agricultural Experimental Station, University of Maryland, College Park, Md. 20742, U.S.A.).

1416 KASHIRAD, A., and MARSCHNER, H., Fe nutrition of sunflower and corn plants in mono and mixed culture, *Plant Soil*, 1974, **41**, 91. (Institut fur Nutzflanzenforschung Pflanzenernahrung, Technische Universitat Berlin, 1 Berlin 33, Lentzeallee 55/57, West Germany).

1417 MOSS, G. I., and HIGGINS, M. L., Mg influences on the fruit quality of sweet orange (*Citrus sinensis* L. Orsbeck), *Plant Soil*, 1974, **41**, 103. (C.S.I.R.O., Div. of Irrigation Research, Griffith, N.S.W., Australia).

1418 SIMAN, A., CRADOCK, F. W., and HUDSON, A. W., The development of Mn toxicity in pasture legumes under extreme climatic conditions, *Plant Soil*, 1974, **41**, 129. (Biological and Chemical Research Institute, Rydalmere, N.S.W. Department of Agriculture, Sydney, N.S.W., Australia).

1419 TYLER, G., Heavy metal pollution and soil enzymatic activity, *Plant Soil*, 1974, **41**, 303. (Dept. of Plant Ecology, O Vallgatan 14, University of Lund, Sweden).

1420 SMILDE, K. W., KOUKOULAKIS, P., and VAN LUIT, B., Crop response to phosphate and lime on acid sandy high in Zn, *Plant Soil*, 1974, **41**, 445. (Institute for Soil Fertility, Haren (Gr)., The Netherlands).

1421 NELSON, L. E.,* and SELBY, R.,† The effect of N sources and Fe levels on the growth and composition of Sitka spruce and Scots pine, *Plant Soil*, 1974, **41**, 573. (*Mississippi State University, Mississippi State, Miss. 39762, U.S.A.; †Rothamsted Experimental Station, Harpenden, Herts. AL5 2JQ, England).

1422 VOLZ, M. G., and JACOBSON, L., A specific Ca requirement for K uptake by excised vetch roots, *Plant Soil*, 1974, **41**, 647. (Dept. of Soils and Plant Nutrition, University of California, Berkeley, Calif. 94720, U.S.A.).

1423 SAFFORD, L. O., Effect of Mn level in nutrient solution on growth and Mg content of *Pinus radiata* seedlings, *Plant Soil*, 1975, **42**, 293. (Yale School of Forestry and Environmental Studies, New Haven, Conn., U.S.A.).

1424 PLANK, C. O., MARTENS, D. C., and HALLOCK, D. L., Effect of soil application of fly-ash on chemical composition and yield of corn (*Zea Mays* L.) and on chemical composition of displaced soil solutions, *Plant Soil*, 1975, **42**, 465. (Dept. of Agronomy, Virginia Polytechnic Institute and State University, Blacksburg, Va., U.S.A.).

1425 WITTWER, R. F., LEAF, A. L., and BICKELHAUPT, D. H., Biomass and chemical composition of fertilized and/or irrigated *Pinus resinosa* Ait. plantations, *Plant Soil*, 1975, **42**, 629. (Dept. of Silviculture and Forest Influences, State University of New York, College of Environmental Science and Forestry, Syracuse, N.Y. 13210, U.S.A.).

1426 NISHITA, H., and HAUG, R. M., Water and NH_4 acetate-extractable Zn, Mn, Cu, Cr, Co and Fe in heated soils, *Soil Sci.*, 1974, **118**, 421. (Laboratory of Nuclear Medicine and Radiation Biology, University of California, Los Angeles, Calif. 90024, U.S.A.).

1427 GOEDERT, W. J., COREY, R. B., and SYERS, J. K., Lime effects on K equiligria in soils of Rio Grande Do Sul, Brazil, *Soil Sci.*, 1975, **120**, 107. (College of Agriculture and Life Sciences, University of Wisconsin, Madison, Wis., U.S.A.).

1428 GRIFFITH, S. M.,* and SCHNITZER, M.,† The isolation and characterization of stable metal-organic complexes from tropical volcanic soils, *Soil Sci.*, 1975, **120**, 126. (*Soil Research Institute, Agriculture Canada, Ottawa, Ont. K1A 0C6, Canada; †University of the West Indies, St. Augustine, Trinidad).

1429 FISKELL, J. G. A.,* and CALVERT, D. V.,† Effects of deep tillage, lime incorporation and drainage on chemical properties of spodosol profiles, *Soil Sci.*, 1975, **120**, 132. (*Soil Science Dept., University of Florida, Gainesville, Fla., U.S.A.; †Agricultural Research Center, Fort Pierce, Fla., U.S.A.).

1430 MISRA, U. K., BLANCHAR, R. K., and UPCHURCH, W. J., Al content of soil extracts as a function of pH and ionic strength, *Soil Sci. Soc. Am., Proc.*, 1974, **38**, 897. (Missouri Agricultural Experimental Station, University of Missouri, Colo. Columbia, Mo. 65201, U.S.A.).

1431 LAHAV, N., and HOCHBERG, M., Kinetics of fixation of Fe and Zn applied as Fe-EDTA, Fe-EDDHA and Zn-EDTA in the soil, *Soil Sci. Soc. Am., Proc.*, 1975, **39**, 55. (Dept. of Soil and Water, Faculty of Agriculture, Rehovat, Hebrew University of Jerusalem, Israel).

1432 HUANG, P. M., Retention of As by hydroxy-Al on surfaces of micaceous mineral colloids, *Soil Sci. Soc. Am., Proc.*, 1975, **39**, 271. (Saskatchewan Institute of Pedology, Dept. of Soil Science, University of Saskatchewan, Sask., Canada).

1433 SUAREZ-HERNANDEZ, A., and HANWAY, J. J., Na availability in non-alkali soils, *Soil Sci. Soc. Am., Proc.*, 1975, **39**, 308. (Iowa Agricultural and Home Economics Experimental Station, Ames, Iowa 50010, U.S.A.).

1434 DANIELS, R. B., GAMBLE, E. E., BUOL, S. W., and BAILEY, H. H., Free Fe sources in an Aquult–Udult sequence from North Carolina, *Soil Sci. Soc. Am., Proc.*, 1975, **39**, 335. (North Carolina Agricultural Experiment Station, Raleigh, N.C. 27607, U.S.A.).

1435 WALLINGFORD, G. W., MURPHY, L. S., POWERS, W. L., and MANGES, H. L., Effects of beef-feedlot manure and lagoon water on Fe, Zn, Mn and Cu content in corn and in DTPA soil extracts, *Soil Sci. Soc. Am., Proc.*, 1975, **39**, 482. (Dept. of Agronomy and Dept. of Agricultural Engineering, Kansas Agricultural Experiment Station, Manhattan, Kan. 66506, U.S.A.).

1436 SINGH, B. R., and STEENBERG, K., Interactions of micronutrients in barley grown on Zn-polluted soils, *Soil Sci. Soc. Am., Proc.*, 1975, **39**, 674. (Institute of Soil Science and Isotope Laboratory, Agricultural University of Norway, Aas, Norway).

1437 YOON, S. K., GILMOUR, J. T., and WELLS, B. R., Micronutrient levels in the rice plant Y leaf as a function of soil solution concentration, *Soil Sci. Soc. Am., Proc.*, 1975, **39**, 685. (Dept. of Agronomy and Rice Branch Experimental Station, University of Arkansas, Agricultural Experimental Station, Fayetteville, Ark. 72701, U.S.A.).

1438 MORTVEDT, J. J., and GIORDANO, P. M., Crop response to Mn sources applied with ortho- and polyphosphate fertilizers, *Soil Sci. Soc. Am., Proc.*, 1975, **39**, 782. (Soils and Fertilizer Research Branch, National Fertilizer Development Center, Tennessee Valley Authority, Muscle Shoals, Ala. 35660, U.S.A.).

1439 IRELAND, M. P., The effect of the earthworm Dendrobaena Rubida on the solubility of Pb, Zn and Ca in heavy-metal contaminated soil in Wales, *J. Soil Sci.*, 1975, **26**, 313. (Dept. of Zoology, University College of Wales, Aberystwyth, Wales).

1440 BOWDEN, D. N.,* and ROBERTS, H. S.,† Analyses and fusion characteristics of New Zealand coal ashes, *N.Z. J. Sci.*, 1975, **18**, 119. (*New Zealand Coal Research Association (Inc.), P.O. Box 3041, Wellington, New Zealand; †Applied Mathematics Div., D.S.I.R., Wellington, New Zealand).

1441 WEISSBERG, B. G., Hg in some New Zealand waters, *N.Z. J. Sci.*, 1975, **18**, 195. (Chemistry Div., D.S.I.R., Private Bag, Petone, New Zealand).

1442 JORDAN, L. D., and HOGAN, D. J., Survey of Pb in Christchurch soils, *N.Z. J. Sci.*, 1975, **18**, 253. (Chemistry Div., D.S.I.R., Christchurch, New Zealand).

1443 WARD, N. I., REEVES, R. D., and BROOKS, R. R., Pb from motor-vehicle exhausts in sweet-corn plants and soils along a highway in Hawke's Bay, New Zealand, *N.Z. J. Sci.*, 1975, **18**, 261. (Address as in ref. 159).

1444 GOH, K. M.,* and WHITTON, J. S.,† Kelp extract as fertilizer. II: Effect on chemical composition and element uptake of white clover, *N.Z. J. Sci.*, 1975, **18**, 391. (*Dept. of Soil Science, Lincoln College, Canterbury, New Zealand; †Soil Bureau, D.S.I.R., Lower Hutt, New Zealand).

1445 CAMPBELL, A. G., COUP, M. R., BISHOP, W. H., and WRIGHT, D. E., Effect of elevated Fe intake on the Cu status of grazing cattle, *N.Z. J. Agric. Res.*, 1974, **17**, 393. (Ruakura Animal Research Station, Private Bag, Hamilton, New Zealand).

The following papers (*) were presented at the New Zealand Institute of Chemistry Annual Conference, 19–22 August, 1975, Palmerston North, New Zealand).

1446* ROBERTS, A. H. C., TURNER, M. A., and SYERS, J. K., Simultaneous extraction and determination of Cd and Zn in waters and aqueous extracts of soils. (Address as in ref. 1427).

1447* PATTERSON, J. E., Rapid heating in flameless AA analysis. (Chemistry Div.,

D.S.I.R., Petone, New Zealand).

The following papers (†) were presented at the 89th Annual Meeting of the Association of Official Analytical Chemists, 13–16 October, 1975, Washington, D.C., U.S.A.

1448† IHNAT, M., Se in foods: evaluation of AAS technique involving hydrogen selenide generation and carbon furnace atomization. (Chemistry and Biology Research Institute, Agriculture Canada, Ottawa, Ont. K1A 0C6, Canada).

1449† MILLER, H. J., ABEL, R. C., and WILLIAMS, J. R., Determination of As and Se in foods by hydride generation/flame AAS. (Food and Drug Administration, 900 Madison Avenue, Baltimore, Md. 21201, U.S.A.).

1450† ANDERSON, P. J., The determination of Sn residues in food by conventional AAS. (Vulcan Laboratories, Pontiac, Mich. 48058, U.S.A.).

1451† SULEK, A. M., and ZINK, E. W., Determination of Pb in evaporated milk by anodic stripping voltammetry and AAS: double-blind refereed collaborative study. (National Canners Association, 1133 20th Street NW, Washington, D.C. 20036, U.S.A.).

1452† HOOVER, W. L., MELTON, J. R., and MORRIS, P. A., Determination of P in feeds and fertilizers by argon plasma ES. (Texas Agriculture Experiment Station, Texas A & M University, College Station, Texas 77843, U.S.A.).

1453† MELTON, J. R., HOOVER, W. L., and MORRIS, P. A., Determination of B in fertilizers by argon plasma ES. (Address as in ref. 1452).

1454† WARD, A. F., Plant tissue analysis using the inductively coupled argon plasma. (Jarrell-Ash Division, Fisher Scientific Co., 590 Lincoln St., Waltham, Mass. 02154, U.S.A.).

1455 SCHRENK, W. G., Analytical AS — Book published by Plenum Press, New York, 1975.

1456 TOROK, T., and ZARAY, G. Y., Experiment with a cooled twin-hollow-cathode interferometer-spectrometer. I, *Spectrochim. Acta, Part B*, 1975, **30***B*, 157. (Address as in ref. 616).

1457 BOUMANS, P. W. J. M., and DE BOER, F. J., Studies of an inductively-coupled h.f. argon plasma for optical ES. II: Compromise conditions for simultaneous multi-element analysis, *Spectrochim. Acta, Part B*, 1975, **30***B*, 309. (Address as in ref. 544).

1458 OMENETTO, N., WINEFORDNER, J. D., and ALKEMADE, C. T. J., An expression for the AF and thermal-emission intensity under conditions of near saturation and arbitrary self-absorption, *Spectrochim. Acta, Part B*, 1975, **30***B*, 335. (Address as in ref. 797).

1459 TSONG, I. S. T., and McLAREN, A. C., An ion beam spectrochemical analyser with application to the analysis of silicate minerals, *Spectrochim. Acta, Part B*, 1975, **30***B*, 343. (Address as in ref. 850).

1460 RADMACHER, H. W., DE SWARDT, M. C., The analysis of steel and cast Fe by means of the Grimm glow discharge lamp, *Spectrochim. Acta, Part B*, 1975, **30***B*, 353. (Address as in ref. 628).

1461 WAGENAAR, H. C., and DE GALAN, L., The influence of line profiles upon analytical curves for Cu and Ag in AAS, *Spectrochim. Acta, Part B*, 1975, **30***B*, 361. (Address as in ref. 207).

1462 MERMET, J. M., Comparison of temperatures and of electronic densities measured on the plasma gas and on some excited elements in an h.f. plasma, *Spectrochim. Acta, Part B*, 1975, **30***B*, 383. (Address as in ref. 182).

1463 RAUTSCHKE, R.,* AMELUNG, G.,* NADA, N.,* BOUMANS, P. W. J. M.,† and MAESSEN, F. J. M. J.,‡ Contribution to the kinetic thermochemical reaction of U compounds in a graphite matrix in the d.c. arc: fundamental considerations on the thermochemical process as partner effect in spectrochemical analysis, *Spectrochim. Acta, Part B*, 1975, **30***B*, 397. (*Martin-Luther-Universitat, Sektion Chemie Weinbergweg, Halle/Saale, East Germany; †Address as in ref. 544; ‡Address as in ref. 206).

1464 GUTSCHE, B., RUDIGER, K., and HERRMANN, R., Method for the determination of F concentration with AA, *Spectrochim. Acta, Part B*, 1975, **30***B*, 441. (Address as in ref. 20).

1465 BOUMANS, P. W. J. M.,* DE BOER, F. J.,* DAHMEN, J.,† HOELZEL, H.,† and MEIER, A.,† A comparative investigation of some analytical performance characteristics of an ICP and a capacitively-coupled microwave plasma for solution analysis by ES, *Spectrochim. Acta, Part B*, 1975, **30***B*, 449. (*Address as in ref. 544; †Address as in ref. 607).

1466 BURRIDGE, J. C., and SCOTT, R. O., A rotating briquetted-disc method for the determination of B and other elements in plant material by ES, *Spectrochim. Acta, Part B*, 1975, **30***B*, 479. (Address as in ref. 174).

The following papers (*) were presented at Svenska Kemistsamfundet, 25–26 November 1975, Gothenburg, Sweden.

1467* MULLINS, C., Peak area measurement of transient signals in AAS. (Instrumentation Laboratory (U.K.) Ltd., Manchester, England).

1468* WELZ, B., Factors affecting accuracy and precision in flameless AAS. (Address as in ref. 803).

1469* PRICE, W. J., Methods of dissolution of siliceous materials for AAS. (Address as in ref. 172).

1470* FRECH, W., Flameless AAS for trace element analysis of metallurgical materials. (Address as in ref. 920).

1471* LUNDGREN, G., Flameless AAS. (Dept. of Analytical Chemistry, University of Umea, Umea, Sweden).

1472* PAUS, P. E., AAS with hydride generation. (Central Institute for Industrial Research, Blindern, Oslo, Norway).

1473* JOHANSSON, L. G., Microwave (EDL) lamps for AAS.

1474* RUDERRUS, H., Preparation of biological material for AAS.

1475* LUNDGREN, G., Analysis of biological material by flameless AAS. (Address as in ref. 1471).

1476* SKUJINS, S., Some examples of sample treatment and analysis using a carbon rod atomizer with and without the addition of hydrogen to the shielding gas. (Varian Techtron, Zug, Switzerland).

1477 JACKSON, K. W., and MITCHELL, D. G., Rapid determination of Cd in biological tissue by microsampling-cup AAS, *Anal. Chim. Acta*, 1975, **80**, 39. (Address as in ref. 80).

1478 NAKAHARA, T., and MUSHA, S., The AAS determination of In in premixed inert gas (entrained air)/hydrogen flames, *Anal. Chim. Acta*, 1975, **80**, 47. (Address as in ref. 186).

1479 AGEMIAN, H., and CHAU, A. S. Y., An AA method for the determination of 20 elements in lake sediments after acid digestion, *Anal. Chim. Acta*, 1975, **80**, 61. (Address as in ref. 185).

1480 KIRK, M., PERRY, E. G., and ARRITT, J. M., The separation and AA measurement of trace amounts of Pb, Ag, Zn, Bi and Cd in high Ni alloys, *Anal. Chim. Acta*, 1975, **80**, 163. (Address as in ref. 760).

1481 CRESSER, M. S., Design and preliminary evaluation of a simple discrete sampler for flame spectrometric analysis, *Anal. Chim. Acta*, 1975, **80**, 170. (Address as in ref. 935).

The following papers (*) were presented at Euroanalysis II, 25–30 August 1975, Budapest, Hungary.

1482* PERMAN, J., What now with visual spectroscopy? (Slovenian Iron and Steel Works, Zelezarna Ravne, Yugoslavia).

1483* HOFFMAN, E., A contribution to modelling of methods in the ES analysis. (Central Institute for Optics and Spectroscopy of the Academy of Sciences, Berlin-Aldershof, West Germany).

1484* SZILVASSY, Z., Excitation with a hollow cathode for the determination of the F and lanthanides content in phosphates. (University of Chemical Engineering, Dept. of Radiochemistry, Verszprem, Hungary).

1485* RYBAROVA, Z., Determination of the spherical distribution of the temperature and electron density in a.c. arc. (Faculty of Metallurgy, Technical University, Kosice, Czechoslovakia).

1486* KRAKOVSKA, E., and MATHERNY, M., Results of applying time-resolution to arc emission spectra. (Address as in ref. 1041).

1487* ZENTAI, P., A method of spectrochemical standardization. (Address as in ref. 599).

1488* IDZIKOWSKI, A., and GADEK, S., LMA-1 laser microspectroanalyser as an excitation source in quantitative spectral analysis. Part II: Determination of the common elements in rocks. (Institute of Inorganic Chemistry, Technical University of Wroclaw, Poland).

1489* ZIVANOVIC-MAGDIC, V., KULOVIC, D., NOVOSEL, V., PRESLOSCAN, M.,

and SOKOLEAN, D., Quantitative spectrographic analysis of nonmetallic inclusions in steel. (Metallurgical Institute, Sisak, Yugoslavia).
1490* FARKAS-UJHIDY, K.,* and DOMBI, A.,† Comparative investigation of ES solution methods in trace analysis of soils. (*Dept. of Radiochemistry; †Dept. of Analytical Chemistry, Veszprem University of Chemical Engineering, Veszprem, Hungary).
1491* SOLLEI, P., The spectrochemical analysis of the nonconducting inorganic materials of the building industry by means of the rotating disc solution method in argon atmosphere. (State Company for the Building Industry in Department Borsod, Miskolc, Hungary).
1492* HAHN-WEINHEIMER, P., and PETER, J., Application of a disc-stabilized d.c. arc to determine trace metals of environmental significance in natural waters. (Research Office for Geochemistry, Technical University, Munich, West Germany).
1493* FLORIAN, K., and PLIESOVSKA, N., Study of the spectrochemical properties of medium-voltage spark discharge. (Address as in ref. 1041).
1494* VECSERNYES, L., and ZARAY, G., Effects of density measuring methods on spectrographic analytical curves. (Address as in ref. 616).
1495* DIMITROV, G.,* DIMOV, D.,† and PANEVA, A.,* Investigation of stripy carbonate formations with the aid of an LMA-1 laser microspectral analyser. (*Dept. of Physics; †Dept. of Geology and Geography, Sofia University, Sofia, Bulgaria).
1496* BURGUDJIEV, Z., The problems of extreme detection limits in spectral analysis using photographic recording. (Dept. of Physics, University of Sofia, Sofia, Bulgaria).
1497* VAN RAAPHORST, J. G., ORDELMAN, J., HAREMAKER, H., and VAN KRALINGEN, N., The ES analysis of water. (Stichting Reactor Centrum, The Netherlands).
1498* NEDYALKOVA, N., and KRASNOBAEVA, N., Effect of easily ionizable additives by the graphite-arc method in a controllable atmosphere. (Address as in ref. 37).
1499* MARINOV, M. I., Investigation of the effect of the basic composition in the spectrographic analysis of microelements in urinary concrements. (Medical Academy, Sofia, Bulgaria).
1500* SKALSKA, S.,* and GLUSZEK, A.,† Spectrographic analysis of Pd powder by d.c. arc technique. (*Institute of Industrial Chemistry; †Institute of Radio Ceramics, Warsaw, Poland).
1501* YUDELEVICH, I. G., LANBINA, T. V., VASILIEVA, A. A., and GINDIN, L. M., AA from non-aqueous solutions in the analytical chemistry of Pt group metals. (Address as in ref. 287).
1502* KOWALCZYK, J., and STRZELBICKA, B., Determination of some rare earths by AAS. (Institute of Inorganic Chemistry and Metallurgy of Rare Elements, Technical University, Wroclaw, Poland).
1503* YUDELEVICH, I. G., BUYANOVA, L. M., BAKHTUROVA, N. F., and KORDA, T. M., AA methods for analysis of semiconductor films. (Address as in ref. 287).
1504* OTWINOWSKI, W., Determination of trace elements in aniline by AAS. (Institute of Industrial Chemistry, Warsaw, Poland).
1505* MICHOTTE, Y., SMEYERS-VERBEKE, J., SEGEBARTH, G., and MASSART, D. L., Matrix interferences in the Massman graphite furnace: analysis of Cu and Mn in biological calcifications. (Address as in ref. 667).
1506* HIRCQ, B., Investigation of the re-partition and transformations of deposit on a graphite filament in AAS. (Commissariat a l'Energie Atomique, Centre d'Etudes de Bruyeres-le-Chatel, 92-Montroug, France).
1507* HALASZ, A., and POLYYAK, K., Investigations on the processes taking place in the graphite tube at the flameless AA determination of Ag. (Institute of Analytical Chemistry, University of Chemical Engineering, Veszprem, Hungary).
1508* SCHREIBER, B., LINDER, H. R., and FREI, R. W., Trace analysis of As, Se, Sb, Sn, Bi and Te by x-ray fluorescence and AAS using the $NaBH_4$ reduction. (Address as in ref. 925).
1509* ORTNER, H. M., and SCHERER, V., A critical comparison between x-ray fluorescence analysis and AA for trace analysis in refractory metals. (Address as in ref. 702).
1510* HAARSMA, J. P. S., VLOGTMAN, J., and AGTERDENBOS, J., Some improvements in the applicability of AFS in practical analysis. (Laboratory for Analytical Chemistry of the State University, Utrecht, The Netherlands).

1511* VAN MONTFORT, P. F. E., VAN SANDWIJK, A., and AGTERDENBOS, J., Trace analysis by microwave excitation of sealed samples. (Address as in ref. 1510).
1512* CSAKOW, J.,* and GLINSKI, J.,† The use of multisource glow discharge lamp in spectral analysis of agricultural materials. (*Address as in ref 626; †Institute of Agrophysics of the Polish Academy of Sciences, Lublin, Poland).
1513* HORVATH, E., PAPP, L., BIRO, Z., and POSTA, J., Investigation of the macro and trace element environmental pollution and its quantitative reflection in the human organism. (Address as in ref. 1320).
1514* TOWNSHEND, A., AKPOFURE, A. K., BELCHER, R., BOGDANSKI, S. L., and KNOWLES, D. J., Determination of S compounds in admixture using MECA. (Address as in ref. 35).
1515* JACKWERTH, E., and MESSERSCHMIDT, J., Preconcentration of trace elements in high-purity Ga by partial dissolution of the matrix. (Address as in ref. 58).
1516 MUSIL, J., Determination of small amounts of Pb or Sn in metals by AAS, *Hutn. Listy,* 1975, **30**, 292. (Kovohute, Minsek pod Brdy, Czechoslovakia).
1517 SHERBURN, F. A., and POOLE, A. G., Determination of Na in gas turbine fuel oil, *J. Inst. Fuel,* 1975, **48**, 21. (Central Electricity Generating Board, S.E. Region, Sumner St., London, S.E.1., England).
1518 CRESSER, M. S., KELIHER, P. N., and WOHLERS, C. C., Aspects of the uses of echelle monochromators in analytical AS, *Lab. Pract.,* 1975, **24**, 335. (Address as in ref. 935).
1519 WILSON, C. A., FERRERO, E. P., and COLEMAN, H. J., Crude-oil spills research: investigation of analytical techniques, *U.S., Bur. Mines, Rep. Invest.,* RI 8024, 1975. (Energy Research Center, Bartlesville, Okla., U.S.A.).
1520 PSZONICKI, L., and ABDALLAH, A. M., Determination of Cr by AAS. I: Role of aliphatic acids, *Chem. Anal. (Warsaw),* 1975, **20**, 473. (Institut Badan Jadrowych, Zakld Chemii Analitycznej, ul. Dorodna 16, 03-195, Warsaw, Poland).
1521 ANON., Non-flame devices in AAS, *Tech. Rep. Water Res. Cent.,* TR 1, 1975. (Water Research Centre, Stevenage Laboratory, Elder Way, Stevenage, Herts., England).
1522 CLINTON, I. E., Curcumin method for B compatible with an AA system of plant analysis, *N.Z. J. Sci.,* 1974, **17**, 445. (Address as in ref. 879).
1523 TARASEVICH, N. I., KOZYREVA, G. V., and PORTUGAL'SKAYA, Z. P., Extraction–AA determination of trace amounts of In, Bi and Pb in silicate rocks and soils, *Vestn. Mosk. Univ., Ser. Khim.,* 1975, **16**, 241. (Address as in ref. 1392).
1524 JOHNSON, J. S., POPE, R., and SANDON, P. T. S., Foliar analysis by direct-reading vacuum ES, *J. Sci. Food Agric.,* 1975, **26**, 441. (Tropical Products Institute, Gray's Inn Road, London, W.C.1., England).
1525 SINHA, R. C. P., and BANERJEE, B. K., Interferences in estimation of trace amounts of Co, Cu and Zn in soils by AAS, *Technology (Sindri, India),* 1974, **11**, 263. (Address as in ref. 722).
1526 SHIMAMURA, A., Application of AAS to routine determination of metallic components in pharmaceutical products, *Flame Notes,* 1975, **7**, 35. (Linden Laboratories, Grosvenor, Los Angeles, Calif. 90066, U.S.A.).
1527 VAN DER PIEPEN, H., Analysis of oils with a geometrically stabilised spark discharge, *Spectrochim. Acta, Part B,* 1975, **30***B*, 179. (Address as in ref. 591).
1528 ZECHEV, D., DYULGEROVA, R., and PACHEVA, Y., The profile of the Fe 372·00 nm line excited in a hollow cathode discharge, *J. Quant. Spectrosc. Radiat. Transfer,* 1975, **15**, 941. (Institute of Solid State Physics, Bulgarian Academy of Science, Sofia, Bulgaria).
1529 HUFFMAN, H. L., and CARUSO, J. A., A preliminary study on the effect of time on apparent Pb content of evaporated milk as determined by non-flame AAS, *Talanta,* 1975, **22**, 871. (Address as in ref. 1097).
1530 GARNYS, V. P., and SMYTHE, L. E., Fundamental studies on improvement of precision and accuracy in flameless AAS using the graphite tube atomizer: Pb in whole blood, *Talanta,* 1975, **22**, 881. (Address as in ref. 159).
1531 KAISER, G., GOTZ, D., SCHOCH, P., and TOLG, G., Determination of nanogram and picogram amounts of elements by microwave induced plasma ES. I: Ultra-sensitive determination of Hg in aqueous solutions, air, organic and inorganic matrices, *Talanta,* 1975, **22**, 889. (Max Planck Institut fur Metallforschung, Laboratorium fur Reinststoffe, Stuttgard und Schevabisch Gmund, West Germany).

1532 VENGSARKAR, B. R., MACHADO, I. J., and MALHOTRA, S. K., Spectrographic analysis of BN for trace impurities, *Talanta,* 1975, **22**, 903. (Spectroscopy Div., Bhabha Atomic Research Centre, Trombay, Bombay 400085, India).

1533 L'VOV, B. V., and ORLOV, N. A., Theory of flame AA analysis. 6: Determination of the temperature of air/acetylene and nitrous oxide/acetylene flames by measuring two-line absorption, *Zh. Anal. Khim.,* 1975, **30**, 1653. (Address as in ref. 348).

1534 L'VOV, B. V., and ORLOV, N. A., Theory of AA analysis. 7: Effect of redox characteristics of flames and chemical form of atomized compounds on the degree of aerosol vaporisation, *Zh. Anal. Khim.,* 1975, **30**, 1661. (Address as in ref. 348).

1535 ZHARNOPOLSKII, A. I., AA spectrophotometer 'Saturn,' *Zh. Anal. Khim.,* 1975, **30**, 1847.

1536 BUKREEV, Y. F., NAGDAEV, V. K., and ZOLOTAVIN, V. L., Study of the effects of acids and cations on the absorbance of Fe atoms with atomisation on a graphite rod, *Zavod. Lab.,* 1975, **41**, 957. (Tambovsk Institute of Chemical Engineering, Tambovsk, U.S.S.R.).

1537 VIKHROV, S. V., The determination of atom concentration in the vapour phase during electron beam fusion, *Zavod. Lab.,* 1975, **41**, 962. (Kiev State University, Kiev, U.S.S.R.).

1538 KATSKOV, D. A., KRUGLIKOVA, L. P., and L'VOV, B. V., Electronic integrator for AAS, *Zavod Lab.,* 1975, **41**, 964. (Address as in ref. 348).

1539 KOLOSOVA, L. P., LISNYANSKAYA, M. G., NADZHINA, L. S., NOVATSKAYA, N. V., and GRINZAID,E. L., Improvement of a fire-assay spectrometric method for the determination of Pt, Pd, Au and Rh in ores and enrichment products, *Zavod. Lab.,* 1975, **41**, 1088. (Address as in ref. 413).

1540 KOROVIN, V. A., KOTENKO, E. F., MASHIREV, L. G., and YUNUSOV, Z. T., Burner for flame-photometric analysis of oil products, *Zavod. Lab.,* 1975, **41**, 1093. (All-Union Scientific and Research Institute for Oil Technology, Moscow, U.S.S.R.).

1541 KUBON, K., and SVARDALA, L., The new vacuum spectrometer E100 Polyvac with computer control, *Hutn. Listy.,* 1975, **30**, 147. (Research and Testing Institute, Ostrava-Kuncice, Czechoslovakia).

1542 SVEHLA, A., and SALCEROVA, M., Study of errors in spectrochemical analysis of ferroalloys, *Hutn. Listy,* 1975, **30**, 206. (Address as in ref. 40).

The following papers (*) were presented at the 2nd Czechoslovak Seminar on AAS, 23–27 April 1975, Reka, Czechoslovakia.

1543* MUSIL, J., and NEHASILOVA, M., Interferences on Si in its determination by AAS. (Address as in ref. 1516).

1544* RUBESKA, I., New interference mechanism in the nitrous oxide/acetylene flame. (Research Institute on Smelting Ceramics, Bratislava, ul. Februarovho Vitastva, Czechoslovakia).

1545* RUSNAKOVA, A., Interferences in the nitrous oxide/acetylene flame. (Address as in ref. 1544).

1546* SLOVAK, Z., and TOMAN, J., Use of a chelating ion exchanger for trace analysis, in particular for the elimination of matrix effects in nonflame AAS. (Research Institute of Pure Reagents, Lachema, Brno, Czechoslovakia).

1547* TOMAN, J., and SLOVAK, Z., Problems and possibilities of nonflame atomization in a graphite furnace for the analysis of pure substances. (Address as in ref. 1546).

1548* LANG, I., WEISER, O., and SYCHRA, V., Notes on the determination of V in fuel samples by AAS. (Address as in ref. 41).

1549* BEK, F., Nonflame atomization in water analysis by AAS. (Institute of Engineering Geology, Prague 4, Na Kovarne 4, Czechoslovakia).

1550 WAHBI, A. M., A critical comment on the curve fitting process, *Fresenius' Z. Anal. Chem.,* 1975, **275**, 203. (Faculty of Pharmacy, University of Alexandria, Alexandria, Egypt).

1551 FRECH, W., Rapid determination of Bi in steels by flameless AA, *Fresenius' Z. Anal. Chem.,* 1975, **275**, 353. (Address as in ref. 920).

1552 WOIDICH, H., and PFANNHAUSER, W., Determination of As in biological material using flame AAS, *Fresenius' Z. Anal. Chem.,* 1975, **276**, 61. (Address as in ref. 72).

1553 SCHMITZ, L., LOOSE, W., and KOCH, K. H., Sample preparation of ferroalloys and sinters by recasting and wet grinding, *Fresenius' Z. Anal. Chem.,* 1975, **276**, 111. (Hoesch Huettenwerke AG, D-4600 Dortmund, Postfach 902, West Germany).

1554 SENSMEIER, M. R., WAGNER, W. F., and CHRISTIAN, G. D., A simple tantalum strip atomizer for the flameless AA determination of trace metals in water, *Fresenius' Z. Anal. Chem.*, 1975, **277**, 19. (Dept. of Chemistry, University of Washington, Seattle, Wash. 98195, U.S.A.).

1555 SHRISTA, I. L., and WEST, T. S., An investigation of the submicroanalysis of metal surfaces by electrography and carbon filament AAS, *Bull. Soc. Chim. Belg.*, 1975, **84**, 549. (Address as in ref. 112).

1556 JAUNIAUX, M., DE MEYER, M., LEJEUNE, W., and LEVERT, J. M., Determination of trace metals in water by AAS after solvent extraction using thiothenoyltrifluoro-acetone and fluorated β-diketones, *Bull. Soc. Chim. Belg.*, 1975, **84**, 565. (Service de Chimie Applique (Analytique et Industrielle), Faculte Polytechnique de Mons, Belgium).

1557 GRIMALDI, R., MEUCCI, A., and RANDI, G., Advances in ferroalloys instrumental analysis, *Biblioteca Tecnica Philips,* 1975. (Laboratorio Centrale, Italsider SpA, Genova, Italy).

1558 BROEKAERT, J. A. C., Analytical methods by ES, *in* Bormans, J., *Editor,* Analyse van Industriele Afvalwaters — Book published by CEBEDOC, Liege, 1975. (Address as in ref. 225).

1559 HERTOGEN-JANSSENS, M. F., AAS: methods and apparatus: application to pollution, *in* Bormans, J., *Editor,* Analyse van Industriele Afalwaters — Book published by CEBEDOC, Liege, 1975. (Institut voor Nucleaire Wetinnhoppen-RUG-Proeftuinlarn, Ghent, Belgium).

1560 DRWIEGA, I., JEDRZEJEWSKA, H., and MALUSECKA, M., The AA method in high purity Al analysis. I: Investigation on the impurities determination in presence of matrix, *Chem. Anal. (Warsaw),* 1975, **20**, 539. (Instytut Metali Niezelaznych, Zaklad Chemii Analitycznej, ul. Sobieskiego 11, 44-100 Gliwice, Poland).

1561 SKORKO-TRYBULA, Z., ROZANSKA, B., and LACHOWICZ, E., AA determination of trace amounts of Ca in spectrally pure Pb oxalate, *Chem. Anal. (Warsaw),* 1975, **20**, 625. (Politechnika Warszawsak, Instytut Chemii Ogolnej i Technologii Nierorganicznej, ul. Noakowskiego 3, 00-664 Warsaw, Poland).

1562 SZOPLIK, J., Spectrographic determination of Cr in layers of Cr diffusion-plated steel with preliminary electrospark sampling, *Chem. Anal. (Warsaw),* 1975, **20**, 647. (Instytut Mechaniki Precyzyjnej, Laboraturium Analiz Chemicznych i Spektralnych, ul. Duchnicka 3, 00-967, Warsaw, Poland).

1563 SOKOLOWSKA, W., Spectrographic determination of metallic impurities in high purity In and Te, *Chem. Anal. (Warsaw),* 1975, **20**, 655. (Osrodek Naukowo-Produkcyjny Materialow Polprzewodnikowych, Zaklad Analiz, ul. Konstruktorska 6, 02-673 Warsaw, Poland).

1564 PSZONICKI, L., and ABDALLAH, A. M., Analysis of Cr by AAS: II. Use of releasing elements, *Chem. Anal. (Warsaw),* 1975, **20**, 683. (Address as in ref. 1520).

1565 DIMITROV, G., and MARINOV, M., Investigation of microdistribution of some elements in urinary concrements by means of laser microprobe analyser LMA-1, *Chem. Anal. (Warsaw),* 1975, **20**, 715. (Address as in ref. 1495).

1566 DIMITROV, G., PETRAKIEV, A., DIMOV, D., and PANEVA, A., Laser microprobe analysis in controlled atmosphere of minerals and mineral formations in the 'Pavel' meteorite and naturally coloured aragonites, *Chem. Anal, (Warsaw),* 1975, **20**, 723. (Address as in ref. 1495).

1567 MARINOV, M., and ALEXIEV, A., Spectrographic determination of microelements in small samples of human skin, *Chem. Anal. (Warsaw),* 1975, **20**, 735. (Address as in ref. 1499).

1568 MALUSECKA, M., JEDRZEJEWSKA, H., and DRWIEGA, I., AA methods in high-purity Al analysis. II: Enrichment of admixtures by extraction and carrier precipitation, *Chem. Anal. (Warsaw),* 1975, **20**, 755. (Address as in ref. 1560).

1569 DABROWSKA, J., Semiquantitative spectrographic analysis of powdered materials, *Chem, Anal. (Warsaw),* 1975, **20**, 855 (Instytut Badan Jadrowych, Zaklad Chemii Analitycznej, ul. Dorodna 16, 03-195, Warsaw, Poland).

1570 ABDALLAH, A. M., and PSZONICKI, L., Analyser of Cr by AAS. III: A new AA method for determination of Cr in industrial Cr bearing materials, *Chem. Anal. (Warsaw),* 1975, **20**, 919. (Address as in ref. 1520).

1571 VAJGAND, V.,* and STOJANOVIC, D.,† Investigation of the possibility of determination of rare-earth elements by AA inhibition titration, *Chem. Anal. (Warsaw),* 1975, **20**, 973. (*Institute of Chemistry, University of Belgrade, Studenski trg. 16, 11000 Belgrade, Yugoslavia; †Institute for the Apllication of Nuclear Energy, Zemun, Yugoslavia).

1572 JAGIELLO-PUCZKA, W., and KLIMECKI, W., Inhomogenity of nonmetallic inclusions in rail steels. II: Laser microprobe spectral analysis, *Chem. Anal. (Warsaw)*, 1975, **20**, 993. (Instytut Metalurgii, Politechnika Czestowchowska, ul. Zawadzkiego 19, 42-201 Czestochowa, Poland).

1573 DANIELSSON, A., SODERMAN, E., and LINDBLOM, P., The IDES system: an image dissector echelle spectrometer for spectrochemical analysis — Paper presented at the XVIII Colloquium Spectroscopicum Internationale, 15–19 September 1975, Grenoble, France. (Address as in ref. 321).

AUTHOR INDEX

SUBJECT INDEX

Methods for the determination of individual elements are not listed in this index as they can be readily obtained from the alphabetical listings in each of the 9 Applications tables.

I—Part I; II—Part II; T—Applications table.